Security in Wireless Communication Networks

Security in Wireless Communication Networks

Yi Qian
University of Nebraska-Lincoln, USA

Feng Ye
University of Dayton, USA

Hsiao-Hwa Chen
National Cheng Kung University, Taiwan

IEEE PRESS

WILEY

Registered Offices
John Wiley & Sons, Inc., 111 River Street, Hoboken, NJ 07030, USA
John Wiley & Sons Ltd, The Atrium, Southern Gate, Chichester, West Sussex, PO19 8SQ, UK

Editorial Office
The Atrium, Southern Gate, Chichester, West Sussex, PO19 8SQ, UK

For details of our global editorial offices, customer services, and more information about Wiley products visit us at www.wiley.com.

Wiley also publishes its books in a variety of electronic formats and by print-on-demand. Some content that appears in standard print versions of this book may not be available in other formats.

Library of Congress Cataloging-in-Publication Data applied for:
ISBN: 9781119244363

Cover design by Wiley
Cover image: © AnuchaCheechang/iStock/Getty Images

Set in 9.5/12.5pt STIXTwoText by Straive, Chennai, India
Printed and bound by CPI Group (UK) Ltd, Croydon, CR0 4YY

C9781119244363_081121

Contents

Preface

We first signed the book contract with Wiley in early 2006. Three years after signing this book contract, we only completed one third of the contents for the whole originally planned book, because of underestimating the challenges on writing such a specialized book. When the first author, Yi Qian, joined the faculty of the Department of Electrical and Computer Engineering at University of Nebraska-Lincoln in August 2009, he created a new course on wireless network security for the department. He has been teaching the wireless network security course in the same department every year in the spring semester since then. When preparing the course materials for wireless network security each year, we feel more and more strongly to have such a comprehensive textbook on wireless network security. In 2015, we signed the revised contract with Wiley and jointly with IEEE Press, and Feng Ye was added as a new co-author when he just received his Ph.D. degree in the Department of Electrical and Computer Engineering at University of Nebraska–Lincoln. We have been improving the contents of the wireless network security course every year, and gradually finished more chapters of the book. Fifteen years after first signing the contract and 12 years after teaching the same course, with over several thousands of hours joint efforts from all the three co-authors, we are very pleased that we have completed the first edition of "Security in Wireless Communication Networks" and published by Wiley/IEEE Press in 2021.

This book intends to be a self-contained and one semester textbook for both undergraduate senior level and graduate level courses. There are five parts with 15 chapters in the book. Part I, Introduction and Mathematics Background, includes the first three chapters on general introduction on computer communication networks and wireless networks, basic concepts on network security, and a brief review of the mathematical background that is needed to understand the rest of the chapters. Part II, Cryptographic Systems, includes the next three chapters on cryptographic techniques for both symmetric and public key crypto systems, as well as message authentication, digital signature, and key management. Part III, Security for Wireless Local Area Networks, includes four chapters on Wi-Fi security, Bluetooth security, Zigbee security, and RFID security. Part IV, Security for Wireless Wide Area Networks, includes three chapters on GSM security, UMTS security, and LTE security. Part V, Security for Next Generation Wireless Networks, includes two chapters on 5G wireless network security, and vehicular communication network security. In the following is a brief introduction for each of the fifteen chapters.

Chapter 1 delivers the general concept of computer networks, highlights the role of wireless communications in the whole picture of networking architecture, and classifies the wireless systems based on coverage, topology, and mobility. This chapter serves as a precursor to the rest of the book by providing the background of different types of wireless networks, including wireless personal area networks (WPAN), wireless local area networks (WLAN), and wireless wide area networks (WWAN). It also explains the security threats in wireless networks and discusses the relationship between network security and wireless security.

Chapter 2 gives an overview on the security concepts used in the rest of this book, including security attacks, security services, and security mechanisms. It first presents the classification of security attacks in terms of passive attacks (e.g. eavesdropping and traffic analysis) and active attacks (e.g. masquerade, replay, modification, and denial of service). It then introduces security services, or the features in system design against possible security attacks, such as confidentiality, integrity, availability, access control, authentication, and non-repudiation. Finally, to provide security service in a system, a list of popular security mechanisms, such as the encipherment, digital signature, etc., is discussed in the remaining part of the chapter.

Chapter 3 goes into the mathematical background related to wireless security, including number theory and modern algebra, modular arithmetic and divisors, finite fields, polynomial arithmetic, Fermat's little theorem, Euler's totient function, Euler's theory, etc. The aforementioned knowledge is critical for the ones to understand cryptography, such as advanced encryption standards and public-key cryptographic systems. In addition, the fundamental principles and exemplary cases are concisely presented from the perspective of mathematics.

After the mathematical background, Chapters 4 and 5 deal with cryptographic techniques. Chapter 4 first introduces several symmetric key cryptographic techniques by illustrating a few classical cryptographic algorithms with substitution and transposition techniques. It then presents the basic concept of modern stream/block cipher as well as Feistel cipher structure. Chapter 5 explains more cryptographic techniques using block ciphers and public key algorithms, including advanced encryption standard, block cipher mode of operations, public key infrastructure, RSA algorithm, etc.

Chapter 6 introduces message authentication and digital signature to protect the integrity of a message and the identity of a sender and a receiver, respectively. First, this chapter discusses MAC and hash functions thoroughly, both widely used to provide message authentication. Then, it goes into the characteristics of digital signature and a series of digital signature standards such as DSA, RSA, and ECDSA. These can protect the sender and receiver against each other. Within the aforementioned mechanisms, key management and distribution play a critical role. The rest of the chapter gives a general idea and some examples of key management schemes. Both symmetric and asymmetric key distributions have been illustrated. The key distribution mechanisms adopt symmetric and public key mechanisms for different purposes. Besides, practical communication systems with massive users need hierarchical key distribution mechanisms. Readers are expected to understand the basic concepts of the cryptographic techniques illustrated in Chapter 5 and Chapter 6. These algorithms will be seen in the wireless systems introduced in the later chapters. The background of the advanced mathematical contents, such as elliptic curve Diffie–Hellman key exchange and elliptic curve digital signature, etc., may be skipped.

The remaining chapters from 7 to 15 focus on the security of specific wireless communication systems, covering different scales of networks and different technologies including WLAN, Bluetooth, ZigBee, RFID, GSM, UMTS, LTE, and 5G. As the emerging vehicle-to-everything (V2X) communications are receiving great attention, the fifteenth chapter especially discusses the security of V2X communications.

Chapter 7 discusses the security of Wireless Local Area Networks (WLAN) or interchangeably Wi-Fi nowadays. It starts with an introduction of WLAN in terms of operating modes and security challenges. WLAN is more vulnerable to attacks than wired connections due to the lack of physical connections. It illustrates a few generations of WLAN security protocols, which evolved from the original Wired Equivalent Privacy defined by the IEEE 802.11, Wi-Fi Protected Access (WPA), WPA2, to the recent WPA3 to improve the security. It also analyzes the implementation details of these security protocols.

Chapter 8 deals with Bluetooth security. Bluetooth is an open standard designed for wireless personal area networks (WPAN). Bluetooth technology enables many wireless devices, such as smartwatches, wireless headphones, wireless keyboards, etc. Bluetooth standard specifies authentication, authorization, and confidentiality for securing data transmission. In this chapter, it analyzes the security mode, trust level, and service level configurations that enable flexibility of Bluetooth security policies and highlight that Bluetooth specifications do not ensure secure connections from all adversary penetrations. If using Bluetooth technology in an organization, it is important to develop security policies to address the use of Bluetooth-enabled devices and the responsibilities of users.

Chapter 9 discusses the security of Zigbee. It first gives an overview of Zigbee standards related to different network layers, and then mainly analyzes the key cryptographic mechanisms. As Zigbee adopts symmetric-key cryptographic mechanisms, it especially emphasizes that the secure storage and distribution of keys is the premise of ensuring the security of Zigbee. In practice, the security provided by Zigbee standards is not enough. For example, if a Zigbee device joins a network, intruders can intercept unprotected keys. Moreover, an attacker may easily get physical access to a Zigbee device and extract privileged information due to the low-cost nature. The security must be carefully considered to provide those applications.

Chapter 10 deals with the security of RFID. It first gives an overview of RFID subsystems, different types of RFID tags, and the frequency bands. It then analyzes the security attacks, risks, and security objectives of RFID systems. RFID systems are vulnerable to some attacks (e.g. counterfeit tag, eavesdropping, and electronic collisions) and privacy risks (e.g. disclosure of location information of users). The security objectives of the RFID system include confidentiality, integrity, non-repudiation, and availability. Due to the low cost and physical constraints of RFID tags, mitigation mechanisms to security risks are limited. The chapter then elaborates on the lightweight cryptographic algorithms, anti-collision algorithms, and physical protection available for RFID. It is imperative to provide security services to RFID systems.

Chapter 11 deals with the security of Global System for Mobile (GSM) Communications. Since the early 1990s, as the most widely used cellular mobile phone system in the world,

GSM can provide services like voice communications, short messaging, etc. This chapter starts with the GSM system architecture and then discusses the network access security features and algorithms. Despite the popularity, the GSM system is exposed to quite a few threats. In the chapter, it mainly discusses the attacks caused by the vulnerability of security algorithms, as well as some possible security improvements. Unfortunately, GSM made very few improvements on these aspects before phasing out recently.

Chapter 12 introduces the security of Universal Mobile Telecommunications System (UMTS). UMTS is a successor of GSM with better security. Several security mechanisms are reused but with modifications. After introducing UMTS architecture, the chapter discusses the security mechanisms of UMTS, like the authentication and key agreement, data confidentiality and integrity, user identity confidentiality. Compared with the GSM, UMTS adds integrity protection. Algorithms f8 and f9 ensure confidentiality and integrity, respectively. Both algorithms are based on block cipher KASUMI. Readers may be interested in some additional security features of UMTS, such as mobile device identification, location services, and user-to-USIM authentication, which are discussed at the end of the chapter.

Chapter 13 illustrates Long-Term Evolution (LTE) security. It starts with the introduction of the LTE system architecture which is based on GSM and UMTS. A key difference with its predecessors is that LTE separates the control plane and user plane, differing LTE security from GSM and UMTS. It then depicts LTE security in terms of security architecture, security mechanisms, and algorithms. LTE covers more keys and security algorithms, such as AES and ZUC, to ensure the security of complex systems. It also highlights the LTE security for interworking with legacy systems as well as non-3GPP access. LTE has strong security implemented comparing with the previous generation system. LTE will continue to serve as an important part of the next-generation wireless system.

Chapter 14 discusses the security of 5th generation (5G) wireless network systems. 5G started large-scale commercial deployment around 2020 and is the next-generation mobile wireless telecommunications beyond 4G/International Mobile Telecommunications (IMT)-Advanced Systems. This chapter illustrates some current development, challenges, and future directions of 5G wireless network security. It especially analyzes several new security requirements and challenges introduced by the advanced features of the 5G wireless network systems. Due to the ongoing development of 5G, the chapter only discusses some present solutions and research results concerning the security of 5G wireless network systems. Quite a few challenges in 5G wireless network security, including new trust models, new security attack models, privacy protection, etc., call for continuous development of 5G security. It briefly analyzes these challenges in the final part of the chapter.

In recent years, as a key component of Intelligent Transportation Systems, vehicle-to-everything (V2X) communications have received great attention. The rapid development of wireless technologies (e.g. DSRC, LTE, and 5G) enables V2X communications in different applications. To integrate the variety of wireless technologies and meet special requirements for V2X communications, security and privacy have become a top priority. Therefore, the last chapter of the book sets off to discuss the security of V2X communications. Standards such as IEEE WAVE and LTE-V2X set a general guideline for V2X security implementations. New cryptography schemes, such as group signature and trust-based schemes, are under development. This chapter covers all these topics. As an emerging type of wireless communication scenario, quite a few unsolved security challenges exist in V2X communications. It discusses several key challenges, including efficient schemes, hardware enhancement, and integration of AI algorithms, etc., at the end of the chapter.

Our teaching philosophy is letting the students learn the basic building blocks that are necessary to design a secure wireless system and learn the security designs of different wireless communication networks from the history to the next generation, also different scales from personal area, local area, to wide area wireless networks, so that the students will be able to handle the new designs of future secure wireless systems.

April 2021

Yi Qian
University of Nebraska-Lincoln, USA

Feng Ye
University of Dayton, USA

Hsiao-Hwa Chen
National Cheng Kung University, Taiwan

Acknowledgments

We would like to thank all the undergraduate and graduate students who have taken the wireless network security classes at the University of Nebraska–Lincoln in the last 12 years, without the interactions with the students and the feedbacks from the students this book would not have been possible. We express our deep appreciation for their enthusiasm and their eagerness of learning the subjects. We especially thank those Ph.D. students who studied the wireless network security course and graduated in the Department of Electrical and Computer Engineering at University of Nebraska–Lincoln or worked as a post doctorate researcher there, for their continuous help with improving the course materials, and adapting our wireless network security course modules in their new schools for teaching, specifically at the University of Wisconsin, University of Dayton, City University of New York, Dakota State University, California Polytechnic State University, University of Central Missouri, and University of Texas.

We are sincerely indebted to Professor David Tipper at the University of Pittsburgh for his advice and help when we first created the wireless network security course at the University of Nebraska-Lincoln 12 years ago. Many thanks to Professor Tipper for his endless advice and support, and the encouragement for completing this book.

We express our thanks to the staff of Wiley, for their continuous support for this undertaking over the last 15 years. We would like to thank Sandra Grayson, Senior Editor at Wiley, for all the support and guidance, as well as for providing the needed extra push to keep us in delivering. We thank Juliet Booker, Managing Editor at Wiley, and several more staff at Wiley, for their patience in dealing with electronic transfer of manuscripts and handling publication issues.

Lastly but not least, we thank our families for their support and patience while we worked on this book over these years.

We believe that we have given our best to ensure the readability, completeness, and accuracy of the book. However, it is possible that some errors and omissions may still have remained undetected. We appreciate any feedback intended to correct such errors.

We are thankful to everyone!

About the Companion Website

This book is accompanied by a companion website:

www.wiley.com/go/qian/sec51

The website includes:

• Lecture Slides

Note: The authors plan to supply additional supplementary resources up to one year after initial publication.

Part I

Introduction and Mathematics Background

1

Introduction

A wireless communication network is a computer network that uses a wireless connection between network nodes. Wireless networking is a method to connect telecommunications networks, and business installations or to connect between various equipment locations, to avoid the costly process of introducing cables. Examples of wireless communication networks include cellular networks, wireless local area networks (WLANs), wireless ad-hoc networks, wireless sensor networks, vehicular communication networks, and satellite communication networks. Wireless communication networks are becoming ubiquitous with the increasing of mobile Internet applications, advances of technological development in radio communications and communication infrastructure backbones, as well as mobile wireless devices and consumer electronics [1]. Over the last three decades, we have witnessed several critical moments for the evolution of next generation wireless communication networks. During the 1990s, we witnessed the popularity of personal computers and Internet access for common households as well as the accessible of 2G cellular wireless communications. During the 2000s, we witnessed the tremendous increasing e-commerce on the Internet and the deployment of 3G cellular wireless communications, as well as WLANs for mobile Internet. Since 2010, we have witnessed increasing bandwidth and quality-of-service for 4G cellular wireless communications with more and more applications on the mobile Internet. The wireless communication technology is continuing to be advanced to the next generation with high capacity, low latency, and low energy consumption, for better implementation of Internet of things and many other new service capabilities. From the beginning, security for wireless communication networks has always been a critical issue. In this chapter, a brief introduction will be given on wireless communication networks and basic concepts on wireless communication network security.

1.1 General Computer Communication Network Architecture

1.1.1 Wired Communication Network Infrastructure

Computer communication networks interconnect a collection of network nodes including computer and communication devices, routers, gateways, and switches [2]. The Internet can be considered as the largest computer network that interconnects billions of autonomous nodes around the globe. Obviously, standalone computer is not the only type

Security in Wireless Communication Networks, First Edition. Yi Qian, Feng Ye, and Hsiao-Hwa Chen.
© 2022 John Wiley & Sons Ltd. Published 2022 by John Wiley & Sons Ltd.
Companion website: www.wiley.com/go/qian/sec51

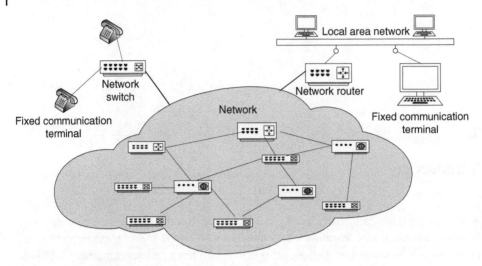

Figure 1.1 Traditional wired networks.

of device that has network access. Smart phones, tablets, smart sensors, vehicles, and many other devices are also connected to computer networks. With the network, data collection and data exchange can be enabled to support further control required by some services. Computer networks have been developed and deployed for many years. In general, computer networks are comprised of wired networks and wireless networks. Although wireless networks are more accessible to regular users in today's communications, the backbone infrastructures still rely significantly on wired networks. Figure 1.1 shows a generic framework of traditional wired networks. User equipment in wired networks is referred to as *fixed communication terminals* due to limited mobility. In the early days, user equipment such as land-line telephones and desktop computers are directly connected to a network switch or a network router through physical network cables. In modern data centers and cloud computing centers, the servers are also hard wired to switches or routers. The core network consists of many switches and routers that are interconnected with physical medium, such as copper wire, Ethernet cable, fiber optics, etc.

1.1.2 Wireless Communication Network Infrastructure

Computer and communication nodes access a wireless communication network through wireless links. However, despite the name, most wireless communication systems only deploy wireless components at the edge of the communication infrastructure, as shown in Figure 1.2. The core network in a general wireless communication infrastructure is a wired network. For example, in a cellular network, its core infrastructure is connected by fiber optic cables and Ethernet. Users are aware of the wireless access only from their user equipment, such as smart phones, tablets, laptops, etc. The wireless access is provided with extra components and resources to the core network infrastructure. The extra components and resources include:

- *Wireless transceivers*: base stations, access point (AP), mobile stations (MSs), etc.
- *Management entities*: mobility management, power management, radio resource management, security management, etc.

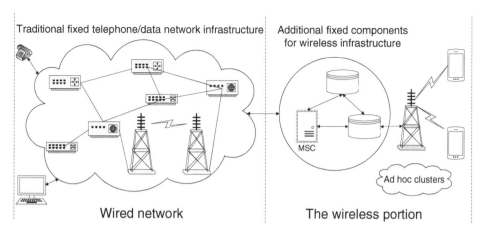

Figure 1.2 Positioning of wireless networks.

- *Spectrum*: radio frequency bands for data transmission and possible air interface.
- *Deployment*: spectrum reuse in communications, wireless network design, etc.

One advantage of wireless communication networks is flexible access from user equipment. Network access can be provided to any user who is within the radio coverage. Therefore, wireless access is more flexible and more convenient compared with wired access. Wireless users would not be restricted by the limited number of Ethernet ports or not long enough cables. The deployment cost of wireless communications is also less than that of wired communications in most cases. For example, a home Wi-Fi network can be established with a single Wi-Fi router, while a traditional Ethernet based home network would require a bulk of Ethernet cables.

1.2 Different Types of Wireless Communication Systems

1.2.1 Classification of Wireless Communication Systems

Wireless communication systems can be classified in several ways, based on *coverage, topology*, or *mobility*, as illustrated in Figure 1.3.

1.2.1.1 Based on Coverage

Wireless communications systems are classified into *wireless personal area networks (WPANs), wireless local area networks (WLANs)*, and *wireless wide area networks (WWANs)*. This classification depends on wireless technology limitations as well as its supporting applications. For example, while both Bluetooth and Wi-Fi can provide a radio coverage large enough for an office, only Wi-Fi is considered as a WLAN. Subtle differences exist due to other classification criteria as well. In some classification, wireless metropolitan area networks may be listed as one type of wireless communication system. Wide area networks in traditional wired computer networks are usually the backbone infrastructure. However, a wireless metropolitan area network has the largest coverage before it is connected to the

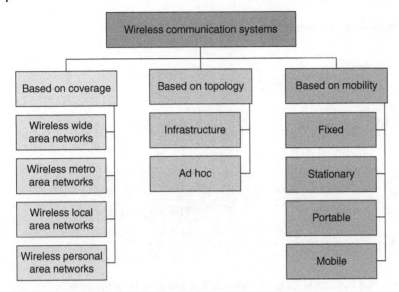

Figure 1.3 Classification of wireless communication systems.

wide area network backbone. Thus, without loss of generality, both wireless metropolitan area networks and WWANs will be considered the same (as WWANs) in this book.

1.2.1.2 Based on Topology

Wireless communication systems are classified into *infrastructure* based and *ad-hoc* based. An infrastructure based wireless communication system requires a fixed backbone communication infrastructure. For example, a cellular network has wireless access from user equipment, but it requires a fixed base station and a backbone network infrastructure. A home Wi-Fi has wireless access from user equipment, but it requires a fixed router that is hard-wired to an Internet service provider. An ad-hoc wireless communication system does not require a fixed infrastructure. For example, a wireless headphone may be connected to a smart phone using Bluetooth technology. In this communication system, data communication between the headphone and the smart phone is wireless based on Bluetooth technology, while a fixed infrastructure is not required for neither end.

1.2.1.3 Based on Mobility

Wireless communication systems are classified into *fixed*, *stationary*, *portable*, and *mobile*. A fixed wireless communication system indicates fixed deployment of equipment. For example, cellular base stations that are micro-wave based only. A stationary wireless communication system indicates a semi-fixed deployment of equipment. For example, a temporary relay vehicle for cellular systems. A portable wireless communication system indicates a more flexible deployment of equipment, with communications enabled when users are not moving fast. For example, users in a home Wi-Fi may have network service with their portable devices. A mobile wireless communication system requires support for services during high speed movement. For example, a general cellular network is a mobile system since services are provided to users, whether moving or not, as long as they are within the radio coverage.

1.2.2 Wireless Personal Area Networks

A WPAN can be used for communications among the personal devices themselves. Therefore, a WPAN usually has an ad-hoc topology. As shown in Figure 1.4, master–slave mode and mesh mode are the two types of ad-hoc networks that can be applied for WPANs. A master–slave ad-hoc network consists of a master node and multiple slave nodes. The master node defines a *cell* or *piconet*. The slave nodes within the piconet connect to the master device. A WPAN based on Bluetooth typically applies master–slave mode. For example, if a wireless headphone is connected to a smart phone using Bluetooth, then the smart phone is the master node where the headphone is a slave node. The user may also connect a Bluetooth keyboard to the same smart phone as a slave node. Some WPANs apply mesh mode, where nodes are interconnected with wireless links without forming a specific cell or piconet, for example, sensor networks, radio-frequency identification (RFID), vehicular ad-hoc networks, etc.

1.2.3 Wireless Local Area Networks

WLANs are infrastructure based wireless communication systems. They are normally built on top of a wired local area network (LAN). One of the typical WLAN settings is a home Wi-Fi, which forms one basic service set (BSS) that includes one AP and multiple user devices. The AP may have extra Ethernet ports to support wired access from servers, desktops, and other devices. As shown in Figure 1.5, a WLAN may have extended service set (ESS) that supports multiple BSSs, similar to a traditional Ethernet based LAN. All APs are interconnected, in most cases through wired connection. A user may be within the radio coverage of multiple APs, nonetheless, each user belongs to one BSS only at a time. That is to say, each user can have access to one AP only in an ESS.

1.2.4 Wireless Wide Area Networks

WWAN has the largest service coverage in all wireless communication systems. As shown in Figure 1.6, a general architecture of WWANs has different components at the radio level, the network level, and the management level.

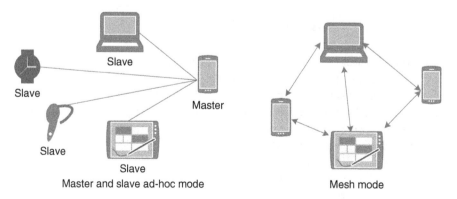

Figure 1.4 Architecture of wireless personal area networks.

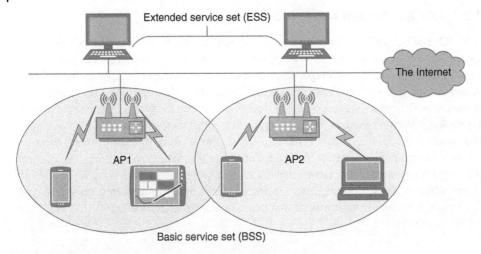

Figure 1.5   Architecture of wireless local area networks.

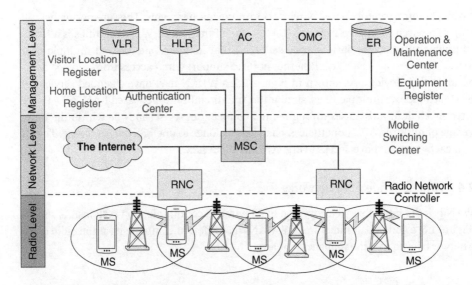

Figure 1.6   Architecture of wireless wide area networks.

The radio level provides wireless access to user equipment, or mobile stations (MSs), which can be a mobile phone, a smart watch, a vehicle, etc. MSs access to WWAN through points of access in the infrastructure. Point of access is the physical radio transceiver. It creates the air interface and communications with MSs. Points of access could be base stations, base transceiver subsystem, mobile data base station, AP, NodeB, eNodeB, etc., depending on the wireless technology it is deployed.

The network level is the backbone infrastructure that connects all switches and routers in the network. A radio network controller (RNC) bridges the radio level and the network level. RNC provides spectrum and power management to base stations, as well as other issues in wireless access. A mobile switching center (MSC) in the network level is a mobile data intermediate system that bridges the network level and the management level in cellular communication systems. MSC manages mobility of devices and keeps track of the location of MSs. MSC also ensures security by using the authentication center and equipment register in the management level to prevent fraudulent devices from using the network.

The management level performs administrative operations of network service providers, such as accounting and billing. In a cellular communication system, the management level includes visitor location register, home location register, authentication center, operation and maintenance center, and equipment register.

1.3 Network Security and Wireless Security

1.3.1 Network Security

Network security is subject to the context in which it is used. Network security is also dictated by the needs of individuals, customs and laws of a region, and policies of an organization. There are different kinds of security breaches. For example, an unauthorized person gets access to confidential records across a network. A malicious user picks up and modifies an authorization file over a network. Or a data file has been received however the sender denies having sent it. All of those examples are security attacks in different ways. In general, *network security is defined as protection of networks and their services from unauthorized modification, destruction, or disclosure*. Network security provides assurance that the network performs its critical functions correctly, with no harmful side-effects [3]. Network security focuses mainly on networks, network protocols, and network applications. It includes all network devices, all applications, an data utilizing a network. For example, routers, switches, smart phones, tablets, etc.

Figure 1.7 illustrates the generic security terminology in a communication network system. As shown in the system, information is usually the target of security attacks. In order to protect the information, requirements and policies are first needed to be specified. Those are the overall and detailed plan for what the potential risks are, and what to protect.

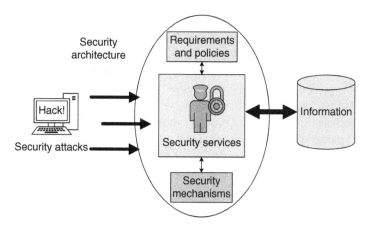

Figure 1.7 Generic security terminology.

This is a statement of what is allowed and what is not. Security services required by a system could be developed based on specific requirements and policies. For example, security services are confidentiality, integrity, availability, etc. Many security mechanisms are developed to provide various security services. Carefully designed security mechanisms detect, prevent, or recover a communication network system from security attacks. In most cases, multiple security mechanisms must be deployed together to provide just one security service. There is no single security mechanism can provide all security services in a communication network system. All the requirements and policies, security services, and security mechanism, form a security architecture of a communication network system.

1.3.2 Security Threats in Wireless Networks

Some security threats are generic in computer networks, for example, hardware sabotage, data leakage, etc. However, wireless networks have unique issues because of the shared transmission medium. Therefore, it is easier for a malicious user to get attached to wireless networks. Even if an access to a wireless communication network system is not granted due to authentication and access control, malicious users may still monitor data traffic by eavesdropping certain radio frequencies. A malicious user may also launch active attacks more easily to a wireless communication network system. For example, a malicious user could continuously send strong signals to jam a radio spectrum. Therefore, vulnerabilities and security problems in wireless communication networks are to be addressed from different aspects.

- Wireless networks suffer from limited coverage and harshness of the radio channels in physical layer. Therefore transmission in wireless networks has relatively high error rates with little to none guarantee of channel quality. Because of that, it is hard to tell denial of service (DoS) attack (an attack to make network resource unavailable to intended users) from channel degradation.
- Wireless networks require decentralized medium access mechanism in medium access control (MAC) layer because of open "broadcast" medium. Fundamental types of medium access mechanisms include frequency division multiple access (FDMA), time division multiple access (TDMA), code division multiple access (CDMA), space division multiple access (SDMA), etc. Besides access control, several other aspects, such as throughput, delay, and quality of service (QoS), also need to be addressed in MAC layer.
- Wireless networks need to deal with mobility of users. On one hand, mobility is a revolutionary advantage of wireless networks. MSs in wireless networks are not restrained to certain deployments; they are free to move within the coverage of the networks. On the other hand, mobility introduces management problems for wireless networks. For example, location tracking and handoff management as MSs move. When the scale of wireless network is large, more issues come to database management.
- Wireless networks need to manage transmission power and radio resources. Generally speaking, raising transmission power level can increase transmission quality for one link. However, interference to other users will be increased thus reducing the transmission quality of other users. Coverage of a wireless network is limited, and it is common that a MS roams from one base station to another one. The process of a MS moving from one

base station to another base station is called handoff. Bear in mind that wireless signals do not have clear boundaries; therefore handoff decision must be carefully made. If a MS moves frequently around the overlap region of two base stations, insufficient handoffs will interrupt transmission, while unnecessary handoffs can increase load to the system.

- Wireless networks are versatile. There is no single type of wireless access available everywhere. Cellular service providers adopt different kinds of wireless technologies. Therefore, very few cell phones can roam across the globe successfully. Even Wi-Fi has different specifications in each AP. For this reason, network design and deployment are to be carefully planned in wireless networks. Besides, spectrum resource is also scarce, therefore coexistence of users and interference among users must be carefully addressed.
- Security concerns in network operations and management need to be addressed in wireless networks. On one hand, network operators need to enable resources and services to MSs safely and privately. On the other hand, network operators also need to authenticate legitimate MSs, especially the roaming ones. Correct accounting and billing for subscribers are based on secure network operations and management.
- Service discovery and data management are problems to be addressed in some wireless networks, e.g. sensor networks and RFIDs. For example, how is data maintained? How to ensure integrity and confidentiality of data? Moreover, a mobile device needs to be lightweight with reasonably long battery life. Therefore, energy efficient designs of software and protocols are unique for wireless networks. While many of these security problems have been studied in wired networks, the solutions proposed there are in general too computationally demanding to work for wireless networks, because mobile devices have limited computational resources and power supply. Communications must also be minimized due to scarce spectrum resource.

1.4 Summary

This chapter gives an introduction on general communication network architectures and wireless communication architectures, as well as security threats in wireless communications networks. The same security objectives that exist in wireline communication networks are also needed for wireless networks. They must be addressed in the context of wireless specific characteristics such as physical layer issues, MAC layer issues, mobility management, radio resource and power management, wireless network design and deployment, wireless network operations and management, wireless application issues, etc. The next chapter provides more security concepts that will be mostly concerned in wireless communication networks. It is recommended to read more on the topics of wireless communication networks for better understanding of security in wireless networks [4–6].

2

Basic Network Security Concepts

Most of the readers are using wireless networks to access communication infrastructure or Internet for work and daily life. Whether you are using the mobile phone or wireless devices through cellular networks, Wi-Fi, or Bluetooth technology, you should have concerns with these networks, and want to make sure that the wireless networks are secure. As discussed in the last chapter, a security architecture consists of requirements and policies, security services, and security mechanisms. In this chapter, we further introduce the concepts that will be used in the rest of this book in the design and operation of a secure communication network in general, and a secure wireless network in specific. These concepts include security attacks, security services, and security mechanisms.

2.1 Security Attacks

Security attacks in communication networks are formally classified as *passive attacks* and *active attacks* in X.800 Recommendation [7] by the International Telecommunication Union, Telecommunication Standardization Sector (ITU-T) and RFC 2828 [8] by the Internet Engineering Task Force (IETF). A high-level description of security attacks is shown in Figure 2.1. Passive attacks do not interfere with a legitimate system directly, leaving users unaware of the attacks. Active attacks intrude a legitimate system by altering original resources, thus causing damages to the system. Both passive and active attacks compromise a legitimate system.

2.1.1 Passive Attacks

Passive attacks include *eavesdropping* and *traffic analysis*, which do not interfere with the legitimate system immediately from current operation point of view.

2.1.1.1 Eavesdropping
Figure 2.2 shows an example with two authorized users and one eavesdropper. The eavesdropper is located somewhere within the transmission range, thus the attacker is able to monitor data transmission and gather unprotected information. The first generation mobile communication system encountered this issue due to its unencrypted analog signal. At that time, anyone can monitor a conversation with necessary equipment. Fortunately, mobile transmissions are encrypted since the second generation mobile communication system;

Security in Wireless Communication Networks, First Edition. Yi Qian, Feng Ye, and Hsiao-Hwa Chen.
© 2022 John Wiley & Sons Ltd. Published 2022 by John Wiley & Sons Ltd.
Companion website: www.wiley.com/go/qian/sec51

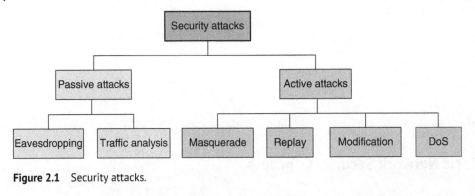

Figure 2.1 Security attacks.

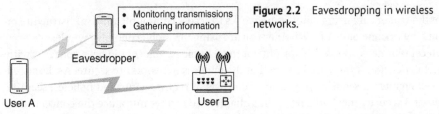

Figure 2.2 Eavesdropping in wireless networks.

therefore the contents of a conversation are free from disclosure to an eavesdropper. Still, much information can be captured by eavesdropping in wireless systems. For example, in a Wi-Fi system, service set identifier (SSID), device MAC address, and some other information are broadcast in clear text. As a security protection, some Wi-Fi users apply white-list to filter unauthorized MAC addresses. However, eavesdroppers can capture MAC addresses by monitoring the traffic with software tools and then update their MAC addresses accordingly to bypass the filter.

2.1.1.2 Traffic Analysis

An attacker doing traffic analysis is essentially an eavesdropper but with other purposes than simply monitoring data traffic contents, as shown in Figure 2.3. With data transmission being encrypted, basic eavesdropping is not very useful in getting information. However, the pattern of transmission can be useful for an attacker. For example, an attacker might be able to locate the legitimate transmitter by measuring signal strength at different locations. Frequency and length of messages might also be revealed to the attacker by tracking the pattern of transmission.

Passive attacks are difficult to detect since there is no data alternation or system manipulation. The system functions as if no security issue to authorized users. Wireless systems

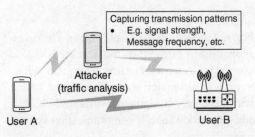

Figure 2.3 Traffic analysis in wireless networks.

usually implement prevention mechanisms rather than detection ones by assuming the existence of passive attacks. For example, encrypting data traffic, sending dumb traffic between real data transmissions, etc.

2.1.2 Active Attacks

Active attacks have direct impact on a legitimate system. There can be modification of the original data stream, injecting false data stream, draining system resources, and several other types of attacks. Generally speaking, active attacks fall into four categories, *masquerade, replay, modification*, and *denial of service* (DoS).

Masquerade: An attacker pretends to be some other entity, usually an authorized user. As illustrated in Figure 2.4, the attacker attempts to establish a connection with user B and pretends to be user A.

Replay: An attacker first captures a message, whether encrypted or not, then replays this message to its designated receiver. As illustrated in Figure 2.5, the attacker first captures the message sent from user A to user B. Later, the attacker replays the message to user B. Capturing the original message in the first step usually involves passive attacks such as eavesdropping.

Modification: An attacker first intercepts and modifies the original message, then forwards the modified message to the legitimate receiver. As illustrated in Figure 2.6, the attacker first intercepts the message sent from user A to user B. Before user B receives

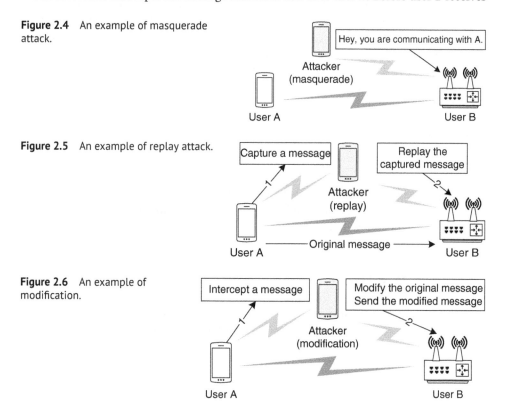

Figure 2.4 An example of masquerade attack.

Figure 2.5 An example of replay attack.

Figure 2.6 An example of modification.

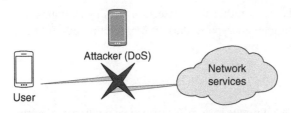

Figure 2.7 An example of DoS.

the original message, the attacker modifies the message and forwards it to user B. Ideally, the attacker should also block user B from receiving the original message from user A, otherwise user B would usually discard the modified message sent from the attacker.

Denial of service: A DoS attack makes system resource unavailable to authorized users. As illustrated in Figure 2.7, a user cannot establish a radio connection for network services because a DoS attack has been lunched. A typical example of DoS attack in the early days of mobile phone system is flooding a legitimate user with constant calls and/or text messages so that the user cannot establish connections with other intended users. Spam text messages still exist in today's mobile phone systems, where a user's device may be paralyzed by tens of thousands of text messages incoming constantly.

Discussion: *Is hacking a Wi-Fi password a passive attack or an active attack? (The attack stops once the password is retrieved.)*

Technically, hacking a Wi-Fi password does not trigger an alarm to the system since the process can be done without interrupting the user or the system. However, eavesdropping or traffic analysis alone is not effective to retrieve the password. The attacker must try some possible passwords with the system. In fact, the attacker is masquerading an authorized user during the process of connection establishment. Moreover, if the attacker launches exhaustive search by trying all possible passwords in high frequency, then the system (i.e. the access point in a Wi-Fi network) resource may be mostly occupied by the attacker, thus causing a DoS attack to intended users. Of course many more attacks, both passive and active, can be launched once the password is revealed to the attacker.

Until now, readers should realize that security attacks are more than hacking some passwords. Many security attacks can be launched without hacking users' passwords. Some security services do not require passwords at all, for example, data integrity.

2.2 Security Services

Security services are the features in a system design against possible security attacks. A wireless system designer and/or provider must evaluate security targets of users and forthcoming threats of a particular system. Corresponding security services are then chosen and implemented. The National Institute of Standards and Technology (NIST) computer security handbook [9] introduces three key security services not only to wireless security, but also to general computer security: *confidentiality, integrity*, and *availability*. In addition, typical security services also include *access control, authentication*, and *non-repudiation*.

2.2.1 Access Control

This service controls access from authorized users to resources in the system. There may be varying levels of access and control, i.e. an authorized user may not have access to all resources in the system. As illustrated in Figure 2.8, user A has access to some resources in the system where user B to some other resources. For example, in a home Wi-Fi setting, the owner may have access to all system resources (e.g. full bandwidth, in-home network attached storage, etc.), whereas a guest only has access to limited resources (e.g. restricted bandwidth, amount of data, etc.).

2.2.2 Authentication

Authentication is a service that verifies the identities of entities in a system. The entities include users connected to the host system and the host system itself. As illustrated in Figure 2.9, user A and user B are two entities connected to the host system. Authentication can be provided between user A and the host system, between user B and the host system, as well as between user A and user B. This example illustrates the concept of mutual authentication. In some cases, only one-way authentication is provided. For example, in

Figure 2.8 Illustration of access control.

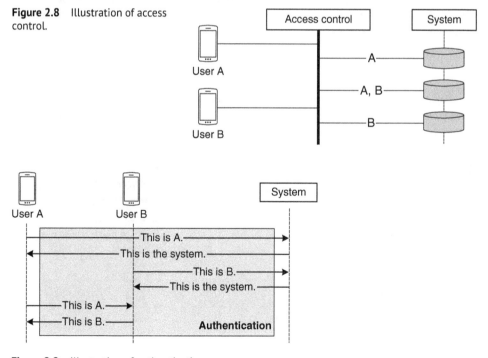

Figure 2.9 Illustration of authentication.

Wi-Fi WEP, only mobile stations are authenticated to the access point. Authentication is mostly provided during the initiation of a connection. It can also be provided during ongoing interactions. Attacks to authentication are usually active, for example, the attacks can be fabrication, masquerade, replay, session hijacking, etc.

Identification is provided in some systems separately as entity authentication. It is the procedure to verify the identity of an entity in a protocol through corroborative evidence. Identification is usually in real-time and associated with access control. Identification must guarantee that honest entities can achieve authentication, where an attacker cannot impersonate a legitimate entity with the observed identification exchanges.

2.2.3 Confidentiality

Confidentiality ensures that information is accessible only to authorized entities. Confidentiality is provided by encryption in many wireless systems. As illustrated in Figure 2.10, data from user A to the host system is first encrypted, then being sent through the network. At the receiver side, the host system needs to decrypt the incoming data. Similarly, confidentiality can be applied to data transmission between user A and user B. To achieve confidentiality, an agreement on the encryption mechanisms and some credentials must be made between the two entities before data transmission. Confidentiality can be broadly applied to all possible data in a system. Or, it can be applied to certain fields or parts of the data. For example, in a Wi-Fi network, confidentiality is provided to data traffic. Nonetheless, some information exchanged during the initial authentication process is sent in clear text. Attacks to confidentiality are usually passive. For example, interception leads to release of contents, and interception leads to traffic analysis.

2.2.4 Integrity

Integrity is a security service also known as *data integrity*. It maintains the accuracy and completeness of data over its entire life cycle. This means that data cannot be modified, reordered, inserted, delayed, or changed in any other way by unauthorized entities.

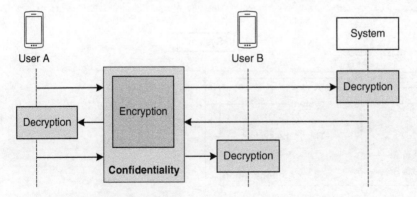

Figure 2.10 Illustration of confidentiality.

Figure 2.11   Illustration of integrity.

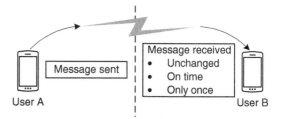

As shown in Figure 2.11, once a message is sent from user A in a transmission, integrity is provided if the receiver (i.e. user B) receives the intended message unchanged, on time, and only once. Attacks to integrity are active. For example, modification and alteration of the original message, and replay of a previous message, etc.

2.2.5   Non-repudiation

Non-repudiation provides proof of the data origin. In other words, non-repudiation guarantees that the sender cannot deny a transmission or its contents. Figure 2.12 illustrates non-repudiation to prove data source from user A. Similarly, non-repudiation service can be applied to the receiver as well. Therefore, there is either source non-repudiation or destination non-repudiation.

The two services *integrity* and *non-repudiation*, while having different security purposes, are often provided together, known as *message authentication*. For example, the *proof* given in Figure 2.12 can be used for both data integrity and source of sender simultaneously.

2.2.6   Availability

Availability is a service that makes information available to authorized parties when needed. In fact, it is an important aspect of reliability and system design rather than a typical security service. However, in addition to natural failures, deliberate attempts to deny access to data and service in information systems may compromise availability. The DoS attack is one example that may fail a system in terms of availability.

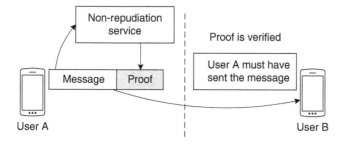

Figure 2.12   Illustration of source non-repudiation.

Discussion: *What are the targets of hacking the password of a home Wi-Fi network? (To keep the problem simple, the attack stops once the password is retrieved.)*

To answer this question, readers need to identify security services required in the system. This particular system is protected by a password. A user who knows the password can access the system at any time. Unauthorized users cannot get access because the password is not known to them. If an attacker is able to hack the password, then the system will grant access to the attacker.

First of all, confidentiality is provided to the Wi-Fi password. Although the confidentiality is not achieved by encryption, a user keeps it as a secret by not sharing to the public. Hacking the Wi-Fi password obviously violates confidentiality of the system. Therefore confidentiality is one security service needed to prevent this attack.

Second, what about privacy? Later, readers may realize that privacy has overlap with confidentiality in security services. Nonetheless, privacy is still different from confidentiality. Revealing you Wi-Fi password to an attacker is a bad thing of losing confidentiality. What's worse is that if the password contains personal information, for example, birthday, bank account password, etc. Hacking such Wi-Fi password violates a user's privacy. Therefore, privacy is not a direct target of this attack. However, it may be compromised if the password is not properly chosen.

Third, is integrity violated? In fact, hacking your Wi-Fi password does not violate its integrity. The password still remains the same even if the password is disclosed to the attacker. If the attacker is able to access the system with a modified password without altering system configurations, then the integrity of the original password is violated. But the problem would be more on the algorithm that uses the password for authentication. This part will be covered later in this book.

Finally, availability of the system is made for authorized users, i.e. users who know the password. Disclosing the password to an attacker does not violate system availability. In other words, authorized users can still type in the password and get access to it.

Security services may have overlapped features. For example, confidentiality and privacy have overlap since both require some information being restricted from unauthorized users. Authenticity and integrity have overlap since data integrity involves message authenticity. A legitimate wireless system may not deploy all the necessary security services. Some of the unprotected security services are caused by improper designs. For example, the wired equivalent privacy used in Wi-Fi does not provide access point authenticity to mobile stations. The others are caused by limited resources that are reserved for more important security targets. For example, data integrity is not provided in the Global System for Mobile Communication (GSM) because real-time phone conversation can tolerate some loss of data integrity. More importantly, there are not enough resources to provide real-time data integrity in GSM.

2.3 Security Mechanisms

Security mechanisms are methods to achieve security services in a system. X.800 defines a list of popular security mechanisms: encipherment, authentication, access control, digital signature, data integrity, traffic padding and routing control, and notarization.

2.3.1 Encipherment

Encipherment can provide confidentiality for either data or traffic flow information and can play a part in or complement a number of other security mechanisms. Encipherment algorithms may be *reversible* or *irreversible*.

Reversible encipherment algorithms are also known as encryption algorithms. There are two general classifications of encipherment algorithms, *symmetric* and *asymmetric* ones. Symmetric encipherment algorithms are also known as *secret-key* encipherment algorithms. A secret-key is pre-shared between the two communication entities for both encryption and decryption. Either communication entity is able to encipher or decipher the same data and produce the same result, thus the name symmetric. Asymmetric encipherment algorithms are also known as public-key encipherment algorithms. Despite its name, each communication entity has a set of two keys, *public key* and *private key*, as opposed to the single private key in symmetric algorithms. A message that is enciphered using the public key can be deciphered using the corresponding private key, and vice versa. That is to say, knowledge of the encipherment key does not imply knowledge of the decipherment key, thus the encipherment is asymmetric.

Irreversible encipherment algorithms may or may not use a key. If a key is used, this key may be public or secret. Irreversible encipherment algorithms cannot be used for encryption because the original data cannot be recovered. They are often applied to other security services, for example, to generate message authentication codes for data integrity.

2.3.2 Authentication

Authentication is applied to authenticate a message or a communication entity. If the mechanism does not succeed in authenticating the entity, this will result in rejection or termination of the connection and may also cause an entry in the security audit trail and/or a report to a security management center. There are different techniques to be applied to authentication exchanges. One of them is to use *authentication information*, such as a password supplied by a sending entity and checked by the receiving entity. Some authentication exchanges rely on *cryptographic techniques*, where successful cryptographic operations at both sides provide authentication. When cryptographic techniques are used, they may be combined with "handshaking" protocols to protect against replay attacks. Moreover, characteristics and/or possessions of the entity may also be used as authentication services, for example, biometrics of human users. The choices of authentication exchange techniques should depend upon requirements of each case. None of the technique provides a universal solution to all authentication services. In some cases, time stamping and synchronized clocks are needed to keep the authentication process fresh. Some cases require only unilateral authentication while others may require mutual authentications. Some cases require more than just authentication techniques. For example, if non-repudiation service is required during an authentication process, then a digital signature mechanism may be required in the process.

2.3.3 Access Control

Access control is applied to determine and enforce the access rights of the entity depending on the authenticated identity of an entity or information about the entity (such as membership in a known set of entities) or capabilities of the entity. If the entity attempts to use an unauthorized resource, or an authorized resource with an improper type of access, then the access control function will reject the attempt. The incident may be further reported as part of a security audit trail. Access control mechanisms may be based on use of one or more of the following:

- Access control information bases, where the access rights of peer entities are maintained. This information may be maintained by authorization centers or by the entity being accessed, and may be in the form of an access control list or matrix of hierarchical or distributed structure. This assumes that peer entity authentication has been assured.
- Authentication information such as passwords and possession of which is evidence of the accessing entity's authorization.
- Capabilities, possession and subsequent presentation of which is evidence of the right to access the entity or resource defined by the capability. (Note: a capability should be unforeseeable and should be conveyed in a trusted manner.)
- Security labels, which when associated with an entity may be used to grant or deny access, usually according to a security policy.
- Time of attempted access, route of attempted access, and duration of access.

An access control mechanism may be applied at either end of a communication association and/or at any intermediate point. Access controls involved at the origin or any intermediate point are used to determine whether the sender is authorized to communicate with the recipient and/or to use the required communication resources. The requirements of the peer level access control mechanisms at the destination side of a connectionless data transmission must be known a priori at the origin, and must be recorded in the security management information base.

2.3.4 Digital Signature

Digital signature is applied to provide certificate of the identity of the origin. A digital signature mechanism defines two procedures, *signing a data unit* and *verifying a signed data unit*.

- **Signing a data unit** uses information that is private (i.e. unique and confidential) to the signer. This process involves either an encipherment of the data unit or the production of a cryptographic check-value of the data unit, using the signer's private information as a private key. This process can be successfully performed by the signer only.

- **Verifying a signed data unit** uses procedures and information that are publicly available but from which private information of the signer cannot be deduced. This process involves using the public procedures and information to determine whether the signature was produced with the signer's private information. This process can be successfully performed by any entity who has the public information of the signer.

The essential characteristic of the signature mechanism is that the signature can only be produced using the private information of the signer. Therefore, when the signature is verified, it can subsequently be proven to any entity at any time that only the original signer could have produced the signature, therefore, the identity of the sender is verified. In addition to the unique private information of the signer, signing a data unit may also use time stamping and synchronized clocks to set a life span.

2.3.5 Data Integrity

Data integrity is applied to provide integrity of the data received or accessed. There are two aspects of data integrity: (i) the integrity of a single data unit or field; and (ii) the integrity of a stream of data units or fields. Those two types of integrity services require different mechanisms. However, provision of the first type of integrity is a prerequisite of the second type. Determining the integrity of a single data unit involves two processes, one at the sending entity and one at the receiving entity.

- At the *sending entity*, it appends to the sending data unit a value that is generated from the data itself. This value may be supplementary information such as a block check code or a cryptographic check value and may itself be enciphered.
- At the *receiving entity*, it generates a corresponding value and compares it with the received one to determine whether the data maintains integrity in transit.

Data integrity mechanism alone is not enough to protect the system against the replay of a single data unit. In a layered network architecture, detection of manipulation may lead to a recovery action (for example, via retransmissions or error correction) at that or a higher layer. For connection-oriented data transfer, protecting the integrity of a sequence of data units (i.e. protecting against mis-ordering, losing, replaying and inserting or modifying data) requires additionally some form of explicit ordering such as sequence numbering, time stamping, or cryptographic chaining. For connectionless data transmission, time stamping may be used to provide a limited form of protection against replay of individual data units.

2.3.6 Traffic Padding and Routing Control

Traffic padding and routing control can be used to provide various levels of protection against traffic analysis. Traffic padding can be effective only if the traffic padding is protected by a confidentiality service. Routing control dynamically chooses routes that use only physically secure subnetworks, relays, or links. Some routing control allows end-systems to instruct the network service provider to establish a connection via a different route, on detection of persistent manipulation attacks. Security policy can be set to forbid data carrying certain security labels to pass through certain subnetworks, relays, or links. The initiator of a connection (or the sender of a data unit) may request to avoid specific sub-networks, links or relays.

2.3.7 Notarization

Notarization can be used to assure data integrity, origin, time, and destination about data communicated between two or more entities. The assurance is provided by a third-party notary, which is trusted by the communicating entities, and which holds the necessary information to provide the required assurance in a verifiable manner. Each instance of communications may use digital signature, encipherment, and integrity mechanisms as appropriate to the service being provided by the notary. When such a notarization mechanism is invoked, the data is communicated between the communicating entities via the protected instances of communications and the notary.

2.4 Other Security Concepts

2.4.1 Levels of Impact

In system design and implementation, what security features should be included? Which security mechanisms should be used to achieve those services? Is it a good idea to provide all security features using the most advanced and perhaps most complicated mechanisms? It depends. A regular house does not need "24/7" armed on-site guards or bullet-proof windows. The level of security to be provided to a communication system shall depend on potential impact of a security breach involving in that particular system. NIST FIPS Publication (PUB) 199 [10] establishes three levels of potential impact, i.e. *low, moderate,* and *high*.

Low level of impact indicates a limited adverse effect on organizational operations, assets, or individuals. Adverse effects on individuals may include, but not limited to, loss of privacy to which individuals are entitled under law. A limited adverse effect means that, for example, the loss of confidentiality, integrity, or availability might: (i) cause a degradation in mission capability to an extent and duration that the organization is able to perform its primary functions, but the effectiveness of the functions is noticeably reduced; (ii) result in minor damage to organizational assets; (iii) result in minor financial loss; or (iv) result in minor harm to individuals.

Moderate level of impact indicates a serious adverse effect on organizational operations, assets, or individuals. A serious adverse effect means that, for example, the loss might: (i) cause a significant degradation in mission capability to an extent and duration that the organization is able to perform its primary functions, but the effectiveness of the functions is significantly reduced; (ii) result in significant damage to organizational assets; (iii) result in significant financial loss; or (iv) result in significant harm to individuals that does not involve loss of life or serious, life-threatening injuries.

High level of impact indicates a severe or catastrophic adverse effect on organizational operations, organizational assets, or individuals. A severe or catastrophic adverse effect means that, for example, the loss might: i) cause a severe degradation in or loss of mission capability to an extent and duration that the organization is not able to perform one or more of its primary functions; ii) result in major damage to organizational assets; iii) result in major financial loss; or iv) result in severe or catastrophic harm to individuals involving loss of life or serious, life-threatening injuries.

NIST Special Publication (SP) 800-53 [9] provides recommendations for minimum management, operational, and technical security controls for information systems based on the FIPS PUB 199 impact categories. The recommendations should be helpful to organizations in identifying controls that are needed to protect system implementations in general.

2.4.2 Cryptographic Protocols

A cryptographic protocol is a sequence of steps precisely specifying the actions required of two or more entities to achieve a specific security objective. Primitives used in cryptographic protocols are the pieces utilized to build the protocol and include encryption algorithms, hash functions, digital signature, random number generators, and other algorithms and functions. Common cryptographic protocols are used for authentication and key establishment.

An **authentication protocol** is a scheme based on the use of cryptographic algorithms designed to authenticate the identity of entities. The cryptographic algorithms may include *symmetric encryption*, *asymmetric encryption*, *data integrity*, and others. Depending on the security requirements, time stamps and/or nonces may also be applied to an authentication protocol.

A **key establishment protocol** is a secure process by which a shared secret key becomes available to two or more parties for subsequent cryptographic use. Key establishment has two ways:

- *Key distribution*: one party chooses the secret key and the secret key is securely transported to the other parties.
- *Key agreement*: two or more parties jointly establish a secret key by communicating over a public channel, e.g. Diffie–Hellman key exchange.

Third-party entities may be involved in some key establishment protocols. For example, a trusted authority, a trusted third party or a key distribution center. Those third-party entities are usually trusted by all entities in the system. They will be involved in all or some of the roles, including key generation, key distribution, certificate generation and verification, etc.

Key management is a set of processes and mechanisms that support key establishment and maintenance of ongoing cryptographic keys in a cryptosystem. Key management deals with new key generations, updates, exchange, storage, and replacement. Key management also deals with trusted third parties, such as assigning roles.

2.5 Summary

This chapter gives an overview on the security concepts that will be used in the rest of this book for security in wireless communication networks. These concepts include security attacks, security services, and security mechanisms.

3

Mathematical Background

Some mathematical concepts, such as finite fields and number theory, are important in developing cryptographic algorithms. Related mathematical background is critical for one to understand cryptography such as Advanced Encryption Standard and public key cryptographic systems. This chapter introduces the basic mathematical background for better understanding cryptography that is applied in wireless security. Section 3.1 gives an overview on basic concepts in modern algebra and number theory. Section 3.2 introduces modular arithmetic and divisors. Section 3.3 studies finite fields. Section 3.4 overviews polynomial arithmetic, followed by Fermat's little theorem, Euler's totient function, Euler's theory in Section 3.5, primality testing in Section 3.6, Chinese remainder theorem in Section 3.7, and discrete logarithm in Section 3.8.

3.1 Basic Concepts in Modern Algebra and Number Theory

In mathematics, number theory is the study of the set of integers and integer-valued functions. Groups, rings, and fields are the basic elements in modern algebra. A finite field is a field that contains a finite number of elements. Finite fields in modern algebra and prime numbers in number theory are two of the most important mathematical concepts applied in modern cryptography.

3.1.1 Group

A *group* denoted as $\{G, \cdot\}$ is an algebraic structure consisting of a set of elements (i.e. $G = \{a, b, c, \dots\}$) together with an operation "$\cdot$" on the elements in the set G that satisfying the four group axioms below this paragraph. The elements can be numbers, polynomials, etc. The operator "$\cdot$" is generic and can refer to any arithmetic operations, such as addition, multiplication, etc., or abstracted mathematical operations. Those four *group axioms*, including *closure*, *associativity*, *identity*, and *invertibility*, are described as follows:

- *Closure*: for any two elements $a, b \in G$, $a \cdot b \in G$.
- *Associativity*: for any three elements $a, b, c \in G$, $(a \cdot b) \cdot c = a \cdot (b \cdot c)$.
- *Identity*: there exists an element $e \in G$, for any element $a \in G$, $e \cdot a = a \cdot e = a$.
- *Invertibility*: for any element $a \in G$, there exists an element $a^{-1} \in G$ such that $a \cdot a^{-1} = e$.

Security in Wireless Communication Networks, First Edition. Yi Qian, Feng Ye, and Hsiao-Hwa Chen.
© 2022 John Wiley & Sons Ltd. Published 2022 by John Wiley & Sons Ltd.
Companion website: www.wiley.com/go/qian/sec51

Example: $\{Z, +\}$ is a group, given Z as the set of integers (i.e., $\ldots - 3, -2, -1, 0, 1, 2, 3 \ldots$), and "+" as the arithmetic addition. A test with group axioms is shown as follows:

- For any integers $a, b \in Z$, $a + b \in Z$. Therefore $\{Z, +\}$ satisfies closure.
- For any integers $a, b, c \in Z$, $(a + b) + c = a + (b + c)$. Therefore $\{Z, +\}$ satisfies associativity.
- For any integer $a \in Z$, it exists $0 \in Z$ such that $0 + a = a + 0 = a$. Thus $e = 0 \in Z$ is the identity element. Therefore $\{Z, +\}$ satisfies identity.
- For any integer $a \in Z$, there exists an integer $b = -a \in Z$ such that $a + b = b + a = 0 = e$. Thus there exists an inverse element $a^{-1} = -a$ for any $a \in Z$ such that $a \cdot a^{-1} = e$. Therefore $\{Z, +\}$ satisfies invertibility.

All the four group axioms are true. It can conclude that $\{Z, +\}$ is a group.

Besides group axioms, there are several other concepts related to groups. In the following, we define an *abelian group* and a *cyclic group*.

3.1.1.1 Abelian Group

An abelian group is a group $\{G, \cdot\}$ that satisfies an additional *commutativity* axioms. *Commutativity* axiom is defined as follows:

- *Commutativity*: $a \cdot b = b \cdot a, \forall a, b \in G$.

Example: Is $(Z, +)$ an abelian group?
One only needs to check if "+" satisfies commutativity due to the fact that $(Z, +)$ is a group. For any integers $a, b \in Z$, $a + b = b + a$, thus the operation "+" satisfies commutativity. Therefore, $(Z, +)$ is an abelian group.

Example: Is $(R, \times)$ an abelian group, where R is the set of real numbers, "$\times$" is the arithmetic multiplication?
It is easy to show that for any $a, b \in R$, $a \times b = b \times a$, thus operation "$\times$" satisfies commutativity. However, $(R, \times)$ is still not an abelian group because it is not a group at all. $(R, \times)$ does not have an identity element e due to element $0 \in R$. Without an identity element, $(R, \times)$ does not satisfy invertibility as well. Therefore, although multiplication satisfies commutativity, $(R, \times)$ is not an abelian group. It is worth noting that the set of real number excluding "0" (i.e. $R' = R \backslash \{0\}$) together with multiplication operator, i.e. $(R', \times)$ forms an abelian group.
  Open question: What is the identity element for $(R', \times)$?

3.1.1.2 Cyclic Group

A cyclic group requires that every element in G is a power of some fixed element $a \in G$. That is to say, for a cyclic group $(G, \cdot)$, any $b \in G$, $b = a^k$ for some integer k. The exponentiation is defined as repeated application of operator "$\cdot$." For example, $a^3 = a \cdot a \cdot a$ and $a^{-3} = a^{-1} \cdot a^{-1} \cdot a^{-1}$, where $a \cdot a^{-1} = a^0 = e \in G$. Note that a cyclic group must satisfy all group axioms, including identity and invertibility.

Example: $(\mathbf{Z}, +)$ is a cyclic group.
There exists an element $a \in \mathbf{Z}$ such that $a^k = a + a + \cdots + a$. Readers may have already found that $a = 1$ satisfies the power operation in $(\mathbf{Z}, +)$. That is to say, for any $n \in \mathbf{Z}$, it calculates as $n = 1^n = 1 + 1 + \cdots + 1$. Therefore $(\mathbf{Z}, +)$ is a cyclic group with $a = 1$.

Note: *A cyclic group is an abelian group*. Because for any $x = a^m, y = a^n$, where $m, n \in \mathbf{Z}$, it satisfies that $x \cdot y = a^m \cdot a^n = a^{m+n} = a^n \cdot a^m = y \cdot x$. Which indicates that commutativity is always satisfied for a cyclic group. Therefore, a cyclic group is an abelian group.

3.1.2 Ring

A *ring* consists of a set R with two operations "+" and "×," and satisfies the three axioms in the following.

(1) $(R, +)$ forms an abelian group.
(2) $(R, \times)$ satisfies conditions of closure and associativity.
(3) Multiplication is distributive with respect to addition, i.e.

$$a \times (b + c) = (a \times b) + (a \times c), \forall a, b, c \in R, \text{and}$$

$$(b + c) \times a = (b \times a) + (c \times a), \forall a, b, c \in R.$$

Example: Integer set $\mathbf{Z}$ with arithmetic operations addition "+" and multiplication "×" forms a ring.
It has been shown that $(\mathbf{Z}, +)$ is an abelian group previously. For any integers $a, b,$ and c, it is multiplication associative, where $(a \times b) \times c = a \times (b \times c)$. Integer set $\mathbf{Z}$ satisfies closure under multiplication. Moreover, any integers $a, b,$ and $c \in \mathbf{Z}$ satisfy multiplicative distribution with respect to addition. Therefore, $\mathbf{Z}$ with operations "+" and "×" forms a ring.

Two additional concepts based on ring are *commutative ring* and *integral domain*, described as follows:

- *Commutative ring*: if multiplication operation is commutative, i.e. for all $a, b \in R, a \times b = b \times a$, then R forms a *commutative ring*.
- *Integral domain*: if multiplication operation has an identity and no zero divisors, then R forms an *integral domain*.

3.1.3 Field

A *field* consists of a set F that forms a nonzero commutative division ring. A ring only requires R to be multiplication associative and has multiplication identity. In addition to be a ring, F is a field if nonzero elements form an abelian group under multiplication. Specifically, for a set F, the conditions are as follows:

(1) $\{F, +\times\}$ forms a ring.
(2) $(F \backslash \{0\}, \times)$ forms an abelian group.

Addition, subtraction, multiplication, and division can be operated in a field without leaving the set F. Division "/" is defined with the following rule: $a/b = ab^{-1}$.

Example of fields: rational numbers, real numbers, and complex numbers with arithmetic operations addition "+" and multiplication "×" are fields.

Question: Is the set of integers **Z** with "+" and "×" a field?
No, because not every element of the set has a multiplicative inverse. In fact, only the elements 1 and −1 have multiplicative inverses in the integers.

3.2 Prime Numbers, Modular Arithmetic, and Divisors

3.2.1 Prime Numbers

Prime numbers are in the central concern to number theory. A *prime number* is a positive integer greater than 1 that has no positive divisors other than 1 and itself.

List of prime number less than 200:
2 3 5 7 11 13 17 19 23 29 31 37 41 43 47 53 59 61 67 71 73 79 83 89 97 101 103 107 109 113 127 131 137 139 149 151 157 163 167 173 179 181 191 193 197 199.

Prime Factorization: A positive integer n can be written as a product of other positive integers, for example, $n = a \times b \times c$. *Factoring a number is relatively hard compared to multiplying the factors together to generate the number. That is to say, it is easy to compute n from a, b, and c. However, it is hard to find a, b, and c given n.* The *prime factorization* of a positive integer n is to write n as a product of prime numbers, such that:

$$n = \prod_{p \in \mathbf{P}} p^{a_p}, \tag{3.1}$$

where **P** is the set of all prime numbers, and a_p is some positive integer.

Example: $91 = 7^1 \times 13^1$; $3600 = 2^4 \times 3^2 \times 5^2$.

3.2.2 Modular Arithmetic

For the set of integers, define modulo operation "$a \bmod n$" as the remainder when the integer a is divided by a positive integer n. Modulo operation is widely applied in cryptography systems. For a positive integer n, two integers a and b are said to be congruent modulo n, if $a \equiv b \bmod n$. The congruence relation is that $(a - b) = kn$ for some integer k. For example, $100 \equiv 34 \bmod 11$. In other words, a and b have the same remainder when divided by n. The same rule holds for negative a and b. For example, $100 \equiv -8 \bmod 11$, $-100 \equiv -1 \bmod 11$. Since we have $a = kn + b$, b is the residue of $a \bmod n$. In general, the smallest positive remainder is chosen as the residue, i.e. $0 \leq b \leq n - 1$. In this case, modular arithmetic operation is "clock arithmetic" such that a finite number of values are results and the values loop back from either end. The modular arithmetic operation is written as a mod $n = b$. For any integers a, b, c, and n, the modular arithmetic exhibits the following properties:

Property 1 $(a + b) \bmod n = [(a \bmod n) + (b \bmod n)] \bmod n$.

> **Example:** $a = 11$, $b = 9$ and $n = 7$.
> $(11 + 9) \bmod 7 = [(11 \bmod 7) + (9 \bmod 7)] \bmod 7 = (4 + 2) \bmod 7 = 6$.

Property 2 $(a - b) \bmod n = [(a \bmod n) - (b \bmod n)] \bmod n$.

> **Example:** $a = 11$, $b = 9$ and $n = 7$.
> $(11 - 9) \bmod 7 = [(11 \bmod 7) - (9 \bmod 7)] \bmod 7 = (4 - 2) \bmod 7 = 2$.

Property 3 $(a \times b) \bmod n = [(a \bmod n) \times (b \bmod n)] \bmod n$.

> **Example:** $a = 11$, $b = 9$ and $n = 7$.
> $(11 \times 9) \bmod 7 = [(11 \bmod 7) \times (9 \bmod 7)] \bmod 7 = (4 \times 2) \bmod 7 = 1$.

Property 4 If $(a + b) \bmod n = (a + c) \bmod n$, then $b \equiv c \bmod n$.

> **Example:** $a = 11$, $b = 9$, $c = 16$, and $n = 7$.
> $(11 + 9) \bmod 7 = (11 + 16) \bmod 7 = 6$, thus $9 \equiv 16 \bmod 7$.

Property 5 If $(a \times b) \bmod n = (a \times c) \bmod n$, then $b \equiv c \bmod n$, if and only if a is relatively prime to n, i.e. $\gcd(a, n) = 1$ *(to be illustrated in the following)*.

> **Example:** $a = 11$, $b = 9$, $c = 72$, and $n = 7$.
> $(11 \times 9) \bmod 7 = (11 \times 72) \bmod 7 = 1$, thus $9 \equiv 72 \bmod 7$; $\gcd(11, 7) = 1$.
> **Example:** $a = 7$, $b = 9$, $c = 10$, and $n = 7$.
> $(7 \times 9) \bmod 7 = (7 \times 10) \bmod 7 = 0$, **but** $9 \neq 10 \bmod 7$; $\gcd(7, 7) = 7$.

Modular arithmetic are normally applied within an integer group Z_n that includes integers from 0 to $n - 1$. For example, $3 + 3 = (3 + 3) \bmod 4 = 2$ in Z_4 and $3 \times 3 = (3 \times 3) \bmod 4 = 1$ in Z_4. The full tables of addition and multiplication in $Z_4 = \{0, 1, 2, 3\}$ are given in Figure 3.1.

3.2.3 Divisors and GCD

If the result of a modular arithmetic is zero, i.e. $a \equiv 0 \bmod b$ (for non-zero integers a and b), then b is defined as a *divisor* of a. It is also said that b divides integer a, denoted as $b|a$.

+	0	1	2	3
0	0	1	2	3
1	1	2	3	0
2	2	3	0	1
3	3	0	1	2

(a)

×	0	1	2	3
0	0	0	0	0
1	0	1	2	3
2	0	2	0	2
3	0	3	2	1

(b)

Figure 3.1 Modulo example of Z_4. (a) Addition modulo 4. (b) Multiplication modulo 4.

For example, all $1, 2, 3, 6$ are divisors of 6 since $1|6$, $2|6$, $3|6$, and $6|6$. Given two non-zero integers a and b, finding their *greatest common divisor* (gcd) is a common problem in number theory. For example, divisors of 60 are $1, 2, 3, 4, 5, 6, 10, 12, 20, 30, 60$; divisors of 24 are $1, 2, 3, 4, 8, 12, 24$. Thus $gcd(60, 24) = 12$. Similarly, $gcd(128, 24) = 8$.

Two integers a and b are *relatively prime* if and only if $gcd(a, b) = 1$. For example, $gcd(8, 15) = 1$ hence 8 and 15 are relatively prime. Readers should be aware that the definition of two numbers be relatively prime does not require either number to be a prime number. One way to find $gcd(a, b)$ is to compare their prime factorizations.

Example: Compute $gcd(18, 300)$.

- Find prime factorization forms of 18 and 300, such that

$$18 = 2^1 \times 3^2, \text{ and } 300 = 2^2 \times 3^1 \times 5^2.$$

- Find common prime factors, i.e. 2, 3, and 5.
- Find the least powers of the common prime factors, i.e. 1 for 2, 1 for 3, and 0 for 5.
- Compute $gcd(18, 300) = 2^1 \times 3^1 \times 5^0 = 6$.

However, finding prime factorization forms of an integer is a hard problem. Therefore finding the greatest common divisor of two integers needs a more efficient approach. The *Euclidean algorithm* is one efficient method to compute $gcd(a, b)$, as detailed in Algorithm 3.1. This algorithm is based on the fact that $gcd(a, b) = gcd(b, a \bmod b)$ where $a > b$.

Algorithm 3.1 Euclidean $gcd(a, b)$

Input: a and b, where $a > b$;
Output: $gcd(a, b)$;
    $A = a; B = b$;
    while $B \neq 0$ **do**
        $temp \leftarrow (A \bmod B)$;
        $A = B$;
        $B = temp$;
    end while
    $gcd(a, b) \leftarrow A$

Example: Use the Euclidean algorithm to compute $gcd(3003, 1440)$.

$$3003 = 2 \times 1440 + 123 \Rightarrow gcd(1440, 123)$$
$$1440 = 11 \times 123 + 87 \Rightarrow gcd(123, 87)$$
$$123 = 1 \times 87 + 36 \Rightarrow gcd(87, 36)$$
$$87 = 2 \times 36 + 15 \Rightarrow gcd(36, 15)$$
$$36 = 2 \times 15 + 6 \Rightarrow gcd(15, 6)$$
$$15 = 2 \times 6 + 3 \Rightarrow gcd(6, 3)$$
$$6 = 2 \times 3 + 0 \Rightarrow gcd(3, 0)$$
$$\text{Return } gcd(3003, 1440) = \boxed{3}.$$

3.2.4 Multiplicative Inverse

A *multiplicative inverse* is defined as follows: given a, b in a finite field, e.g. $MOD(p)$, if $a \times b = 1$ in $MOD(p)$, then b is a multiplicative inverse of a in $MOD(p)$. It is also denoted as $a \cdot b \equiv 1 \bmod p$, or $b = a^{-1} \bmod p$. Back to the example shown in Figure 3.4c, the multiplicative inverse of 1 in $MOD(2)$ is also 1.

Question: Does a multiplicative inverse always exist for any element a in $MOD(p)$? *No, a multiplicative inverse is not guaranteed for any given a. As shown in the example of MOD(2), element 0 does not have a multiplicative inverse. Only element 1 has a multiplicative inverse, which is 1 itself.*

How to determine if a multiplicative inverse b exists for an element a in $MOD(p)$? If it exists, how to find b? For example, does multiplicative inverse of 3 exist in $MOD(20)$? If so, what is it? In this example, it is easy to find that $3 \times 7 = 21 \equiv 1 \bmod 20$, thus the answer is 7. What if a is a very large number, e.g. a binary number in 512 bits? In theory, the answers can be found through exhaustive search given unlimited time and computing resources. Nonetheless, it is necessary to have a much more efficient way to determine if a multiplicative inverse exists and locates the exact value. Fortunately, finding the multiplicative inverse b for a in $MOD(p)$ is a simple process, by applying the *extended Euclidean Algorithm*. Note that the algorithm is based on the Euclidean algorithm, which computes remainder iteratively. The extended Euclidean algorithm adds a few more steps, as detailed in Algorithm 2. Earlier, the multiplicative inverse of 3 in $MOD(20)$ was found by trying $3 \times 7 \equiv 1 \bmod 20$. Figure 3.2a shows the process of finding 7 using the extended Euclidean algorithm. In this example, the extended Euclidean algorithm took three iterations to stop. In the final iteration, B_3 returns 1, which indicates that the multiplicative inverse exists for 3 in $MOD(20)$. B_2 in the final iteration is the multiplicative inverse. If the multiplicative inverse does not exist,

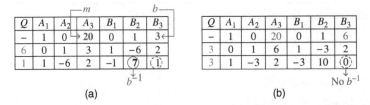

Q	A_1	A_2	A_3	B_1	B_2	B_3
–	1	0	20	0	1	3
6	0	1	3	1	–6	2
1	1	–6	2	–1	7	1

(a)

Q	A_1	A_2	A_3	B_1	B_2	B_3
–	1	0	20	0	1	6
3	0	1	6	1	–3	2
3	1	–3	2	–3	10	0

No b^{-1}

(b)

Figure 3.2 Applying the extended Euclidean algorithm to find multiplicative inverse. (a) Multiplicative inverse of 3 in MOD (20). (b) Multiplicative inverse of 6 in MOD (20).

for example, 6 in $MOD(20)$ has no multiplicative inverse. In this case, the final iteration of the extended Euclidean algorithm returns 0. Detailed process is given in Figure 3.2b.

Algorithm 3.2 Extended Euclidean (p, a)

Input: p and a
Output: a^{-1} in $MOD(p)$;
    $(A_1, A_2, A_3) = (1, 0, p); (B_1, B_2, B_3) = (0, 1, a);$
    while $B_3 \neq 0$ or 1 **do**
      $Q = \lfloor A_3/B_3 \rfloor;$
      $(T_1, T_2, T_3) = (A_1 - QB_1, A_2 - QB_2, A_3 - QB_3);$
      $(A_1, A_2, A_3) = (B_1, B_2, B_3);$
      $(B_1, B_2, B_3) = (T_1, T_2, T_3)$
    end while
    if $B_3 = 1$ **then**
      $gcd(p, a) \leftarrow B_3;$
      $a^{-1} \leftarrow B_2;$
    else if $B_3 = 0$ **then**
      $gcd(p, a) \leftarrow A_3;$
      a^{-1} does not exist;
    end if

3.3 Finite Field and Galois Field

Finite fields are used in a variety of cryptography algorithms such as the AES encryption algorithm and public key cryptography. A finite field is a field with finite number of elements. Please note that the fields of rational numbers, real numbers, and complex numbers all have infinite number of elements, therefore, not finite fields. A typical example of finite field is a ring of integers modulo p, or $MOD(p) = \{0, 1, \ldots, p-1\}$ where p is a prime number, which can perform addition, subtraction and multiplication. The results of addition and multiplication in $MOD(7)$ are shown in Figure 3.3.

For a finite field, the number of elements is called its *order*. A special case of Galois field is a finite field with an order $q = p^n$, where p is a prime number and n is some positive integer. A typical Galois field is commonly denoted as $GF(p^n)$, e.g. $GF(2^{64})$, $GF(3^{128})$, etc. Two of the most interested finite fields used in cryptography are $GF(p)$ and $GF(2^n)$. Known from its definition, $GF(p)$ is the set $\{0, 1, \ldots, p-1\}$ with arithmetic operations

+	0	1	2	3	4	5	6
0	0	1	2	3	4	5	6
1	1	2	3	4	5	6	0
2	2	3	4	5	6	0	1
3	3	4	5	6	0	1	2
4	4	5	6	0	1	2	3
5	5	6	0	1	2	3	4
6	6	0	1	2	3	4	5

(a)

×	0	1	2	3	4	5	6
0	0	0	0	0	0	0	0
1	0	1	2	3	4	5	6
2	0	2	4	6	1	3	5
3	0	3	6	2	5	1	4
4	0	4	1	5	2	6	3
5	0	5	3	1	6	4	2
6	0	6	5	4	3	2	1

(b)

Figure 3.3 Addition and multiplication in $MOD(7)$. (a) Addition in MOD (7). (b) Multiplication in MOD (7)

+	0	1
0	0	1
1	1	0

(a)

×	0	1
0	0	0
1	0	1

(b)

w	w^{-1}
0	–
1	1

(c)

Figure 3.4 Operations in $GF(2)$. (a) Addition in $GF(2)$. (b) Multiplication in $GF(2)$. (c) Multiplicative inverse in GF (2).

modulo prime number p. The order of $GF(p)$ is also p. Within $GF(p)$, the operations of addition, multiplication, subtraction, and division are performed without leaving the field. The simplest form of $GF(p)$ is $GF(2) = \{0, 1\}$. In $GF(2)$, both addition and subtraction are equivalent to exclusive-OR (XOR, $\oplus$) operation. For example, $1 + 1 = 1 - 1 = 1 \oplus 1 = 0$ in $GF(2)$. Multiplication in $GF(2)$ is equivalent to logical AND ($\wedge$) operation. For example, $1 \times 1 = 1 \wedge 1 = 0$ in $GF(2)$. The addition and multiplication operations are summarized in Figures 3.4a and 3.4b, respectively. *Multiplicative inverse* of all elements in $GF(2)$ is illustrated in Figure 3.4c. More discussion on multiplicative inverse will be given in Section 3.4.3. For other finite fields, addition, subtraction, and multiplication cannot be converted to simple XOR or AND operations.

3.4 Polynomial Arithmetic

Besides integer arithmetic, polynomial arithmetic is also widely applied in many cryptographic algorithms. A polynomial $f(x)$ is defined as:

$$f(x) = a_n x^n + a_{n-1} x^{n-1} + \cdots + a_1 x + a_0 = \sum_{i=0}^{n} a_i x^i,$$

where a_n are the coefficients in the polynomial, and x is the indeterminate.

3.4.1 Ordinary Polynomial Arithmetic

Polynomial arithmetic is to apply arithmetic operations to polynomials, for example, addition, subtraction, multiplication, and division. For two polynomials $f(x) = \sum_{i=0}^{n} a_i x^i$ and

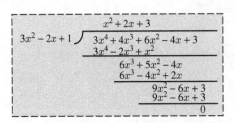

(a)

$$x^2 + 2x + 3$$
$$\times \quad 3x^2 - 2x + 1$$
$$x^2 + 2x + 3$$
$$-2x^3 - 4x^2 - 6x$$
$$3x^4 + 6x^3 + 9x^2$$
$$3x^4 + 4x^3 + 6x^2 - 4x + 3$$

(b)

Figure 3.5 Ordinary polynomial addition and multiplication. (a) $f(x) + g(x)$. (b) $f(x) \times g(x)$.

$$x^2 + 2x + 3$$
$$3x^2 - 2x + 1 \,\big)\, \overline{\,3x^4 + 4x^3 + 6x^2 - 4x + 3\,}$$
$$3x^4 - 2x^3 + x^2$$
$$\overline{\quad 6x^3 + 5x^2 - 4x}$$
$$6x^3 - 4x^2 + 2x$$
$$\overline{\quad 9x^2 - 6x + 3}$$
$$9x^2 - 6x + 3$$
$$\overline{\quad\quad 0}$$

Figure 3.6 An example of polynomial division in ordinary polynomial arithmetic.

$g(x) = \sum_{i=0}^{m} b_i x^i$, the ordinary polynomial addition is defined as follows:

$$f(x) + g(x) = \sum_{i=0}^{m} (a_i + b_i)x^i, \quad \text{for } m \geq n. \tag{3.2}$$

The polynomial multiplication is defined as follows:

$$f(x) \times g(x) = \sum_{i=0}^{n+m} c_i x^i, \tag{3.3}$$

where $c_i = \sum_{k=0}^{i} a_k b_{i-k}$. For example, given $f(x) = x^2 + 2x + 3$ and $g(x) = 3x^2 - 2x + 1$, the polynomial addition and polynomial multiplication are as follows:

$$\begin{cases} f(x) + g(x) = 4x^2 + 4, \\ f(x) \times g(x) = 3x^4 + 4x^3 + 6x^2 - 4x + 3. \end{cases}$$

Figure 3.5a and 3.5b shows the process of addition and multiplication in ordinary polynomial arithmetic.

Polynomial division is usually calculated using long division. Let $f(x) = 3x^4 + 4x^3 + 6x^2 - 4x + 3$ and $g(x) = 3x^2 - 2x + 1$, the polynomial division is calculated as follows:

$$\frac{f(x)}{g(x)} = \frac{3x^4 + 4x^3 + 6x^2 - 4x + 3}{3x^2 - 2x + 1} = x^2 + 2x + 3.$$

The polynomial division is illustrated in Figure 3.6.

3.4.2 Polynomial Arithmetic in Finite Fields

In Section 3.4.1, the ordinary polynomial arithmetic has been illustrated, i.e. the coefficients are in the set of integers. In cryptography, modular arithmetic is usually applied to coefficients in polynomial arithmetic, known as polynomial with coefficients in $GF(p)$. That is to say, for $f(x) = \sum_{i=0}^{n} a_i x^i$, the coefficients are in $GF(p)$, i.e. $a_i \in GF(p)$. The polynomial addition with coefficients in $GF(p)$ is defined as:

$$f(x) + g(x) = \sum_{i=0}^{m} [(a_i + b_i) \bmod p]x^i, \quad \text{for } m \geq n. \tag{3.4}$$

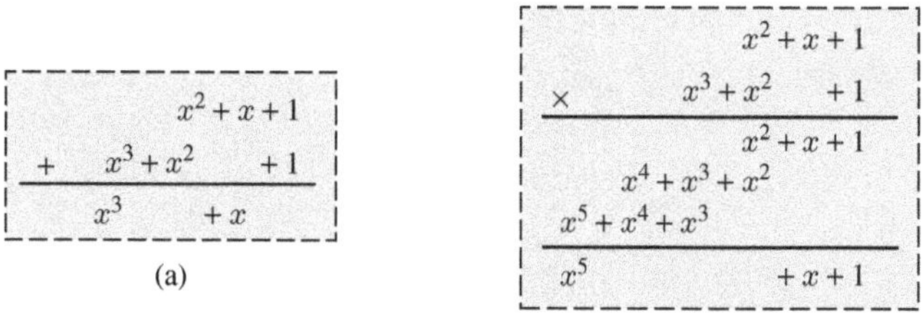

$$x^2 + x + 1$$
$$+ \quad x^3 + x^2 \qquad + 1$$
$$\overline{x^3 \qquad + x}$$

(a)

$$x^2 + x + 1$$
$$\times \quad x^3 + x^2 \qquad + 1$$
$$\overline{x^2 + x + 1}$$
$$x^4 + x^3 + x^2$$
$$x^5 + x^4 + x^3$$
$$\overline{x^5 \qquad\qquad + x + 1}$$

(b)

Figure 3.7 Ordinary polynomial addition and multiplication. (a) $f(x) + g(x)$ with coefficients in $GF(2)$. (b) $f(x) \times g(x)$ with coefficients in $GF(2)$.

Figure 3.8 $f(x)/g(x)$ with coefficients in $GF(2)$.

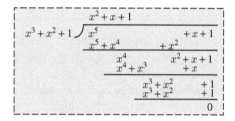

The polynomial multiplication with coefficients in $GF(p)$ is defined as:

$$f(x) \times g(x) = \sum_{i=0}^{n+m} c_i x^i, \tag{3.5}$$

where $c_i = \left(\sum_{k=0}^{i} a_k b_{i-k} \bmod p \right)$. For instance, let the polynomial arithmetic have coefficients in $GF(2)$, then the same coefficients used in the example in Section 3.4.1 cannot be applied. In this new example, let $f(x) = x^2 + x + 1$ and $g(x) = x^3 + x^2 + 1$, then the polynomial addition is computed as:

$$f(x) + g(x) = x^3 + (2 \bmod 2)x^2 + x + (2 \bmod 2) = x^3 + x.$$

Detail is shown in Figure 3.7a. The polynomial multiplication of $f(x) \times g(x)$ is:

$$f(x) \times g(x) = x^5 + (2 \bmod 2)x^4 + x + (2 \bmod 2)x^3 + (2 \bmod 2)x^2 + x + 1$$
$$= x^5 + x + 1.$$

Detail is shown in Figure 3.7b.

The polynomial division $f(x)/g(x)$ with coefficients in $GF(p)$ is to apply long division where the operations of coefficients are in $GF(p)$. For $f(x) = x^5 + x + 1$ and $g(x) = x^3 + x^2 + 1$, the polynomial division of $f(x)/g(x)$ in $GF(2)$ is computed as:

$$\frac{f(x)}{g(x)} = \frac{x^5 + x + 1}{x^3 + x^2 + 1} = x^2 + x + 1.$$

Detail of the calculation is shown in Figure 3.8.

3.4.3 Modular Polynomial Arithmetic

Polynomial can apply modular operations similar to integers. A polynomial $f(x)$ can be expressed as:

$$f(x) = q(x)c(x) + r(x),$$

where $r(x)$ can be interpreted as a polynomial residue from $f(x)$ divided by $c(x)$. By applying modular arithmetic, it is expressed as:

$$f(x) \bmod c(x) = r(x),$$

or

$$f(x) \equiv r(x) \bmod c(x).$$

Example: given $f(x) = 4x^3 + 6x^2 + 3x + 2$, $c(x) = x^2 + 1$, find $r(x)$.

$$4x^3 + 6x^2 + 5x + 8 = (4x + 6)(x^2 + 1) + (x + 2)$$

Thus $r(x) = x + 2$.

Similar to integer arithmetic, if residue $r(x) = 0$, then $c(x)$ divides $f(x)$, denoted as $c(x)|f(x)$. Equivalently speaking, $c(x)$ is a polynomial divisor of $f(x)$. For example, $c(x) = x + 1$ divides $f(x) = x^2 + 2x + 1$. If $c(x)$ is also a polynomial divisor of another polynomial $g(x)$, then $c(x)$ is a common divisor of $f(x)$ and $g(x)$. Now, if this particular polynomial $c(x)$ is the polynomial of greatest degree which divides both $f(x)$ and $g(x)$, then $c(x)$ is the greatest common divisor of $f(x)$ and $g(x)$, denoted as $c(x) = gcd(f(x), g(x))$. The Euclidean algorithm can be modified to find $c(x)$, as illustrated in Algorithm 3.3.

Algorithm 3.3 Euclidean $gcd(f(x), g(x))$

Input: $f(x)$ and $g(x)$
Output: $gcd(f(x), g(x))$;
  $A(x) = f(x)$; $B(x) = g(x)$;
  while $B(x) \neq 0$ **do**
    $temp(x) \leftarrow (A(x) \bmod B(x))$;
    $A(x) = B(x)$;
    $B(x) = temp(x)$;
  end while
  $gcd(f(x), g(x)) \leftarrow A(x)$

In some cryptographic algorithms, modular polynomial arithmetic computes in finite field $GF(2^n)$ whose degree is less than n, with coefficients modulo p (i.e. coefficients are in $GF(p)$). Readers should note that there are two moduli involved in a polynomial modular arithmetic, (i) a polynomial modulus that calculates the polynomial remainder, and (ii) an integer modulus that restricts coefficients to $GF(p)$.

Example: Given $f(x) = x^5 + x^3 + 1$ and $g(x) = x^3 + x^2 + 1$ with coefficients in $GF(2)$, find $f(x) \bmod g(x)$.

One can apply long division to find the polynomial quotient $q(x)$ and polynomial remainder $r(x)$ such that

$$f(x) = q(x) \cdot g(x) + r(x), \Leftrightarrow$$
$$x^5 + x^3 + 1 = (x^2 + x)(x^3 + x^2 + 1) + (x^2 + x + 1).$$

Figure 3.9 An example of polynomial modulus.

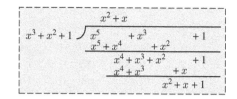

Thus $q(x) = x^2 + x$ and $r(x) = x^2 + x + 1$. Details of the calculation are shown in Figure 3.9. Note that all coefficients are in $GF(2)$, thus subtraction applied in the long division is the same as XOR operation.

Multiplicative inverse can be applied to polynomials over a finite field. For a polynomial $f(x)$, if a multiplicative inverse $g(x)$ exists in $GF(2^n)$, then $f(x) \cdot g(x) = 1$ in $GF(2^n)$, where $g(x) = f^{-1}(x)$. Similar to finding multiplicative inverse for integers, the extended Euclidean algorithm can be applied to find a polynomial multiplicative inverse $f^{-1}(x)$ in $GF(2^n)$. If a polynomial $g(x)$ has no divisor other than itself or 1, then it $g(x)$ is an *irreducible polynomial*, or *prime polynomial*. Arithmetic modulo an irreducible polynomial forms a finite field.

Example: $g(x) = x + 1$ is an irreducible polynomial in $GF(2)$.

Example: $g(x) = x^4 + 1$ is not an irreducible polynomial in $GF(2)$.
$$x^4 + 1 = (x + 1)(x^3 + x^2 + x + 1) = (x + 1)(x + 1)(x^2 + 1) = (x + 1)^4.$$

3.4.4 Computational Considerations

A common modular polynomial arithmetic used in cryptography is in $GF(2)$ for coefficients, thus the coefficients are either 0 or 1. For this reason, a polynomial $f(x)$ can be represented as a bit string with 0's and 1's. For example, in $GF(2^3)$, $(x^2 + 1)$ can be represented as $(101)_2$, $(x^2 + x + 1)$ can be represented as $(111)_2$. Polynomial addition can be computed as XOR function of strings. For example,

$$(x^2 + 1) + (x^2 + x + 1) = x \Leftrightarrow 101 \oplus 111 = (010)_2.$$

Polynomial multiplication is computed as shift and XOR functions. For example,

$$(x + 1) \cdot (x^2 + 1) = x \cdot (x^2 + 1) + 1 \cdot (x^2 + 1)$$
$$= x^3 + x^2 + x + 1$$
$$\Leftrightarrow 011 \cdot 101 = 101 \ll 1 \oplus 101 \ll 0 = 1010 \oplus 101 = (1111)_2.$$

Figure 3.10 shows the results of modular polynomial addition in $GF(2^3)$ over irreducible polynomial $f(x) = x^3 + x + 1$. Figure 3.11 shows the results of modular polynomial multiplication in the same finite field. For simplicity, the subscript $(\cdot)_2$ for binary strings is omitted in the illustration.

+	000	001	010	011	100	101	110	111
000	000	001	010	011	100	101	110	111
001	001	000	011	010	101	100	111	110
010	010	011	000	001	110	111	100	101
011	011	010	001	000	111	110	101	100
100	100	101	110	111	000	001	010	011
101	101	100	111	110	001	000	011	010
110	110	111	100	101	010	011	000	001
111	111	110	101	100	011	010	001	000

Figure 3.10 Polynomial addition modulo ($x^3 + x + 1$).

×	000	001	010	011	100	101	110	111
000	000	000	000	000	000	000	000	000
001	001	001	010	011	100	101	110	111
010	000	010	100	110	011	001	111	101
011	000	011	110	101	111	100	001	010
100	000	100	011	111	110	010	101	001
101	000	101	001	100	010	111	011	110
110	000	110	111	001	101	011	010	100
111	000	111	101	010	001	110	100	011

Figure 3.11 Polynomial multiplication modulo ($x^3 + x + 1$).

3.4.5 Generating a Finite Field with a Generator

A generator g may be applied to define a specific finite field F (e.g. $GF(2^n)$) of order q defined over an irreducible polynomial $f(x)$, i.e. $F = \{0, g^0, g, \dots, g^{q-2}\}$. The generator g can be found as $f(g) = 0$.

For example, given the finite field $GF(2^3)$ defined over the irreducible polynomial $f(x) = x^3 + x + 1$, generator g is found by setting

$$f(g) = 0 \Leftrightarrow g^3 + g + 1 = 0,$$
$$\Leftrightarrow g^3 = -g - 1 = g + 1 \quad \text{(the coefficients are in } GF(2)\text{)}.$$

Thus, g^3 is represented with g and g^0. For g^k where $k > 3$, they are generated as follows:

$$g^4 = g(g^3) = g(g + 1) = g^2 + g$$
$$g^5 = g(g^4) = g(g^2 + g) = g^3 + g^2 = g^2 + g + 1$$
$$g^6 = g(g^5) = g(g^2 + g + 1) = g^3 + g^2 + g = g^2 + 1$$
$$g^7 = g(g^6) = g(g^2 + 1) = g^3 + g = 1 = g^0$$
$$g^8 = g(g^7) = g(g^0) = g$$
$$\vdots$$
$$g^k = g^{k \bmod 7}, \text{ for any integer } k.$$

Therefore, generator g defines the entire finite field $GF(2^3)$ over $f(x) = x^3 + x + 1$. The power representation, polynomial and binary representation of generator for $GF(2^3)$ using $f(x) = x^3 + x + 1$, is shown in Table 3.1.

In general, for any finite field $GF(2^n)$, if its generator is g, then all the elements of the finite field can be found by calculating $g^k = g^{k \bmod (2^n - 1)}$ for any integer k.

Table 3.1 Generator for $GF(2^3)$ using $f(x) = x^3 + x + 1$.

Power of generator g	Polynomial	Binary
0	0	000
g^0	1	001
g^1	g	010
g^2	g^2	100
g^3	$g + 1$	011
g^4	$g^2 + g$	110
g^5	$g^2 + g + 1$	111
g^6	$g^2 + 1$	101

3.5 Fermat's Little Theorem, Euler's Totient Function, and Euler's Theorem

3.5.1 Fermat's Little Theorem

Fermat's little theorem is applied in some public key cryptography and primality testing. The theorem is described as follows.

Theorem 3.1 *Fermat's Little Theorem: Given p as a prime number, and a as an integer that is not divisible by p, i.e. GCD(a, p) = 1, Fermat's little theorem gives:*

$$a^{p-1} \equiv 1 \ mod \ p. \tag{3.6}$$

Proof: First create a sequence of numbers $(a, 2a, 3a, \dots, (p-1)a)$. Given $gcd(a, p) = 1$, if $m \cdot a \equiv n \cdot a \ mod \ p$, then we should have $m = n \ mod \ p$. Since the $(p-1)$ multipliers are distinct and nonzero, the $p-1$ multiples of $a \ mod \ p$ are distinct and nonzero. Therefore, multiply all the congruence together,

$$a \times 2a \times 3a \times \dots \times (p-1)a \equiv 1 \times 2 \times 3 \dots \times (p-1) \ mod \ p$$
$$\Rightarrow (p-1)! a^{p-1} \equiv (p-1)! \ mod \ p.$$

Divides $(p-1)!$ from both sides,

$$a^{p-1} \equiv 1 \ mod \ p.$$

That completes the proof. #

Example: Given $p = 7$ and $a = 2$, verify Fermat's little theorem. It can be seen that:

$$a^{p-1} \ mod \ p \Rightarrow 2^{7-1} \ mod \ 7 = 64 \ mod \ 7 = 1 \Rightarrow 2^{7-1} \equiv 1 \ mod \ 7.$$

Table 3.2 List of $\phi(n)$ for $1 \leq n \leq 30$.

n	1	2	3	4	5	6	7	8	9	10
$\phi(n)$	1	1	2	2	4	2	6	4	6	4
n	11	12	13	14	15	16	17	18	19	20
$\phi(n)$	10	4	12	6	8	8	16	6	18	8
n	21	22	23	24	25	26	27	28	29	30
$\phi(n)$	12	10	22	8	20	12	18	12	28	8

Multiply a to both sides of Eq. (3.6), Fermat's little theorem can be alternatively presented as follows:

$$a^p \equiv a \bmod p, \forall a \in \mathbb{Z}^+. \tag{3.7}$$

Note that for the original Fermat's little theorem in Eq. (3.6), a is relatively prime to p, while the alternatively form in Eq. (3.7) can be applied to any integers. For example, given $p = 6$ and $a = 4$ ($gcd(a, p) = 2$), it can be seen that $4^6 = 4096 = 6 \times 682 + 4 \equiv 4 \bmod 6$.

3.5.2 Euler Totient Function $\phi(n)$

Before giving the definition of Euler totient function, readers need to understand the concept of a complete set of remainders and a reduced set. When doing arithmetic modulo n, where n is a positive integer, the *complete set of remainders* is $R = \{0, \ldots, n - 1\}$. The reduced set $R' \subset R$ includes all the elements that are relatively prime to n. For example, given $n = 10$, $R = \{0, 1, 2, \ldots, 9\}$, and $R' = \{1, 3, 7, 9\}$. It is now ready to give the definition of Euler totient function. The *Euler totient function* $\phi(n)$ is defined as the number of elements in R', i.e. $\phi(n) = |R'|$. For example, $\phi(10) = 4$, since $R' = \{1, 3, 7, 9\}$ has four elements. Table 3.2 lists $\phi(n)$ for the first 30 positive integers.

What would be a good way to calculate $\phi(n)$? In general, one can list R' and count $|R'|$ to find $\phi(n)$. However, it is not efficient and not even applicable if n is a large number. On the other hand, a few techniques can be applied to compute $\phi(n)$, listed as follows:

(1) $\phi(p) = p - 1$, if p is a prime number;
(2) $\phi(mn) = \phi(m)\phi(n)$, if $gcd(m, n) = 1$;
(3) $\phi(p^k) = p^k - p^{k-1} = p^k(1 - 1/p)$, if p is a prime number and k is a positive integer.

Example: Find $\phi(37)$, $\phi(10)$, and $\phi(25)$.

$\phi(37) = 37 - 1 = 36,$

$\phi(10) = \phi(2 \times 5) = \phi(2)\phi(5) = (2 - 1)(5 - 1) = 4,$

$\phi(25) = \phi(5^2) = 5^2 - 5^{2-1} = 20.$

With the three techniques, it is able to derive a general formula to compute $\phi(n)$ efficiently. Recall the prime factorization of an integer, i.e. $n = \prod_{p_i \in \mathbf{P}} p_i^{k_i}$. Note that all p_i are relatively prime to each other, therefore $\phi(n)$ can be calculated as follows:

$$\phi(n) = \phi\left(\prod_{p_i \in \mathbf{P}} p_i^{k_i}\right) = \phi\left(p_1^{k_1}\right) \phi\left(p_2^{k_2}\right) \phi\left(p_3^{k_3}\right) \dots.$$

According to the three techniques mentioned earlier, $\phi(n)$ can be further calculated as follows:

$$\begin{aligned}
\phi(n) &= \phi\left(\prod_{p_i \in \mathbf{P}} p_i^{k_i}\right) \\
&= \phi\left(p_1^{k_1}\right) \phi\left(p_2^{k_2}\right) \dots \\
&= p_1^{k_1}\left(1 - \frac{1}{p_1}\right) p_2^{k_2}\left(1 - \frac{1}{p_2}\right) \dots \\
&= p_1^{k_1} p_1^{k_2} \dots \left(1 - \frac{1}{p_1}\right)\left(1 - \frac{1}{p_2}\right) \dots \\
&= n\left(1 - \frac{1}{p_1}\right)\left(1 - \frac{1}{p_2}\right) \dots .
\end{aligned} \tag{3.8}$$

However, in order to find $\phi(n)$ according to Eq. (3.8), prime factorization has to be available. Although finding all the prime divisors is a hard problem, it is less stressful than listing all elements in R'.

3.5.3 Euler's Theorem

If the Euler totient function is applied to Fermat's little theorem, it becomes Euler's theorem, stated as follows:

Theorem 3.2 *Euler's Theorem: If a and n are relatively prime, then*

$$a^{\phi(n)} \equiv 1 \bmod n. \tag{3.9}$$

Proof: Let $R' = \{x_1, x_2, \dots, x_{\phi(n)}\}$ be the set of reduced residues of $MOD(n)$, where $gcd(x_i, n) = 1$, $\forall x_i \in R'$. Thus $aR' = \{ax_1, ax_2, \dots, ax_{\phi(n)}\}$. Given $gcd(a, n) = 1$, if $ax_i \equiv ax_k \bmod n$, then $x_i = x_k$. Now, let

$$S = \{ax_1 \bmod n, ax_2 \bmod n, \dots, ax_{\phi(n)} \bmod n\}.$$

Since $x_1, x_2, \dots, x_{\phi(n)}$ are distinct and nonzero, S is a permutation of $\{x_1, x_2, \dots, x_{\phi(n)}\}$, which is R'. Therefore

$$\prod_{i=1}^{\phi(n)} x_i \equiv \prod_{i=1}^{\phi(n)} ax_i \bmod n$$

$$\Rightarrow \prod_{i=1}^{\phi(n)} x_i \equiv a^{\phi(n)} \prod_{i=1}^{\phi(n)} x_i \bmod n.$$

Divides $\prod_{i=1}^{\phi(n)} x_i$ from both sides,

$$a^{\phi(n)} \equiv 1 \text{ mod } n.$$

That completes the proof. #

Example: Given $a = 3$ and $n = 10$; $a = 2$ and $n = 11$, demonstrate Euler's theorem.
For $n = 10$, it has $\phi(n) = \phi(10) = 4$. Then it can be seen that

$$3^{\phi(10)} = 3^4 = 81 = 8 \times 10 + 1 \equiv 1 \text{ mod } 10.$$

For $n = 11$, it has $\phi(11) = 10$. Then it can be seen that

$$2^{\phi(11)} = 2^{10} = 1024 = 11 \times 93 + 1 \equiv 1 \text{ mod } 11.$$

Multiply a to both sides of Eq. (3.9), Euler's theorem can be relaxed as follows:

$$a^{\phi(n)+1} \equiv a \text{ mod } n. \tag{3.10}$$

Similar to the case with Fermat's little theorem, the original form of Euler's theorem requires that a be relatively prime to n, but the alternative form does not.

3.6 Primality Testing

Prime numbers play an important role in cryptography. However, given an integer n, how to verify the primality of it? If the number is relatively small, it is possible to have an exhaustive search, by dividing with all prime numbers that are smaller than $\sqrt{n}$. However, prime numbers used in cryptography are usually very large, e.g. 128-bit long. In that case, how to test its primality? Fortunately, prime numbers have a few properties, therefore statistical primality tests can be applied. Two properties of prime numbers are given as follows:

(1) If p is a prime number and a is a positive integer less than p, then $a^2 \text{ mod } p = 1$ if and only if either

$$a \text{ mod } p = 1,$$

or

$$a \text{ mod } p = -1 \equiv p - 1 \text{ mod } p.$$

Since

$$ab \text{ mod } p = (a \text{ mod } p)(b \text{ mod } p) \text{ mod } p,$$

The two possibilities can be rewritten into

$$a^2 \text{ mod } p = (a \text{ mod } p)^2 = 1.$$

(2) Let p be a prime number greater than 2, p can be written as $p = 2^k q + 1$ for some integer $k > 0$ and odd integer q. Let a be an integer such that $1 < a < p - 1$, then one of the two following conditions is true.

(a) $a^q \equiv 1 \bmod p$.

(b) There exists an integer j, for $0 < j < k - 1$, where $a^{2^j q} \equiv p - 1 \bmod p$.

The Miller–Rabin algorithm is a primality testing algorithm based on the two properties of prime numbers. The algorithm is shown in details in Algorithm 3.4.

Algorithm 3.4 Miller–Rabin Algorithm

Input: An odd integer n for primality testing;

Output: "Composite" if n is composite, otherwise "Probably Prime";

Find integer k, q where $k > 0$, q is odd, and $(n - 1) = 2^k q$;

$a \leftarrow$ Random integer, where $2 \le a \le n - 1$;

if $a^q \bmod n = 1$ **then**

Return "Probably Prime";

else

  for $j = 0$ to $k - 1$ **do**

    if $a^{2^j q} \bmod n = n - 1$ **then**

      Return "Probably Prime";

    end if

  end for

  Return "Composite";

end if

Example: Apply the Miller–Rabin algorithm to test the primality of $n = 37$. Given $n = 37$, it has $n - 1 = 36 = 2^2 \times 9 = 2^k q$, thus $k = 2$ and $q = 9$. Randomly choose $a = 3$, one can find $3^9 \bmod 37 = 36 = -1$, and continue the "For Loop," and computes $3^{2^0 9} \bmod 37 = 36$, which returns "Probably Prime."

Example: Apply the Miller–Rabin algorithm to test the primality of $n = 35$. Given $n = 35$, we have $n - 1 = 34 = 2^1 \times 17$, thus $k = 1$ and $q = 17$. Randomly choose $a = 3$, it will find $3^{17} \bmod 35 = 33$ that is neither 1 nor 34, thus the algorithm returns "Composite."

If the Miller–Rabin algorithm returns "composite" then the testing number is definitely composite. However, if the Miller–Rabin algorithm returns "probably prime" then the testing number is not guaranteed to be prime. One possibility is *pseudo-prime number*. A pseudo-prime number is a composite number that also satisfies the properties. For example, given a composite number $n = 221 = 13 \times 17$, then $n - 1 = 220 = 2^2 \times 55$, thus $k = 2$ and $q = 55$. If the Miller–Rabin algorithm starts with $a = 21$, then it computes $21^5 5 \bmod 221 = 200$, which returns "Probably Prime." The algorithm then computes $21^{2 \times 55} \bmod 221 = 220$, which also returns "Probably Prime." However, $221 = 13 \times 17$ is a composite number thus the primality testing fails to detect it with $a = 21$. However, it does not make the Miller–Rabin algorithm useless. The probability that the Miller–Rabin algorithm returns "Probably Prime" given a pseudo-prime number is less than $1/4$. In other words, the Miller–Rabin algorithm can still be trusted with more than 75% confidence is it returns "Probably Prime." In order to further reduce the probability of false positive output, a user may run the Miller–Rabin algorithm multiple times with different input a. If the Miller–Rabin algorithm is repeated with t different a values, then the probability of giving false positive result is reduced to 4^{-t}. Thus the probability of a correct primality testing result of n after t times is $1 - 4^{-t}$. For example, if running the Miller–Rabin algorithm 10 times with all results "Probably Prime," then the probability of getting a real prime is more than $1 - 4^{-10} > 0.99999$. That is to say, after testing for 10 times with different initials, if all the return results are "Probably Prime," one can trust that n is prime with 99.999% confidence.

Prime Distribution: Prime number theorem states that a prime number occurs roughly every $(\ln n)$ integers. Since all the even numbers (other than 2) can be ignored, only $(0.5 \ln n)$ numbers of size n need to be tested to locate a prime number. For example, if a prime on the order of magnitude of 2^{100}, then about $0.5 \ln(2^{100}) \approx 69$ trials would be needed to locate a prime number. Readers need to note that $(0.5 \ln n)$ only indicates an average value. Two prime numbers are sometimes close to each other, and far apart other times.

3.7 Chinese Remainder Theorem

One of the most useful results in number theory is the Chinese remainder theory (CRT), so called because it is believed to have been discovered by an ancient Chinese mathematician. It is useful in speeding up computations in the public key crypto algorithms such as RSA.

Theorem 3.3 *Chinese Remainder Theorem (CRT): If an integer M can be expressed as*

$$M = \prod_{i=1}^{k} m_i, \tag{3.11}$$

where m_i are pairwise relatively prime, i.e. for all $i, j \in [1, k]$, $i \neq j$ and $\gcd(m_i, m_j) = 1$. Then every integer in the set $Z_M = \{0, 1, \ldots, M - 1\}$ can be reconstructed from residues w.r.t to those k numbers.

CRT makes two assertions about any integer A, described as follows:

(1) For any given integer A where $0 \leq A \leq M$, there is a unique sequence of integers $a_1, a_2, \ldots, a_k$ that represents it. For any given sequence $a_1, a_2, \ldots, a_k$, there is a unique integer $A \in Z_M$.

(2) Operations performed on the elements of Z_M can be equivalently performed on the corresponding sequence $a_1, a_2, \ldots, a_k$ by performing the operation independently in each coordinate position in the appropriate system.

According to CRT, for any integer $A \in Z_M$, the exact sequence of integers $a_1, a_2, \ldots, a_k$ are calculated as follows:

$$a_i \equiv A \bmod m_i, \text{ for } 1 \leq i \leq k. \tag{3.12}$$

With a given sequence $a_1, a_2, \dots, a_k$, the integer A is reconstructed in the following steps:

(1) First, calculate M_i such that

$$M_i = \frac{M}{m_i}, \quad \text{for } 1 \leq i \leq k. \tag{3.13}$$

(2) Since $gcd(M_i, m_i) = 1$, there exists a multiplicative inverse M_i^{-1} for M_i modulo m_i. Find M_i^{-1} for all M_i and calculate c_i as follows:

$$c_i = M_i \times (M_i^{-1} \bmod m_i), \text{ for } 1 \leq i \leq k. \tag{3.14}$$

(3) Finally, reconstruct A as follows:

$$A = \left(\sum_{i=1}^{k} a_i \times c_i \right) \bmod M. \tag{3.15}$$

As it shows, CRT provides a way to manipulate numbers modulo M in terms of a sequence of smaller numbers. However, readers may note that the factorization of M is needed beforehand. Assuming that two large integers A and B have been represented using their unique sequence of integers, such that:

$$A \leftrightarrow a_1, \dots, a_k, \text{ and}$$
$$B \leftrightarrow b_1, \dots, b_k,$$

CTR enables us to operate on the small integers instead of A and B directly. For example, $(A \pm B) \bmod M$ can be calculated as follows:

$$(A \pm B) \bmod M \leftrightarrow \left((a_1 + b_1) \bmod m_1, \dots, (a_k + b_k) \bmod m_k \right), \tag{3.16}$$

and $(A \times B) \bmod M$ can be calculated as follows:

$$(A \times B) \bmod M \leftrightarrow \left((a_1 \times b_1) \bmod m_1, \dots, (a_k \times b_k) \bmod m_k \right). \tag{3.17}$$

For example, given $A = 12345$, $B = 23456$, $M = 6873$ and the factorization such that $6873 = 79 \times 87$, compute $(A + B) \bmod M$. CRT will be applied first and then a solution from direct computation will be used for verification. First, $m_1 = 79$ and $m_2 = 87$ are given, thus M_1 and M_2 will be calculated as follows:

$$\begin{cases} M_1 = \dfrac{M}{m_1} = 87, \\[2mm] M_2 = \dfrac{M}{m_2} = 79. \end{cases}$$

Second, the multiplicative inverse M_1^{-1} and M_2^{-1} are found by applying the extended Euclidean algorithm. The values are as follows:

$$\begin{cases} M_1^{-1} \bmod m_1 = 10, \\ M_2^{-1} \bmod m_2 = 76. \end{cases}$$

One can easily verify that $87 \times 10 \equiv 1 \bmod 79$, and $79 \times 76 \equiv 1 \bmod 87$. Before computing $(A + B) \bmod M$, two unique integer sequences need to be established for A and B respectively. For A, its unique integer sequence is calculated as follows:

$$\begin{cases} a_1 = A \bmod m_1 = 21, \\ a_2 = A \bmod m_2 = 78. \end{cases}$$

And for B, its unique integer sequence is calculated as follows:

$$\begin{cases} b_1 = B \bmod m_1 = 72, \\ b_2 = B \bmod m_2 = 53. \end{cases}$$

With the two integer sequences (a_1, a_2) and (b_1, b_2), it is ready to compute $(A + B) \bmod M$. Assuming the result is C with unique integer sequence (c_1, c_2), which are computed as follows:

$$\begin{cases} c_1 = (a_1 + b_1) \bmod m_1 = (21 + 72) \bmod 79 = 14, \\ c_2 = (a_2 + b_2) \bmod m_2 = (78 + 53) \bmod 87 = 44. \end{cases}$$

To reconstruct C from (c_1, c_2), it needs to be calculated as follows:

$$C = (c_1 \times M_1 \times M_1^{-1} + c_2 \times M_2 \times M^{-1}) \bmod M,$$

$$= (14 \times 87 \times 10 + 44 \times 79 \times 76) \bmod 6873,$$

$$= 1436.$$

Now, verify the result by compute $(A + B) \bmod M$ directly, such that:

$$(A + B) \bmod M = (12345 + 23456) \bmod 6873 = 1436.$$

The results are the same. In this example, since the integers are relatively small, CRT seems less efficient than computing directly. However, if the integers are very large (e.g. in the order of 512 bits), CRT can significantly speed up the modular operations.

3.8 Discrete Logarithm

One last concept to introduce in this chapter is *discrete logarithm*. Before getting into it, one needs to understand what is a *primitive root modulo a positive number n*. Euler's theorem gives that $a^{\phi(n)} \equiv 1 \bmod n$ if $\gcd(a, n) = 1$. Thus for $a^m \equiv 1 \bmod n$, there must exist $m = \phi(n)$ that satisfies the requirement. Note that m may also be smaller than $\phi(n)$. If $m = \phi(m)$ is the smallest solution, then a is called a *primitive root modulo n*. If n is a prime number, then successive powers of a generate the group of units modulo n. For example, $a = 3$ is a primitive root modulo $n = 7$, because

$$3^0 \equiv 1 \bmod 7,$$
$$3^1 \equiv 3 \bmod 7,$$
$$3^2 \equiv 2 \bmod 7,$$
$$3^3 \equiv 6 \bmod 7,$$
$$3^4 \equiv 4 \bmod 7,$$
$$3^5 \equiv 5 \bmod 7,$$
$$3^6 \equiv 1 \bmod 7.$$

It is obvious that the group of units modulo 7 (i.e. $\{1, 2, 3, 4, 5, 6\}$) is generated by powers of 3. Therefore, 3 is a primitive root modulo 7. Primitive roots are useful in cryptography; however, they are relatively hard to find especially for large prime numbers.

After knowing the concept of primitive root, it is ready to introduce the discrete logarithm. The discrete logarithm is the finite-group-theoretic analogue of ordinary logarithms. For real numbers x, y, and z, one can define $y = \log_x z$ if $x^y = z$, with positive $x \neq 1$. For integers x, y, z, and n, if

$$x^y \equiv z \bmod n, \tag{3.18}$$

then, one can define discrete logarithm as

$$y = d \log_{x,n} z, \tag{3.19}$$

where d log indicates discrete logarithm. If x is a primitive root modulo n, then y always exists. Recall the example that 3 is a primitive root modulo 7. Then, with given

$$3^y \equiv 4 \bmod 7.$$

One can find that

$$y = d \log_{3,7} 4 = 4.$$

In fact, one can find all d $\log_{3,7} z$ for $1 \leq z \leq 6$ as follows:

$$y = d \log_{3,7} 1 = 0.$$
$$y = d \log_{3,7} 2 = 2.$$
$$y = d \log_{3,7} 3 = 1.$$
$$y = d \log_{3,7} 4 = 4.$$
$$y = d \log_{3,7} 5 = 5.$$
$$y = d \log_{3,7} 6 = 3.$$

If x is not a primitive root modulo n, then a unique discrete logarithm modulo n may not exist. It is also worth noting that while exponentiation is relatively easy, finding a discrete logarithm is generally a hard problem.

3.9 Summary

In this chapter, several basic concepts in number theory and modern algebra are introduced. Prime numbers, finite fields as well as polynomial arithmetic operations are widely applied in security algorithms. The mathematical concepts given in this chapter will be used in future chapters of this book. Readers are encouraged to have further study in those concepts if necessary [11–13].

Part II

Cryptographic Systems

4

Cryptographic Techniques

4.1 Symmetric Encryption

Symmetric encryption was the only type of encryption in use before the development of public-key encryption in the 1970s. Symmetric encryption is also referred to a conventional encryption or single-key encryption [14, 15]. It remains the most widely used of the two types of encryption. For simplicity, the basic terminology of cryptographic techniques is listed in Table 4.1. The symmetric cipher model is shown in Figure 4.1. A *single key* is pre-shared between the sender and the receiver. The sender and the receiver can either encrypt or decrypt messages using the pre-shared key. Normally, the sender runs an encryption algorithm that translates the input plaintext to its corresponding ciphertext. The ciphertext is then transmitted to the receiver side. The receiver runs a decryption algorithm, which is the inverse of the encryption algorithm, with the same pre-shared key to decrypt the ciphertext into plaintext. If the pre-shared key is K, the plaintext is X, the ciphertext is Y, the encryption/decryption algorithms are E/D, and then the encryption and decryption processes are

$$E_K(X) = Y, \tag{4.1}$$

and

$$D_K(Y) = X. \tag{4.2}$$

A symmetric cipher model requires that the encryption algorithm is strong enough to provide security. It also requires that the pre-shared key is known only to the sender and the receiver. Moreover, the encryption and decryption algorithms must be known to the sender and the receiver beforehand. Finally, a symmetric cipher requires a secure method to distribute the pre-shared key.

4.2 Classical Cryptographic Schemes

In this section, we introduce a group of traditional cryptographic schemes. A study of these techniques helps to illustrate the basic approaches to symmetric encryption used today

Security in Wireless Communication Networks, First Edition. Yi Qian, Feng Ye, and Hsiao-Hwa Chen.
© 2022 John Wiley & Sons Ltd. Published 2022 by John Wiley & Sons Ltd.
Companion website: www.wiley.com/go/qian/sec51

Table 4.1 Basic terminology of cryptographic techniques.

Term	Definition
Plaintext	Original message
Ciphertext	Coded message
Cipher	Algorithm for transforming plaintext to ciphertext
Key	Information used in cipher known only to sender/receiver
Encipher (encrypt)	Converting plaintext to ciphertext
Decipher (decrypt)	Recovering ciphertext to plaintext
Cryptography	Study of encryption principles/methods
Cryptanalysis	Study of principles/methods of deciphering ciphertext without knowing key
Cryptology	Field of both cryptography and cryptanalysis

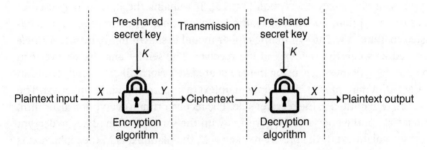

Figure 4.1 Symmetric cipher model.

and the types of cryptanalytic attacks that must be anticipated. All classical cryptographic schemes are symmetric encryption algorithms with techniques of substitution and transposition. A few classical cryptographic algorithms will be presented in details, such as Caesar cipher, monoalphabetic cipher, the Playfair cipher, polyalphabetic cipher, Vigenère cipher, etc.

4.2.1 Classical Substitution Ciphers

The two basic building blocks of all encryption technique are *substitution* and *transposition*. In substitution based ciphers, characters of plaintext are replaced by other letters, numbers, or symbols. If plaintext is viewed as a sequence of bits in binary format, then substitution involves replacing plaintext bit patterns with ciphertext bit patterns.

4.2.1.1 Caesar Cipher
Caesar cipher is one of the earliest known substitution ciphers [16]. The original Caesar cipher replaces each letter with the third letter after. Nonetheless, any cipher using a simple

letter shift can be called a Caesar cipher now. In the original Caesar cipher, the transformation is defined in Table 4.2.

Given a plaintext consists of English letters, its corresponding ciphertext can be found by substituting the letters, for example,

Plaintext:	see	you	in	the	midnight
Ciphertext:	VHH	BRX	LQ	WKH	PLGQLJKW

Alternatively, Caesar cipher can be defined as modulo arithmetic mathematically. Each letter or character can be mapped to a number, such that $a = 0, b = 1, ..., z = 25$. Let k be the number of letters to shift (i.e. the symmetric key), then the encryption and decryption of Caesar cipher are calculated as:

$$c = E_k(p) = (p + k) \bmod 26, \tag{4.3}$$

and

$$p = D_k(c) = (c - k) \bmod 26, \tag{4.4}$$

where p and c are plaintext and ciphertext, respectively. $E_k(\cdot)$ and $D_k(\cdot)$ are encryption and decryption functions, respectively.

Cryptanalysis of Caesar cipher Caesar cipher has only 26 possible keys, of which only 25 are of any use, since mapping "a" to "A" (i.e. $k = 0$) does not obscure the message. Therefore, a brute force search can be applied to try all possible keys (shifts) in turn until the original message can be recognized. For example, given a ciphertext "ZIVC IEWC XS JMRH," we can find that the key (shift) is "4" and the original message is "very easy to find." An original message (i.e. English or other language) is usually easy for humans to recognize. However, it is harder to computers to detect valid messages. Furthermore, if the original message is compressed data, it would be much harder to recognize.

4.2.1.2 Monoalphabetic Cipher

Caesar cipher is far from secure with only 25 possible keys. Monoalphabetic cipher maps each plaintext letter to a distinct ciphertext letter. Hence the key of monoalphabetic cipher is 26-letter long. In comparison, Monoalphabetic cipher achieves a dramatic increase in key space. For example, with the key given in Table 4.3, a plaintext can be encrypted by substituting the letters as follows:

Plaintext:	if	we	wish	to	replace	letters
Ciphertext:	OY	CT	COLI	MG	KTHSZET	STMMTKL

Table 4.2 Transformation of Caesar cipher.

Plaintext	a	b	c	d	e	f	g	h	i	j	k	l	m	n	o	p	q	r	s	t	u	v	w	x	y	z
Ciphertext	D	E	F	G	H	I	J	K	L	M	N	O	P	Q	R	S	T	U	V	W	X	Y	Z	A	B	C

Table 4.3 An example key of monoalphabetic cipher.

Plaintext	a b c d e f g h i j k l m n o p q r s t u v w x y z
Ciphertext	Z A E R T Y U I O P Q S D F G H J K L M W X C V B N

Cryptanalysis of monoalphabetic cipher monoalphabetic cipher has a total of $26! - 1 \approx 4 \times 10^{26}$ keys. With so many keys, brute force is not feasible to crack the key in general. Does this mean that monoalphabetic cipher is secure? Unfortunately the answer is NO. The weakness is due to *language characteristics*. Human languages are redundant (English is discussed as an example here). We don't actually need all the letters in order to understand written English text. For example, "ths s nt dffclt" ("this is not difficult") can be understood although the vowels were removed. The redundancy is also the reason we can compress text files. The computer can derive a more compact encoding without losing any information. The basic idea is to count the relative frequencies of letters, and calculate the frequency of each letter. The frequency of each letter can reveal the letter itself, because letters are not equally commonly used. In English, "E" is by far the most common letter, followed by "T," "R," "N," and so on. Some letters (e.g. "Z," "J," etc.) are rarely used. Figure 4.2 shows the frequency distribution of English letter [17].

For example, given ciphertext:

> Z COKTSTLL EGDDWFOEZMOGF FTMCGKQ OL Z EGDHWMTK
> FTMCGKQ MIZM WLTL Z COKTSTLL EGFFTEMOGF WLWZSSB
> KZROG EGDDWFOEZMOGF ATMCTTF FTMCGKQ FGRTL.

By counting the frequency of each letter, one may find that "T," "M," and "Z" has appeared 14, 10, and 8 times, respectively. Therefore, guess that "T"→"e," "M"→"t", and "Z"→"a".

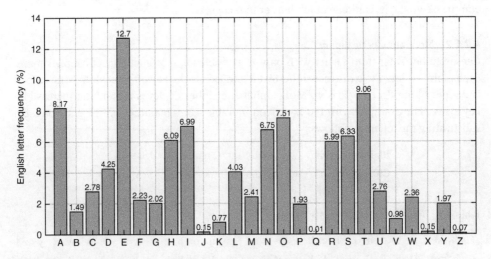

Figure 4.2 English letter frequency distribution. Source: Modified from Lewand [17] .

Then "MIZM"→"T?AT" reveals that "I"→"h". One may keep proceeding with trial and error for the rest of the analysis. Shortly, the original message is revealed:

> a wireless communication network is a computer
> network that uses a wireless connection usually
> radio communication between network nodes

Monoalphabetic cipher maps one letter to another, thus the frequency distribution of each letter is only shuffled. In order to reduce the "spikiness" of natural language text, one approach is to encrypt more than one letter at once.

4.2.1.3 Playfair Cipher

Playfair cipher treats two letters in the plaintext as a single unit and translates the unit into ciphertext letters. Playfair cipher was invented by Charles Wheatstone in 1854, but was named after his friend Baron Playfair [18]. Playfair encryption/decryption is based on the use of a 5×5 matrix of letters constructed using a keyword. The keyword is used as the key. The rules for filling in this 5×5 Playfair matrix are listed in the following.

- Left to right.
- Top to bottom.
- First with keyword after duplicate letters have been removed.
- Then with the remaining letters in alphabetic order.
- "I/J" are used as a single letter.

For example, given keyword "DECEMBER," the 5×5 matrix is the one in the following:

D	E	C	M	B
R	A	F	G	H
I/J	K	L	N	O
P	Q	S	T	U
V	W	X	Y	Z

The encryption of Playfair cipher is performed two letters at a time using the 5×5 matrix according to the rules as shown in the following:

- If a pair is a repeated letter, insert a filler like "X," e.g. "ballon" is encrypted as "ba lx lo on."
- If both letters fall in the same row of the matrix, replace each with letter to its right (wrapping back to start from end).
- If both letters fall in the same column of the matrix, replace each with the letter below it (again wrapping to top from bottom).
- Otherwise, each letter is replaced by the one in its row in the column of the other letter of the pair.

For instance, using the matrix illustrated earlier, "mb" → "BD," "dv" → "RD", "at" → "GQ," and "od" → "IB" or "JB." Decryption in Playfair cipher works exactly in reverse. Readers shall be able to decrypt the example pairs on your own.

Cryptanalysis of Playfair cipher Security is much improved over monoalphabetic cipher since it has a total of $26 \times 26 = 676$ diagrams. It would need a 676 entry frequency table to analyze verses 26 for a monoalphabetic. Playfair cipher was widely used for many years by US and British military until World War I. However, Playfair cipher can be broken given just a few hundred letters because the ciphertext still has much plaintext structure.

4.2.1.4 Polyalphabetic Cipher

Polyalphabetic cipher improves security by using multiple cipher alphabets. It makes cryptanalysis harder with more alphabets to guess and flatters frequency distribution. Polyalphabetic cipher uses a key to select which alphabet is used for each letter of the message. It uses each alphabet in turn and repeats from start after the key is used up. Vigenère cipher is the simplest polyalphabetic substitution cipher. It effectively uses multiple Caesar ciphers. The key of Vigenère cipher is multiple letters long, s.t. $K = k_1 k_2, \ldots, k_d$. The *ith* letter specifies the *ith* alphabet to use. After the first d letters in a message, it repeats from the start. Since Vigenère cipher is a symmetric cipher, its decryption simply works in reverse. For example, given keyword "wireless," we then have the key $K = 22, 8, 17, 4, 11, 4, 18, 18, \ldots$, A complete example is give in the following.

Key:	w	i	r	e	l	e	s	s	w	i	r	e	l	e
Plaintext:	p	o	l	y	a	l	p	h	a	b	e	t	i	c
Ciphertext:	L	W	C	C	L	P	H	Z	W	J	V	X	T	G

Key:	s	s	w	i	r	e	l	e	s	s	w	i	r	e
Plaintext:	c	i	p	h	e	r	i	m	p	r	o	v	e	s
Ciphertext:	U	A	L	P	V	V	T	Q	H	J	K	D	V	W

Implementing a polyalphabetic cipher by hand can be tedious. Various aids were devised to assist the process. *Saint-Cyr Slide* is a simple manual aid that was popularized and named by Jean Kerckhoffs, who published a famous early text "La Cryptographie Militaire" (Military Cryptography) in 1883. He named the slide after the French National Military Academy where the methods were taught. Saint-Cyr Slide is a slide with repeated alphabet. It lines up plaintext "A" with key letter, e.g. "C." Then, it reads off any mapping for key letter. The slide can also be bended round into a cipher disk or expanded into a Vigenère tableau, as shown in Figure 4.3.

Cryptanalysis of polyalphabetic cipher The Vigenère cipher had been *le chiffre indéchiffrable* (the unbreakable cipher) for several centuries. As a result of a challenge, it was broken by the British cryptographer Charles Babbage in 1854. However, the cracking technique was not published during that time. In fact, the technique was finally made public in the 1970s. The Vigenère and related polyalphabetic ciphers have multiple ciphertext letters for each plaintext letter, thus letter frequencies are obscured. However, it does not completely obscure the underlying language characteristics. The key to breaking them was to identify the number of translation alphabets, and then attack each separately.

	A	B	C	D	E	F	G	H	I	J	K	L	M	N	O	P	Q	R	S	T	U	V	W	X	Y	Z
A	A	B	C	D	E	F	G	H	I	J	K	L	M	N	O	P	Q	R	S	T	U	V	W	X	Y	Z
B	B	C	D	E	F	G	H	I	J	K	L	M	N	O	P	Q	R	S	T	U	V	W	X	Y	Z	A
C	C	D	E	F	G	H	I	J	K	L	M	N	O	P	Q	R	S	T	U	V	W	X	Y	Z	A	B
D	D	E	F	G	H	I	J	K	L	M	N	O	P	Q	R	S	T	U	V	W	X	Y	Z	A	B	C
E	E	F	G	H	I	J	K	L	M	N	O	P	Q	R	S	T	U	V	W	X	Y	Z	A	B	C	D
F	F	G	H	I	J	K	L	M	N	O	P	Q	R	S	T	U	V	W	X	Y	Z	A	B	C	D	E
G	G	H	I	J	K	L	M	N	O	P	Q	R	S	T	U	V	W	X	Y	Z	A	B	C	D	E	F
H	H	I	J	K	L	M	N	O	P	Q	R	S	T	U	V	W	X	Y	Z	A	B	C	D	E	F	G
I	I	J	K	L	M	N	O	P	Q	R	S	T	U	V	W	X	Y	Z	A	B	C	D	E	F	G	H
J	J	K	L	M	N	O	P	Q	R	S	T	U	V	W	X	Y	Z	A	B	C	D	E	F	G	H	I
K	K	L	M	N	O	P	Q	R	S	T	U	V	W	X	Y	Z	A	B	C	D	E	F	G	H	I	J
L	L	M	N	O	P	Q	R	S	T	U	V	W	X	Y	Z	A	B	C	D	E	F	G	H	I	J	K
M	M	N	O	P	Q	R	S	T	U	V	W	X	Y	Z	A	B	C	D	E	F	G	H	I	J	K	L
N	N	O	P	Q	R	S	T	U	V	W	X	Y	Z	A	B	C	D	E	F	G	H	I	J	K	L	M
O	O	P	Q	R	S	T	U	V	W	X	Y	Z	A	B	C	D	E	F	G	H	I	J	K	L	M	N
P	P	Q	R	S	T	U	V	W	X	Y	Z	A	B	C	D	E	F	G	H	I	J	K	L	M	N	O
Q	Q	R	S	T	U	V	W	X	Y	Z	A	B	C	D	E	F	G	H	I	J	K	L	M	N	O	P
R	R	S	T	U	V	W	X	Y	Z	A	B	C	D	E	F	G	H	I	J	K	L	M	N	O	P	Q
S	S	T	U	V	W	X	Y	Z	A	B	C	D	E	F	G	H	I	J	K	L	M	N	O	P	Q	R
T	T	U	V	W	X	Y	Z	A	B	C	D	E	F	G	H	I	J	K	L	M	N	O	P	Q	R	S
U	U	V	W	X	Y	Z	A	B	C	D	E	F	G	H	I	J	K	L	M	N	O	P	Q	R	S	T
V	V	W	X	Y	Z	A	B	C	D	E	F	G	H	I	J	K	L	M	N	O	P	Q	R	S	T	U
W	W	X	Y	Z	A	B	C	D	E	F	G	H	I	J	K	L	M	N	O	P	Q	R	S	T	U	V
X	X	Y	Z	A	B	C	D	E	F	G	H	I	J	K	L	M	N	O	P	Q	R	S	T	U	V	W
Y	Y	Z	A	B	C	D	E	F	G	H	I	J	K	L	M	N	O	P	Q	R	S	T	U	V	W	X
Z	Z	A	B	C	D	E	F	G	H	I	J	K	L	M	N	O	P	Q	R	S	T	U	V	W	X	Y

Figure 4.3 Vigenère tableau.

4.2.1.5 Autokey Cipher

Proposed by Vigenère, the *autokey cipher* took the polyalphabetic idea nearly to the extreme since it wanted as many different translation alphabets as letters in the message being sent [19]. In the Autokey cipher, the keyword is prefixed to message as the key. Knowing the keyword can recover the first few letters that will be used in turn on the rest of the message. For example, given a keyword "WIRELESS" and a message "autokey cipher example." First, the keyword is prefixed to as much of the message as needed. When deciphering, one should recover the first eight letters using the keyword "WIRELESS." Then, instead of repeating the keyword, the recovered letters from the message "AUTOKEYC" are used to recover the next eight letters, and so on. The complete encryption is shown as follows:

Key:	W	I	R	E	L	E	S	S	A	U
Plaintext:	a	u	t	o	k	e	y	c	i	p
Ciphertext:	W	C	K	S	V	I	Q	U	I	J
Key:	T	O	K	E	Y	C	I	P	H	E
Plaintext:	h	e	r	e	x	a	m	p	l	e
Ciphertext:	A	S	B	I	V	C	U	E	S	I

The problem of the Autokey cipher is that the same language characteristics are used by the key as the message. That is, a key of "E" will be used more often than "T," etc. Hence an "E" encrypted with a key of "E" occurs with probability $0.1275^2 = 0.01663$, about twice as often as a "T" encrypted with a key of "T." Cryptanalysis has to use a larger frequency table. However, given sufficient ciphertext, the autokey cipher can still be broken.

4.2.1.6 One-Time Pad

One-time pad is an evolution of the Vernham cipher, which was invented by Gilbert Vernham in 1918. The Vernham cipher used a long tape of random letters to encrypt the message. An Army Signal Corp officer, Joseph Mauborgne, proposed an improvement using a random key that was as long as the message with no repetitions, which can totally obscure the original message. One-time pad produces a random output that bears no statistical relationship to the plaintext. Because the ciphertext contains no information about the plaintext, there is no way to break the code since any plaintext can be mapped to any ciphertext given some key. Although one-time pad offers complete security, it has two fundamental difficulties in practice. One difficulty is the practical problem of making large quantities of random keys. The other is the problem of key distribution and protection, where for every message to be sent, a key of equal length is needed by both sender and receiver. Because of these difficulties, one-time pad is of limited usage, and is only useful for low-bandwidth channels requiring very high security.

4.2.2 Classical Transposition Ciphers

All the techniques introduced in Section 4.2.1 involve the substitution of a ciphertext symbol for a plaintext symbol. Another type of mapping is achieved by performing some sort of permutation on the plaintext letters. This technique is referred to as a transposition cipher, and forms the second basic building block of ciphers. The core idea is to rearrange the order of basic units (letters/bytes/bits) without altering their actual values.

4.2.2.1 Rail Fence Cipher

The simplest transposition cipher is the rail fence cipher. The technique is to write down the plaintext as a sequence of diagonals and then read off as a sequence of rows. For example, to encrypt 'rail fence cipher example' with a rail fence of depth 2, first write message letters out diagonally over a number of rows, i.e. 2 in this example.

```
r   i   f   n   e   i   h   r   x   m     l
  a   l   e   c   c   p   e   e   a   p   e
```

Thus, the ciphertext is "RIFNEIHRXMLALECCPEEAPE." A transposition cipher is easy to be recognized because it has the same letter frequencies as the original plaintext. For the type of transposition just shown, cryptanalysis is fairly straightforward by laying out the ciphertext in a matrix and playing around with column positions.

4.2.2.2 Row Transposition Cipher

Row transposition cipher is a more complex transposition cipher. It writes the letters of a message in rows over a specified number of columns. Then the columns are reordered column by column according to some key before reading off. For example, given a key "3,1,4,2,7,6,5" (meaning seven columns), the plaintext "row transposition cipher test" is rewritten into

Key:	3	1	4	2	7	6	5
Plaintext:	r	o	w	t	r	a	n
	s	p	o	s	i	t	i
	o	n	c	i	p	h	e
	r	t	e	s	t	x	y

Note that "x,y" are attached to fill the columns. The ciphertext is column by column according to the sequence of the key, i.e.

Ciphertext: OPNT TSIS RSOR WOCE NIEY ATHX RIPT'.

To decipher the ciphertext, simply rebuild the columns and the plaintext will be the row rewriting of it.

4.2.2.3 Product Cipher

Ciphers based solely on either substitutions or transpositions are not secure. They can be attacked because they do not sufficiently obscure the underlying language structure. A solution to enhance the security is to use several ciphers in succession. For instance, two substitutions make a more complex substitution. Two transpositions make a more complex transposition. Furthermore, a substitution followed by a transposition makes a new and much harder cipher. *Product cipher* was a concept brought by Claude Shannon, who presented his idea in "Communication Theory of Secrecy Systems" [20]. Product cipher uses a combination of a substitution followed by a transposition. It is a much more secure cipher compared with the other classical ciphers discussed earlier. Product cipher forms the bridge to modern ciphers. It has become the basic building blocks for several encryption standards such as the Data Encryption Standard (DES) and the Advanced Encryption Standard (AES).

4.2.3 More Advanced Classical Ciphers

The next major advance in ciphers deployed mechanical devices that enabled for complex varying substitutions. Before modern ciphers, rotor machines were most common complex ciphers in use.

4.2.3.1 Rotor Machines

A rotor machine consists of a set of independently rotating cylinders through which electrical pulses can flow. Each cylinder has 26 input pins and 26 output pins, with internal wiring that connects each input pin to a unique output pin. If we associate each input and output pin with a letter of the alphabet, then a single cylinder defines a monoalphabetic substitution. After each input key is depressed, the cylinder rotates one position, so that the internal connections are shifted accordingly. The power of the rotor machine is in the use of multiple cylinders, in which the output pins of one cylinder are connected to the input pins of the next, and with the cylinders rotating like an "odometer," leading to a very large number of substitution alphabets being used. Rotor machines were extensively used in World War II, and the history of their use and analysis is one of the great stories from World War II.

4.2.3.2 Steganography

Steganography is an alternative to encryption that hides the very existence of a message by some means. There are a large range of techniques for doing this. Steganography has a number of drawbacks when compared with encryption. It requires a lot of overhead to hide a relatively few bits of information. Also, once the system is discovered, it becomes virtually worthless, although a message can be first encrypted and then hidden using steganography. An example of steganography is shown in Figure 4.4, can you find the original message?
    (Hint: check the last word of each line).

Wireless communications have been much improved in our daily life. Most households are now equipped with Wi-Fi routers. The default password implemented in the router is usually a random value. Therefore, it is hard to be hacked without physical access. Many users would prefer to update with an easy to remember password.

Figure 4.4 An example of steganography.

4.3 Stream Cipher

A stream cipher is a symmetric key cipher. Each plaintext digit (in bit or byte) is encrypted one at a time with a corresponding digit from the keystream. As shown in Figure 4.5, the encryption is generally XOR function, where a digit of the plaintext is XORed with the corresponding digit of the keystream. The keystream is usually a pseudo-random number generated by an algorithm with a (pseudo) random seed value. Because a stream cipher is a symmetric key cipher, the seed value is synchronized between a transmitter and a receiver to serve as the cryptographic key. The keystream is generated at both side instead of being transmitted.

4.3.1 Rivest Cipher 4

Rivest Cipher 4 (RC4) is one of the typical and widely used stream ciphers. It was designed by Ron Rivest of RSA Security in 1987. Because of its speed and simplicity, RC4 has been applied in many applications for years, e.g. Wired Equivalent Privacy (WEP) and Wireless Protected Access (WPA) in wireless local area networks, and Transport Layer Security (TLS) for Internet security. There are two elements in the RC4 algorithm: *Key Scheduling Algorithm (KSA)* and *Pseudo-Random Generation Algorithm (PRGA)*.

KSA is the key scheduling algorithm. First, it has the key initialization process that takes input the *l*-byte secret key K and generates a random 256-value state array S. Detailed process of KSA is illustrated in Algorithm 4.1. The array S is initialized from 0 to 255 and it is shuffled based on the input key. PRGA is the pseudo-random generation algorithm that runs after KSA. As shown in Figure 4.6, two states in the array S are first chosen, for example, $S[i]$ and $S[j]$. Swap $S[i]$ and $S[j]$. Then, a new index is calculated as $(S[i] + S[j])$ mod 256. Lastly, the element in array S from this index is selected as the keystream byte k. PRGA is summarized in Algorithm 4.2.

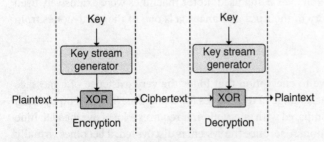

Figure 4.5 Structure of stream cipher.

Algorithm 4.1 Key Scheduling Algorithm (KSA) for RC4

Input: K, l;
Output: state array S;
  for $i = 0$ to 255 **do**
    $S[i] \leftarrow i$;
  end for
  $j \leftarrow 0$;
  for $i = 0$ to 255 **do**
    $j \leftarrow (j + S[i] + K[i \bmod l]) \bmod 256$;
    SWAP($S[i], S[j]$);
  end for

Figure 4.6 The overview of the keystream generation.

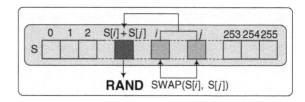

Algorithm 4.2 Pseudo-Random Generation Algorithm (PRGA) for RC4

Input: K, l, S;
Output: k;
  $i = 0, j = 0$;
  while Generating output **do**
    $i \leftarrow (i + 1) \bmod 256$;
    $j \leftarrow (j + S[i]) \bmod 256$;
    SWAP($S[i], S[j]$);
    $k \leftarrow S[(S[i] + S[j]) \bmod 256]$;
  end while

The keystream k generated from PRGA algorithm is XORed with the plaintext byte to produce the ciphertext byte in the sender side. In the receiver side, the same keystream k generated from the same PRGA algorithm is XORed with the received ciphertext byte to get the original plaintext.

4.4 Modern Block Ciphers

4.4.1 Overview of Modern Block Ciphers

Modern block ciphers are some of the most widely used types of cryptographic algorithms. They provide encryption of quantities of information (secrecy), and/or a cryptographic checksum to ensure that the contents have not been altered (authentication). In this section, we present the basics of modern block cipher design and DES. AES and different modes of block cipher operations are introduced in the next chapter. Different from stream ciphers, block ciphers operate on large blocks of digits with a fixed transformation. However, this distinction is not always clear-cut: in some modes of operation (which will be discussed later in the next chapter), a block cipher primitive is used in such a way that it acts effectively as a stream cipher. Compared with stream ciphers, block ciphers typically execute at a slower speed and have higher hardware complexity. However, stream ciphers can be susceptible to serious security problems if used incorrectly. In particular, the same starting state (seed) must never be used twice. Therefore, block ciphers have a broader range of applications. A block cipher operates on a plaintext block of n bits to produce a ciphertext block of n bits. An arbitrary reversible substitution cipher for a large block size is not practical, however, from an implementation and performance point of view. In general, for an n-bit substitution block cipher, the length of the key is $n \times 2^n$. For example, a 64-bit block, which is a desirable length to thwart statistical attacks, has a key length of $64 \times 2^{64} = 2^{70} \approx 10^{21}$ bits. Several symmetric block encryption algorithms in current use are based on a structure referred to as a *Feistel block cipher*.

4.4.2 Feistel Block Cipher

4.4.2.1 Ideal Block Cipher

Before heading into the Feistel block cipher system, let us discuss some of the basics for block ciphers. First, for block size $n = 2$, there are $2^2 = 4$ possible different plaintext blocks. In order to make the encryption reversible (i.e. for decryption to be possible), each plaintext block must produce a unique ciphertext block. Also, a block cipher operates on a plaintext block of n bits to produce a ciphertext block of n bits. The mapping shown in Figure 4.7a is reversible. The mapping shown in Figure 4.7b is irreversible since both plaintexts "01" and "10" are mapped into "01." Given "01" as the ciphertext, it is not possible to recover a unique plaintext. A block cipher must be a reversible mapping. For an n-bit general substitution ideal block cipher, it allows for the maximum number of possible encryption mappings from a plaintext block to a ciphertext block. For example, a 4-bit input produces one of 16 possible input states, which is mapped by the substitution cipher into a unique one of 16 possible output states, each of which is represented by 4 bits ciphertext. The encryption and decryption mappings can be defined by two tables. Figure 4.8 shows a set of encryption and decryption tables of a 4-bit substitution block cipher.

In Claude Shannon's 1949 paper, he proposed the key ideas that lead to the development of modern block ciphers [20]. It was the technique of layering groups of substitution boxes (S-boxes) separated by a larger permutation box (P-box) to form the S–P (substitution and

Plaintext	Ciphertext
00	10
01	11
10	01
11	00

(a)

Plaintext	Ciphertext
00	11
01	01
10	01
11	10

(b)

Figure 4.7 Examples of reversible and irreversible mappings for $n = 2$. (a) Reversible mapping. (b) Irreversible mapping.

Plaintext	Ciphertext
0000	1101
0001	0100
0010	1110
0011	0010
0100	0001
0101	1111
0110	1011
0111	0011
1000	1000
1001	1010
1010	1100
1011	0110
1100	1001
1101	0101
1110	0111
1111	0000

(a)

Ciphertext	Plaintext
0000	1111
0001	0100
0010	0011
0011	0111
0100	0001
0101	1101
0110	1011
0111	1110
1000	1000
1001	1100
1010	1001
1011	0110
1100	1010
1101	0000
1110	0010
1111	0101

(b)

Figure 4.8   Encryption and decryption tables for a 4-bit substitution cipher.

permutation) network, a complex form of a product cipher. He also introduced the ideas of confusion and diffusion of messages and keys, notionally provided by S-boxes and P-boxes. Every block cipher involves a transformation of a block of plaintext into a block of ciphertext, where the transformation depends on the key. The mechanism of diffusion seeks to make the statistical relationship between the plaintext and ciphertext as complex as possible in order to thwart attempts to deduce the key. Confusion seeks to make the relationship between the statistics of the ciphertext and the value of the encryption key as complex as possible, again to thwart attempts to discover the key. Due to the success of diffusion and confusion in capturing the essence of the desired attributes of a block cipher, they have become the cornerstone of modern block cipher design.

4.4.2.2   Feistel Cipher Structure

Feistel cipher structure was devised by Horst Feistel. One of Feistel's main contributions was the invention of a suitable structure, which adapted Shannon's S–P network in an easily inverted structure. Figure 4.9 illustrates the classical Feistel cipher structure. As shown in Figure 4.9a, the input data in the encryption process is split in two halves (i.e. L_0 and R_0) and processed through a number of rounds. In each round, a substitution is performed on the left half (i.e. L_{i-1} in the ith round) using the output of the round function $f(\cdot)$. Round function $f(\cdot)$ performs a series of substitution and permutation on right half (i.e. R_{i-1} in the ith round) and a subkey (i.e. k_i in the ith round). The final stage of a round is a permutation function, which swaps the two halves. The output of the left half is input into the right side of the next round, where the output of the right half is input into to left side of the next round, such that

$$L_i = R_{i-1},$$
$$R_i = L_{i-1} \oplus f(R_{i-1}, k_i).$$

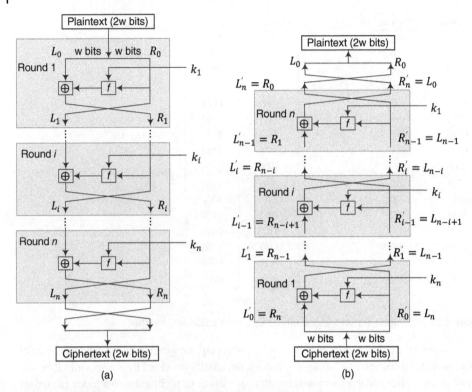

Figure 4.9 Encryption and decryption structures of Feistel cipher.

The output of the final round gets swapped such that $R_n \| L_n$ as the ciphertext.

The decryption process with a Feistel cipher is essentially the same as the encryption process. As illustrated in Figure 4.9b, the input to the ith decryption round is $L'_{i-1} \| R'_{i-1}$, or equivalently, $R_{n-i+1} \| L_{n-i+1}$. The subkey k_i is applied in reverse order, that is, k_n is used in the first round, k_{n-1} is used in the second round, and so on until k_1 is used in the last round. The decryption process can be presented as follows:

$$L'_0 = R_n,$$
$$R'_0 = L_n,$$
$$L'_i = R'_{i-1} = R_{n-i},$$
$$R'_i = L'_{i-1} \oplus f(R'_{i-1}, k_i) = L_{n-i},$$
$$\vdots$$
$$L'_n = R_0,$$
$$R'_n = L_0.$$

A final swap recovers the original plaintext $R'_n \| L'_n = L_0 \| R_0$, demonstrating the validity of the Feistel decryption process. Note that f does not have to be a reversible function. As you may see, the Feistel cipher has a very nice feature that only one algorithm is needed for both encryption and decryption. Essentially the same hardware or software is used for both encryption and decryption, with just a slight change in how the keys are used. The

Table 4.4 Parameters and design features of Feistel cipher structure.

Parameter and features	Description
Block size	Increasing size improves security, but slows cipher
Key size	Increasing size improves security, makes exhaustive key searching harder, but may slow cipher
Number of rounds	Increasing number improves security, but slows cipher
Subkey generation	Greater complexity can make analysis harder, but slows cipher
Round function	Greater complexity can make analysis harder, but slows cipher
Software implementation	More recent concern for practical use
Ease of analysis	For easier validation and testing of strength

exact realization of a Feistel cipher depends on the choice of block size, key size, number of rounds, subkey generation algorithm, round function, etc. Details of the parameters and design features are listed in Table 4.4.

4.4.3 Block Cipher Design

The cryptographic strength of a Feistel cipher derives from three aspects of the design: the number of rounds, the function $f(\cdot)$, and the key schedule algorithm. The greater the number of rounds, the more difficult it is to perform cryptanalysis, even for a relatively weak $f(\cdot)$. In general, the criterion should be that the number of rounds is chosen so that known cryptanalytic attacks require greater effort than a simple brute-force key search attack. This criterion makes it easy to judge the strength of an algorithm and to compare different algorithms. The function $f(\cdot)$ provides the element of confusion in a Feistel cipher, want it to be difficult to "unscramble" the substitution performed by $f(\cdot)$. One obvious criterion is that $f(\cdot)$ be nonlinear. The more nonlinear $f(\cdot)$, the more difficult any type of cryptanalysis will be. It is expected to have good avalanche properties, or even the strict avalanche criterion. Another criterion is the bit independence criterion. One of the most intense areas of research in the field of symmetric block ciphers is that of S-box design. Would like any change to the input vector to an S-box to result in random-looking changes to the output. The relationship should be nonlinear and difficult to approximate with linear functions. A final area of block cipher design, and one that has received less attention than S-box design, is the key schedule algorithm. With any Feistel block cipher, the key schedule is used to generate a subkey for each round. Would like to select subkeys to maximize the difficulty of deducing individual subkeys and the difficulty of working back to the main key. The key schedule should guarantee key/ciphertext strict avalanche criterion and bit independence criterion.

4.5 Data Encryption Standards (DES)

4.5.1 Overview of DES

In 1973, the National Bureau of Standards (NBS) issued a request for proposals for a national cipher standard. IBM submitted the modified *LUCIFER*, which was originally designed for

IBM mainframe computers. It was by far the best algorithm proposed and was adopted in 1977 by the National Bureau of Standards as Federal Information Processing Standard 46 (FIPS PUB 46), also known as the DES. Despite its popularity as a symmetric key block cipher, the DES has been the subject of much controversy its security. Before its adoption as a standard, the proposed DES was subjected to intense and continuing criticism over the size of its key and the classified design criteria. Subsequent analysis showed that despite this controversy, DES is well designed. DES is theoretically broken using Differential or Linear Cryptanalysis but in practice is unlikely to be a problem yet. Also rapid advances in computing speed though have rendered the 56 bit key susceptible to exhaustive key search. DES has flourished and is widely used, especially in financial applications. It is still standardized for legacy systems, with either AES or triple DES for new applications.

The overall scheme for DES encryption is illustrated in Figure 4.10. It takes input of 64-bit data and a 56-bit key. The basic process for enciphering a 64-bit data block consists of an Initial Permutation (IP) of the 64 bit plaintext, and a Permuted Choice 1 (PC1) of the key, which selects 56 bits out of the 64-bit input in two 28-bit halves; 16 rounds of the same function that involves both permutation and substitution functions, with a generated 48-bit subkey in each of the 16 rounds, and a final permutation (i.e. the inverse IP) before the left and right swap of the 32-bits after the round 16. As you may have observed, with the exception of the initial and final permutations, DES has the exact structure of a Feistel cipher.

4.5.2 Initial Permutation (IP)

The initial permutation (IP) is the first step of the data operation. It reorders the input data bits B to output data bits X, i.e. $X = IP(B)$. The IP is defined in Table 4.5. The input table

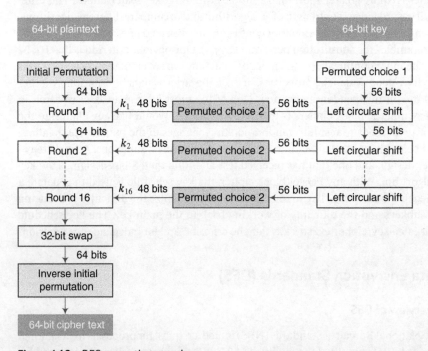

Figure 4.10 DES encryption overview.

Table 4.5 Initial permutation (*IP*).

58	50	42	34	26	18	10	2
60	52	44	36	28	20	12	4
62	54	46	38	30	22	14	6
64	56	48	40	32	24	16	8
57	49	41	33	25	17	9	1
59	51	43	35	27	19	11	3
61	53	45	37	29	21	13	5
63	55	47	39	31	23	15	7

consists of 64 bits numbered left to right from 1 to 64. The 64 entries in the permutation table contain a permutation of the numbers from 1 to 64. Each entry in the permutation table indicates the position of a numbered input in the output, which also consists of 64 bits. For example, given an entry (*E71A69475E5A6B5E*) in hexadecimal (HEX), we first fill in the 64-bit table (from left to right, e.g. $B_1 B_2 B_3 B_4 = 1110$).

$$B = \begin{matrix} B_1 & B_2 & B_3 & B_4 & B_5 & B_6 & B_7 & B_8 \\ B_9 & B_{10} & B_{11} & B_{12} & B_{13} & B_{14} & B_{15} & B_{16} \\ B_{17} & B_{18} & B_{19} & B_{20} & B_{21} & B_{22} & B_{23} & B_{24} \\ B_{25} & B_{26} & B_{27} & B_{28} & B_{29} & B_{30} & B_{31} & B_{32} \\ B_{33} & B_{34} & B_{35} & B_{36} & B_{37} & B_{38} & B_{39} & B_{40} \\ B_{41} & B_{42} & B_{43} & B_{44} & B_{45} & B_{46} & B_{47} & B_{48} \\ B_{49} & B_{50} & B_{51} & B_{52} & B_{53} & B_{54} & B_{55} & B_{56} \\ B_{57} & B_{58} & B_{59} & B_{60} & B_{61} & B_{62} & B_{63} & B_{64} \end{matrix} = \begin{matrix} 11100111 \\ 00011010 \\ 01101001 \\ 01000111 \\ 01011110 \\ 01011010 \\ 01101011 \\ 01011110 \end{matrix}$$

Then we permute the input according to the IP table (Table 4.5), such that

$$X = \begin{matrix} B_{58} & B_{50} & B_{42} & B_{34} & B_{26} & B_{18} & B_{10} & B_2 \\ B_{60} & B_{52} & B_{44} & B_{36} & B_{28} & B_{20} & B_{12} & B_4 \\ B_{62} & B_{54} & B_{46} & B_{38} & B_{30} & B_{22} & B_{14} & B_6 \\ B_{64} & B_{56} & B_{48} & B_{40} & B_{32} & B_{24} & B_{16} & B_8 \\ B_{57} & B_{49} & B_{41} & B_{33} & B_{25} & B_{17} & B_9 & B_1 \\ B_{59} & B_{51} & B_{43} & B_{35} & B_{27} & B_{19} & B_{11} & B_3 \\ B_{61} & B_{53} & B_{45} & B_{37} & B_{29} & B_{21} & B_{13} & B_5 \\ B_{63} & B_{55} & B_{47} & B_{39} & B_{31} & B_{23} & B_{15} & B_7 \end{matrix} = \begin{matrix} 11111101 \\ 10110010 \\ 10011001 \\ 01001101 \\ 00000001 \\ 01000101 \\ 11110110 \\ 11111011 \end{matrix}$$

After the permutation, the output is (*FDB2994D0145F6FB*) in HEX.

During the decryption process, an inverse permutation table IP^{-1} is applied, as shown in Table 4.6.

4.5.3 DES Round Function

As shown in Figure 4.11, each DES round follows the classic structure for a Feistel cipher, where $L_i = R_{i-1}$ and $R_i = L_{i-1} \oplus F(R_{i-1}, K_i)$. The round function $f(\cdot)$ of DES takes in 32-bit right half (R) and a 48-bit subkey. The round function first processes the 32-bit R through Expansion Permutation $E(\cdot)$.

Table 4.6 Inverse initial permutation (IP^{-1}).

40	8	48	16	56	24	64	32
39	7	47	15	55	23	63	31
38	6	46	14	54	22	62	30
37	5	45	13	53	21	61	29
36	4	44	12	52	20	60	28
35	3	43	11	51	19	59	27
34	2	42	10	50	18	58	26
33	1	41	9	49	17	57	25

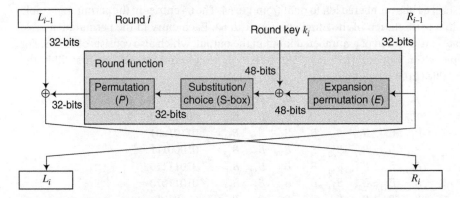

Figure 4.11 DES round structure.

As shown in Table 4.7, the input to $E(\cdot)$ is written in a 8×4 table. Each entry in the left most column expands to its left column by the entry to its left in the original matrix. Note that the first entry expands with the last entry (i.e. 32 is expanded to the left of 1). Each entry in the right most column expands to its right column by the entry to its right in the original matrix. Note that the last entry expands with the first entry (i.e. 1 is expanded to the right of 32). The output of E is 48 bits (8×6).

After the expansion, the 48-bit subkey is added to the 48-bit output using XOR. The results are split into eight 6-bit sequences, and get passed through 8 substitution boxes (S-boxes) respectively. The output of each S-box is 4-bit. The overall output of S-boxes is 32-bit.

Table 4.7 Expansion permutation (E).

32	1	2	3	4	5
4	5	6	7	8	9
8	9	10	11	12	13
12	13	14	15	16	17
16	17	18	19	20	21
20	21	22	23	24	25
24	25	26	27	28	29
28	29	30	31	32	1

4.5.3.1 DES S-Boxes

DES has eight S-boxes, as defined in Table 4.8. The S-boxes are designed to provide non-linearity, resistance to differential cryptanalysis and good confusion. Each of the eight S-boxes accepts 6 bits as input and produces 4 bits as output. For example, $S_1(011000) = 0101$. The input of an S-box is 6-bit, e.g. $b_0b_1b_2b_3b_4b_5$. The first and last bits of the input form a 2-bit binary number (i.e. b_0b_5), which is used to select one of four substitution defined by the four rows in the table (00, 01, 10, 11). The four bits in the

Table 4.8 DES S-boxes.

	14	4	13	1	2	15	11	8	3	10	6	12	5	9	0	7
S_1	0	15	7	4	14	2	13	1	10	6	12	11	9	5	3	8
	4	1	14	8	13	6	2	11	15	12	9	7	3	10	5	0
	15	12	8	2	4	9	1	7	5	11	3	14	10	0	6	13
	15	1	8	14	6	11	3	4	9	7	2	13	12	0	5	10
S_2	3	13	4	7	15	2	8	14	12	0	1	10	6	9	11	5
	0	14	7	11	10	4	13	1	5	8	12	6	9	3	2	15
	13	8	10	1	3	15	4	2	11	6	7	12	0	5	14	9
	10	0	9	14	6	3	15	5	1	13	12	7	11	4	2	8
S_3	13	7	0	9	3	4	6	10	2	8	5	14	12	11	15	1
	13	6	4	9	8	15	3	0	11	1	2	12	5	10	14	7
	1	10	13	0	6	9	8	7	4	15	14	3	11	5	2	12
	7	13	14	3	0	6	9	10	1	2	8	5	11	12	4	15
S_4	13	8	11	5	6	15	0	3	4	7	2	12	1	10	14	9
	10	6	9	0	12	11	7	13	15	1	3	14	5	2	8	4
	3	15	0	6	10	1	13	8	9	4	5	11	12	7	2	14
	2	12	4	1	7	10	11	6	8	5	3	15	13	0	14	9
S_5	14	11	2	12	4	7	13	1	5	0	15	10	3	9	8	6
	4	2	1	11	10	13	7	8	15	9	12	5	6	3	0	14
	11	8	12	7	1	14	2	13	6	15	0	9	10	4	5	3
	12	1	10	15	9	2	6	8	0	13	3	4	14	7	5	11
S_6	10	15	4	2	7	12	9	5	6	1	13	14	0	11	3	8
	9	14	15	5	2	8	12	3	7	0	4	10	1	13	11	6
	4	3	2	12	9	5	15	10	11	14	1	7	6	0	8	13
	4	11	2	14	15	0	8	13	3	12	9	7	5	10	6	1
S_7	13	0	11	7	4	9	1	10	14	3	5	12	2	15	8	6
	1	4	11	13	12	3	7	14	10	15	6	8	0	5	9	2
	6	11	13	8	1	4	10	7	9	5	0	15	14	2	3	12
	13	2	8	4	6	15	11	1	10	9	3	14	5	0	12	7
S_8	1	15	13	8	10	3	7	4	12	5	6	11	0	14	9	2
	7	11	4	1	9	12	14	2	0	6	10	13	15	3	5	8
	2	1	14	7	4	10	8	13	15	12	9	0	3	5	6	11

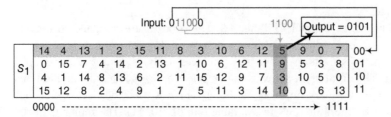

Figure 4.12 Illustration of S-box process.

middle (i.e. $b_1b_2b_3b_4$) determine one of the 16 columns (from 0000 to 1111). Note that both the rows and the columns are indexed starting from 0. The decimal value in the cell selected by the row and column is then converted to its 4-bit representation to produce the output. The process of $S_1(011000)$ is shown in Figure 4.12. The row is 00 (the first row) and the column is 0110 (the 13th column). The value is 5 in decimal. Thus the output is 0101 by converting 5 into binary format.

Question: Given an input of S-boxes as a sequence of 48 bits, or equivalently in hexadecimal:

$$S = (27, 12, 03, 3d, 21, 37, 15, 24).$$

Can you find the corresponding output? (Solution: $(2, 7, 7, 2, B, 7, 5, 4)$ in hexadecimal).

4.5.3.2 DES Permutation Function

The output of S-boxes is finally processed by permutation function P. The detailed permutation table is shown in Table 4.9. Note that the S-boxes provide the "confusion" of data and key values, while the permutation $P(\cdot)$ spreads this as widely as possible, so each S-box output affects as many S-box inputs in the next round as possible to increase "diffusion."

4.5.4 DES Key Schedule

In each of the 16 rounds, a subkey is also part of the input to round function $f(\cdot)$. The subkeys are generated from the 56-bit key for each data encryption round. In practice, the input of the symmetric key K is 64-bit. The extra 8 bits can be parity bits or simply arbitrary. The overall subkey generation process is shown in Figure 4.13. Note that the input of the key schedule is a 64-bit symmetric key K.

The first operation of subkey generation is Permutation Choice 1 (*PC1*). Given an input key as shown in Table 4.10, the output of *PC1* is shown in Table 4.11. The output of *PC1*

Table 4.9 Permutation function (*P*).

16	7	20	21	29	12	28	17
1	15	23	26	5	18	31	10
2	8	24	14	32	27	3	9
19	13	30	6	22	11	4	25

Figure 4.13 The key schedule of DES.

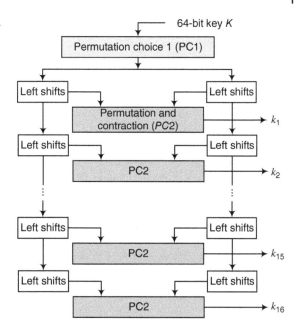

Table 4.10 Indexes of the input key K.

1	2	3	4	5	6	7	8
9	10	11	12	13	14	15	16
17	18	19	20	21	22	23	24
25	26	27	28	29	30	31	32
33	34	35	36	37	38	39	40
41	42	43	44	45	46	47	48
49	50	51	52	53	54	55	56
57	58	59	60	61	62	63	64

Table 4.11 Permutation choice 1 ($PC1$).

57	49	41	33	25	17	9
1	58	50	42	34	26	18
10	2	59	51	43	35	27
19	11	3	60	52	44	36
63	55	47	39	31	23	15
7	62	54	46	38	30	22
14	6	61	53	45	37	29
21	13	5	28	20	12	4

Table 4.12 Schedule of left shifts.

Round #	1	2	3	4	5	6	7	8	9	10	11	12	13	14	15	16
Bits rotated	1	1	2	2	2	2	2	2	1	2	2	2	2	2	2	1

is 56-bit, where $(8, 16, 24, 32, 40, 48, 56, 64)$ are discarded during the process because those 8 bits are not used in the 56-bit key input to the DES key schedule. The output of *PC*1 is treated as two 28-bit quantities (i.e. the upper 4 rows and the lower 4 rows). After each round, the two halves are separately subjected to a circular left shift of 1 or 2 bits according to the schedule listed in Table 4.12.

The shifted values are then input into Permutation Choice 2 (*PC*2). Detailed permutation of *PC*2 is illustrated in Table 4.13. The output of *PC*2 is the 48-bit subkey. Note that the values in *PC*2 are re-ordered from left shifts, starting from 1 to 56, with the original bits in 9, 18, 22, 25, 35, 38, 43, 54 being eliminated, so that the output only has 48 bits.

> **DES decryption**: as a feature of Feistel block cipher, DES decryption uses the same algorithm as encryption except that the subkeys derived are used in reverse order from k_{16} to k_1.

4.5.5 DES Security

The analysis of DES security can be done in several aspects:

- *Avalanche effect*: It is a desirable property for an encryption algorithm. A small change in either the plaintext or the key should produce a significant change in the ciphertext. A change in one bit of the plaintext or one bit of the key should produce a change in many bits of the ciphertext. It is proved that DES exhibits a strong avalanche effect.
- *Key size*: DES has a key size of 56 bits. There are $2^{56} \approx 7.2 \times 10^{16}$ keys. A brute-force attack appeared impractical (in a reasonable amount of time) to DES when proposed. However, DES could be broken within a day as shown in 1999 with the capability of an average computer and some assumptions. The time is much shortened nowadays. Unless known plaintext is provided, the analyst must be able to recognize plaintext.
- *Analytic attacks*: Cryptanalysis is possible by exploiting the characteristics of DES. The focus has been on the eight S-boxes that are used in each iteration. These techniques utilize some deep structure of the cipher by gathering information about encryptions so that eventually you can recover some/all of the sub-key bits, and then exhaustively search for

Table 4.13 Permutation choice 2 (*PC*2).

14	17	11	24	1	5	3	28
15	6	21	10	23	19	12	4
26	8	16	7	27	20	13	2
41	52	31	37	47	55	30	40
51	45	33	48	44	49	39	56
34	53	46	42	50	36	29	32

the rest if necessary. Generally these are statistical attacks that depend on the amount of information gathered for their likelihood of success. Attacks of this form include differential cryptanalysis, linear cryptanalysis, and related key attacks.

- *Timing attacks*: A timing attack is one in which information about the key or the plaintext is obtained by observing how long it takes a given implementation to perform decryptions on various ciphertexts. A timing attack exploits the fact that an encryption or decryption algorithm often takes slightly different amounts of time on different inputs. DES appears to be fairly resistant to a successful timing attack.

Despite its popularity in the past, DES is generally considered insecure in today's increasing computing power. Alternatives must be found to replace DES. The most important of the alternatives are AES and triple DES. Triple DES is to run DES three times (E–D–E as encryption, and D—E–D as decryption).

4.5.6 Multiple Encryption and DES

As discussed before, DES is vulnerable to a brute-force attack. Therefore, there has been considerable interest in finding an alternative. One approach is to design a completely new algorithm, of which AES is a prime example. Another alternative, which would preserve the existing investment in software and equipment, is to use multiple encryption with DES and multiple keys.

Double DES (2-DES): The simplest form of multiple encryption has two encryption stages and two keys, i.e. *2-DES* or *double DES*. Two keys (e.g. K_1 and K_2) are applied to two DES encryptions. The encryption and decryption processes are shown in Figure 4.14. A ciphertext is produced as:

$$C = E_{K_2}(E_{K_1}(P)), \tag{4.5}$$

and the decryption is:

$$P = D_{K_1}(D_{K_2}(C)). \tag{4.6}$$

With two keys K_1 and K_2, 2-DES increases the effective key size to 112 bits. However, "meet-in-the-middle" attack makes 2-DES vulnerable. The "meet-in-the-middle" attack was first described by Whitfield Diffie and Martin Hellman in 1977 [21]. It is a known plaintext attack (i.e. with a pair of known plaintext and ciphertext (P, C), attempting to find by trial-and-error a value X in the "middle" of the double-DES encryption of this pair, s.t., $X = E_{K_1}(P) = D_{K_2}(C)$. Chances of this are much better at $O(2^{56})$ than exhaustive search at $O(2^{112})$. Therefore, 2-DES is not a secure alternative to DES.

Triple-DES (3-DES) with two keys: 3-DES is another alternative to DES. 3-DES with two keys is a popular implementation. With two keys (e.g. K_1 and K_2), the encryption

Figure 4.14 Double-DES (2-DES).

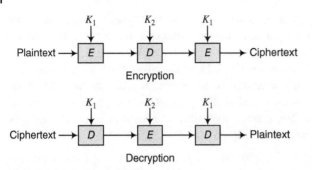

Figure 4.15 Triple-DES (3-DES).

and decryption processes are shown in Figure 4.15. Mathematically, the encryption is as follows:

$$C = E_{K_1}(D_{K_2}(E_{K_1}(P))), \tag{4.7}$$

and the decryption is as follows:

$$P = D_{K_1}(E_{K_2}(D_{K_1}(C))). \tag{4.8}$$

The uses of encryption and decryption stages are equivalent, but the chosen structure allows for compatibility with single-DES implementations. If $K_1 = K_2$, then $E_{K_1}(D_{K_2}(E_{K_1}(P))) = E_{K_1}(P)$. Currently, there are no practical cryptanalytic attacks on 3-DES. The cost of a brute-force key search on 3-DES is on the order of $2^{112} \approx 5 \times 10^{33}$. It estimates that the cost of differential cryptanalysis suffers an exponential growth, compared with single DES, exceeding $O(10^{52})$. However, 3-DES suffers from being three times slower to run.

3-DES with three keys: Although there are no practical attacks on two-key triple-DES, anyone using two-key triple-DES may feel some concern of possible attacks on 2-key triple-DES. Therefore, triple-DES with three keys (168 bits) is being used in some applications, including PGP and S/MIME on the Internet, for greater security. The encryption of 3-key triple-DES is as follows:

$$C = E_{K_3}(D_{K_2}(E_{K_1}(P))), \tag{4.9}$$

and the decryption of 3-key triple-DES is as follows:

$$P = D_{K_1}(E_{K_2}(D_{K_3}(C))). \tag{4.10}$$

4.6 Summary

In this chapter, several symmetric key cryptographic techniques were presented. A few classical cryptographic algorithms with substitution and transposition techniques were illustrated for the understanding of basics of symmetric key algorithms. Modern stream cipher and block cipher were also presented. The design of modern block ciphers based on Feistel cipher structure was also demonstrated. More cryptographic techniques using block ciphers and public key algorithms will be given in the next chapter.

5

More on Cryptographic Techniques

This chapter continues to introduce more on modern block ciphers, including Advanced Encryption Standard, block cipher mode of operations, public key infrastructure, RSA algorithm, and Diffie–Hellman (D–H) key exchange protocol.

5.1 Advanced Encryption Standards

The Advanced Encryption Standard (AES) was standardized by National Institute of Standards and Technology (NIST) in 2001 [22]. While 3-DES is considered secure and well understood, it is too slow in software implementation. NIST solicited a new symmetric block cipher AES to replace DES for a wide range of applications. The Rijndael proposal was finally selected as the AES standard in November 2001 [23]. The two researchers who developed and submitted Rijndael for the AES are both cryptographers from Belgium, Joan Daemen and Vincent Rijmen. The chosen AES cipher and other candidates for the latest generation of block ciphers have a significant increase in the sizes of block and keys.

5.1.1 The AES Cipher: Rijndael

The Rijndael proposal for AES defined a cipher in which the block size is 128 bits and the key length can be independently set to 128, 192, or 256 bits. Depending on different key lengths, AES has three parameter configurations, as depicted in Table 5.1. Although the AES standard uses three key size alternatives, it limits the block length to 128 bits for all configurations. AES is an iterative cipher (instead of a Feistel cipher) that operates on the entire data block (i.e. 4 × 4 bytes) in every round. Note that Feistel cipher operates on halves of data block at a time. AES is designed to have characteristics as follows: resistance against all known attacks; speed and code compactness on a wide range of platforms; and design simplicity.

5.1.2 AES Data Structure

The overall structure of AES with a key size of 128 bits is illustrated in Figure 5.1. The input to the AES encryption and decryption algorithms is a single 128-bit block as a square matrix of bytes, depicted in the Federal Information Processing Standards Publication

Security in Wireless Communication Networks, First Edition. Yi Qian, Feng Ye, and Hsiao-Hwa Chen.
© 2022 John Wiley & Sons Ltd. Published 2022 by John Wiley & Sons Ltd.
Companion website: www.wiley.com/go/qian/sec51

Table 5.1 Parameters for different AES configurations.

Parameter	AES-128	AES-192	AES-256
Key size	128 bits	192 bits	256 bits
Plaintext block size	128 bits	128 bits	128 bits
Number of rounds	10	12	14
Round key size	128 bits	128 bits	128 bits
Expanded key size	176 bytes	208 bytes	240 bytes

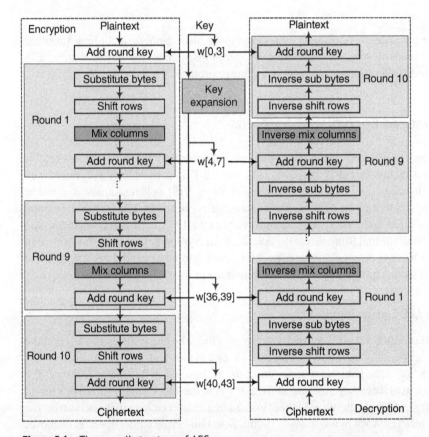

Figure 5.1 The overall structure of AES.

(FIPS PUB) 197. The other two AES configurations have the similar structure with more rounds.

As shown in Figure 5.2a, the input block is copied into the state array, which is modified at each stage of encryption or decryption. After the final stage, the state is copied to an output. As shown in Figure 5.2b, the input key is expanded into 44/52/60 (depending on the original key size) words with 32-bit in each word, and four words used in each round. The data computation consists of an *add round key* step, followed by 9/11/13 (depending

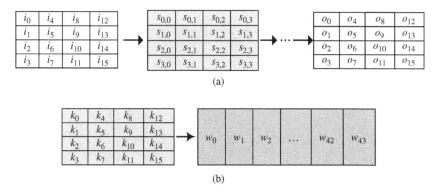

Figure 5.2 AES data structure. (a) Input, state array, and output block. (b) Input key and expanded key.

on the original key size) rounds with all four steps in each round, and a final round with three steps. The four steps before the last round are: *substitute bytes*, *shift rows*, *mix columns*, and *add round key*. The three steps in the last round are: *substitute bytes*, *shift rows*, and *add round key*. All the steps are easily reversed, and can be efficiently implemented using XOR's and table lookups for both AES encryption and AES decryption.

5.1.3 Details in Each Round

The detailed illustration will be given with AES-128, which has a key size of 128 bits. The other two AES implementations are similar. In AES-128, there are ten rounds in both encryption and decryption algorithm. For the first nine rounds, one permutation and three substitutions are applied as follows:

- *Substitute bytes*: Uses an S-box to perform a byte-by-byte substitution of the block.
- *Shift rows*: A simple permutation.
- *Mix columns*: A substitution that makes use of arithmetic over $GF(2^8)$, with the irreducible polynomial $m(x) = x^8 + x^4 + x^3 + x + 1$.
- *Add round key*: A simple bitwise XOR of the current block with a portion of the expanded key.

Only the *add round key* stage makes use of the round key. For this reason, the cipher begins and ends with an *add round key* stage. Any other stage, applied at the beginning or end, is reversible without knowledge of the key and so would add no security. The tenth round (the final round) of both encryption and decryption consists of only three stages (i.e. substitute bytes, shift rows, and add round key).

5.1.3.1 Substitute Bytes

The *substitute bytes* stage uses a 16×16 S-box to perform a byte-by-byte substitution of the block. This S-box is a permutation of all 256 8-bit values, constructed using a transformation that treats the values as polynomials in $GF(2^8)$. Decryption requires the inverse of the table. The two tables are fixed, as given in Tables 5.2 and 5.3.

Both the S-box and the inverse S-box were designed to be resistant to known cryptanalytic attacks, as required by NIST. Specifically, the design guarantees a low correlation between

Table 5.2 AES *S*-box for encryption.

	0	1	2	3	4	5	6	7	8	9	A	B	C	D	E	F
0	63	7C	77	7B	F2	6B	6F	C5	30	01	67	2B	FE	D7	AB	76
1	CA	82	C9	7D	FA	59	47	F0	AD	D4	A2	AF	9C	A4	72	C0
2	B7	FD	93	26	36	3F	F7	CC	34	A5	E5	F1	71	D8	31	15
3	04	C7	23	C3	18	96	05	9A	07	12	80	E2	EB	27	B2	75
4	09	83	2C	1A	1B	6E	5A	A0	52	3B	D6	B3	29	E3	2F	84
5	53	D1	00	ED	20	FC	B1	5B	6A	CB	BE	39	4A	4C	58	CF
6	D0	EF	AA	FB	43	4D	33	85	45	F9	02	7F	50	3C	9F	A8
7	51	A3	40	8F	92	9D	38	F5	BC	B6	DA	21	10	FF	F3	D2
8	CD	0C	13	EC	5F	97	44	17	C4	A7	7E	3D	64	5D	19	73
9	60	81	4F	DC	22	2A	90	88	46	EE	B8	14	DE	5E	0B	DB
A	E0	32	3A	0A	49	06	24	5C	C2	D3	AC	62	91	95	E4	79
B	E7	C8	37	6D	8D	D5	4E	A9	6C	56	F4	EA	65	7A	AE	08
C	BA	78	25	2E	1C	A6	B4	C6	E8	DD	74	1F	4B	BD	8B	BA
D	70	3E	B5	66	48	03	F6	0E	61	35	57	B9	86	C1	1D	9E
E	E1	F8	98	11	69	D9	8E	94	9B	1E	87	E9	CE	55	28	DF
F	8C	A1	89	0D	BF	E6	42	68	41	99	2D	0F	B0	54	BB	16

Table 5.3 AES inverse *S*-box for decryption.

	0	1	2	3	4	5	6	7	8	9	A	B	C	D	E	F
0	52	09	6A	D5	30	36	A5	38	BF	40	A3	9E	81	F3	D7	FB
1	7C	E3	39	82	9B	2F	FF	87	34	8E	43	44	C4	DE	E9	CB
2	54	7B	94	32	A6	C2	23	3D	EE	4C	95	0B	42	FA	C3	4E
3	08	2E	A1	66	28	D9	24	B2	76	5B	A2	49	6D	8B	D1	25
4	72	F8	F6	64	86	68	98	16	D4	A4	5C	CC	5D	65	B6	92
5	6C	70	48	50	FD	ED	B9	DA	5E	15	46	57	A7	8D	9D	84
6	90	D8	AB	00	8C	BC	D3	0A	F7	E4	58	05	B8	B3	45	06
7	D0	2C	1E	8F	CA	3F	0F	02	C1	AF	BD	03	01	13	8A	6B
8	3A	91	11	41	4F	67	DC	EA	97	F2	CF	CE	F0	B4	E6	73
9	96	AC	74	22	E7	AD	35	85	E2	F9	37	E8	1C	75	DF	6E
A	47	F1	1A	71	1D	29	C5	89	6F	B7	62	0E	AA	18	BE	1B
B	FC	56	3E	4B	C6	D2	79	20	9A	DB	C0	FE	78	CD	5A	F4
C	1F	DD	A8	33	88	07	C7	31	B1	12	10	59	27	80	EC	5F
D	60	51	7F	A9	19	B5	4A	0D	2D	E5	7A	9F	93	C9	9C	EF
E	A0	E0	3B	4D	AE	2A	F5	B0	C8	EB	BB	3C	83	53	99	61
F	17	2B	04	7E	BA	77	D6	26	E1	69	14	63	55	21	0C	7D

input bits and output bits with the property that the output cannot be described as a simple mathematical function of the input. When applying *S*-box (or inverse *S*-box) substitution, each byte of state is replaced by byte indexed by row (left 4 bits) and column (right 4 bits), for example, byte "58" is replaced by byte in row 5 and column 8, which is "6A." For example, a complete process of *substitute bytes* is given as follows:

89	85	2D	CB		A7	97	D8	1F
----	----	----	----		----	----	----	----
D8	5A	18	12	*S-box* $\rightarrow$	61	BE	AD	C9
10	CE	43	8F		CA	8B	1A	73
E8	68	D8	E4		9B	45	61	69

5.1.3.2 Shift Rows

The *shift rows* stage shifts three of the rows of the state array to provide a simple "permutation" of the data. This stage also provides for diffusion of values between columns. In particular, *shift rows* performs a circular rotate on each row as follows:

- First row is unchanged.
- Second row does 1 byte circular shift to left.
- Third row does 2 byte circular shift to left.
- Fourth row does 3 byte circular shift to left.

The process is shown in Figure 5.3. For example, given a current state block in the left below, the *shift rows* is performed as follows:

A7	97	D8	1F		A7	97	D8	1F
61	BE	AD	C9	*Shift rows*	BE	AD	C9	61
CA	8B	1A	73	→	1A	73	CA	8B
9B	45	61	69		69	9B	45	61

$s_{0,0}$	$s_{0,1}$	$s_{0,2}$	$s_{0,3}$		$s_{0,0}$	$s_{0,1}$	$s_{0,2}$	$s_{0,3}$
$s_{1,0}$	$s_{1,1}$	$s_{1,2}$	$s_{1,3}$	*Shift rows*	$s_{1,1}$	$s_{1,2}$	$s_{1,3}$	$s_{1,0}$
$s_{2,0}$	$s_{2,1}$	$s_{2,2}$	$s_{2,3}$	→	$s_{2,2}$	$s_{2,3}$	$s_{2,0}$	$s_{2,1}$
$s_{3,0}$	$s_{3,1}$	$s_{3,2}$	$s_{3,3}$		$s_{3,3}$	$s_{3,0}$	$s_{3,1}$	$s_{3,2}$

Figure 5.3 Illustration of *shift rows*.

In AES decryption, the *inverse shift rows* stage performs the circular shifts in the opposite direction for each row.

5.1.3.3 Mix Columns

The *mix columns* stage is a substitution that makes use of arithmetic over $GF(2^8)$. In this stage, each byte in the block is mapped into a new value based on all values in that column. The mapping function is a polynomial matrix multiplication in $GF(2^8)$, with the irreducible polynomial $m(x) = x^8 + x^4 + x^3 + x + 1$. The chosen constants mix all four bytes in the same column with maximal distance between code words. Specifically, the transformation is defined by the matrix multiplication as follows:

$$[\text{Mix Column}] = \begin{bmatrix} 02 & 03 & 01 & 01 \\ 01 & 02 & 03 & 01 \\ 01 & 01 & 02 & 03 \\ 03 & 01 & 01 & 02 \end{bmatrix} \times [\text{Current State}]. \tag{5.1}$$

Continue with the output of *shift rows*, the results of the mix column can be found as follows:

A7	97	D8	1F		FF	31	64	77
BE	AD	C9	61	*Mix columns*	87	D8	51	3A
1A	73	CA	8B	→	96	6A	51	D0
69	9B	45	61		84	51	FA	09

AES decryption applies the *inverse mix columns* stage, where the matrix multiplication are given as follows:

$$[\text{Inverse Mix Column}] = \begin{bmatrix} 0E & 0B & 0D & 09 \\ 09 & 0E & 0B & 0D \\ 0D & 09 & 0E & 0B \\ 0B & 0D & 09 & 0E \end{bmatrix} \times [\text{Current State}]. \qquad (5.2)$$

Since *mix columns* are the matrix multiplication, it can be implemented as four equations to map all 4 bytes in the same column. Therefore, software implementation relies only on shifts, XORs and conditional XORs.

5.1.3.4 Add Round Key

Add round key is the first operation in both AES encryption and AES decryption. It is also the last stage in each round of AES encryption. The stage is simply XORing the current state block with the block round key, illustrated as follows:

$s_{0,0}$	$s_{0,1}$	$s_{0,2}$	$s_{0,3}$
$s_{1,0}$	$s_{1,1}$	$s_{1,2}$	$s_{1,3}$
$s_{2,0}$	$s_{2,1}$	$s_{2,2}$	$s_{2,3}$
$s_{3,0}$	$s_{3,1}$	$s_{3,2}$	$s_{3,3}$

$\oplus\ [w_i | w_{i+1} | w_{i+2} | w_{i+3}].$

The *add round key* stage is simple thus it does not provide too much security in terms of confusion, diffusion, and non-linearity. Security is mainly provided by the other three stages in each round. Therefore, the *add round key* stage is applied at the end of each round, in which case the confusion, diffusion, and non-linearity provided from other stages can be kept in AES encryption. The inverse add round key is identical to the forward add round key, because the XOR operation is its own inverse.

Example: Given the current state block and the round key block, the *add round key* stage is operated as follows:

FF	31	64	77
87	D8	51	3A
96	6A	51	D0
84	51	FA	09

$\oplus$

B6	64	BE	68
92	3D	9B	30
CF	BD	C5	B3
0B	F1	00	FE

where the first matrix is current state, the second matrix is the round key. The result is the output of *add round key* as follows:

49	55	DA	1F
15	E5	CA	0A
59	D9	94	63
8F	A0	FA	F7

5.1.3.5 AES Key Expansion

The AES key expansion algorithm takes as input the secret key (16 bytes in AES-128) and produces an array of words used as round keys in *add round key* stages. The AES key expansion algorithm processes as follows:

- Firstly, it copies the initial key into the first group of four words of four round keys, i.e. $W[0]$, $W[1]$, $W[2]$, $W[3]$, with 4 bytes in each word.
- Secondly, it constructs subsequent groups of four round keys. The first word in each group of 4 (e.g. $W[4]$) is performed with *rotate word*, S-box, and XOR constant on the previous word. Rotate words performs a one-byte circular left shift on a word, i.e. $[B_0, B_1, B_2, B_3] \rightarrow [B_1, B_2, B_3, B_0]$.
- Thirdly, the output is substituted using the S-box in Table 5.2.
- After substitution, the first byte of the results are XORed with a round constant $RC[i]$. The constant is determined in the following:

Round #	1	2	3	4	5	6	7	8	9	10
$RC[i]$	01	02	04	08	10	20	40	80	1B	36

Note that $RC[i]$ is only performed on the first byte of a word, the next three bytes are not XORed, i.e. $(S_0 \oplus RC[i], S_1, S_2, S_3)$.

- Finally, the output is XORed with the 4*th* back words (i.e. $W[i-4]$). For example, if the output of XOR with RC is $W'[4]$, then $W[4] = W'[4] \oplus W[0]$.
- The next three words in a group (e.g. $W[5]$, $W[6]$, and $W[7]$) are simply based on the values of the previous and the 4*th* back words as follows:

$$W[i] = W[i-1] \oplus W[i-4]. \tag{5.3}$$

The detailed AES key expansion algorithm is summarized in Algorithm 5.1.

Algorithm 5.1 AES key expansion

Input: AES key *byte key*[16];
Output: Expanded key *word w*[44];
    word *temp*;
    for $i = 0; i < 4; i + +$ **do**
      $w[i] = (key[4i], key[4i + 1], key[4i + 2], key[4i + 3]);$
    end for
    for $i = 4; i < 44; i + +$ **do**
      $temp = w[i - 1];$
      if $i \bmod 4 == 0$ **then**
        $temp = SubWord(RotWord(temp)) \oplus RC[i/4];$
      end if
      $w[i] = w[i - 4] \oplus temp;$
    end for

Key expansion rationale: The algorithm is designed to be resistant to known cryptanalytic attacks. Specifically, given part of the key, it would be insufficient to find the rest of the key; the key expansion must be an invertible transformation thus decryption can be performed; the algorithm needs to use round constants to break symmetry; the algorithm needs to provide enough non-linearity to hinder analysis; the algorithm must be simple and easy to implement on a wide range of CPUs.

Example: Given the round key for the second round as:

$$(W[8], W[9], W[10], W[11])$$

$$=(B692CF0B, \quad 643DBDF1, \quad BE30B3FE, \quad 6830B3FE).$$

The first four bytes (first column) of the round key for round 3 (i.e. $W[12]$) are calculated as follows:

- i=12;
- $temp = W[i-1] = W[11] = 6830B3FE$;
- After *RotWord* → 30B3FE68;
- After *SubWord* → 046DBB45;
- After ⊕ with $RC[3] = 006DBB45$;
- $W[i-4] = W[8] = B692CF0B$;
- $W[i] = W[12] = temp \oplus W[i-4] = B6FF744E$.

Then, we can find the rest three words of round 3 as follows:

$$W[13] = W[12] \oplus W[9] = D2C2C9BF,$$

$$W[14] = W[13] \oplus W[10] = 6C590CBF,$$

$$W[15] = W[14] \oplus W[11] = 0469BF41.$$

5.1.3.6 AES Decryption
The AES decryption cipher has a similar 9-and-1 round structure where an *add round key* stage is at the beginning of entire process. However, it is not the exact reverse of the encryption cipher. In each round, the four stages are applied with their inverse functions in a different order: *inverse shift rows, inverse substitute bytes, add round key*, and *inverse mix columns*. The *add round key* stage is the same as it in the encryption cipher. Note that the *add round key* stage is applied neither the first nor the last in the decryption cipher, which may break the security provided by other stages. Fortunately, the purpose of the decryption cipher is to reveal the encrypted message instead of making it harder to understand. Due to the different stage functions, AES encryption and decryption need two separate software or firmware modules.

5.1.3.7 AES Implementation Aspects
Due to byte operation design, AES is easy to implement on an 8-bit processor. *Add round key* is a byte wise XOR operation. *Shift rows* is a simple byte shifting operation. *Substitute bytes* operates at the byte level and only requires a lookup of a 256-byte table *S*. *Mix columns* is matrix multiplication that can be implemented as byte XOR's and table lookups with a second 256 byte table. AES can also be very efficiently implemented on 32-bit and 64-bit processors, by rewriting the stage transformation to use 4/8 table lookups and 4/8 XOR's per column of state. These tables can be computed in advance that needs 4/8 Kbytes to store.

5.2   Block Cipher Modes of Operation

A block cipher forms a basic building block which encrypts or decrypts a fixed sized block of data. For example, DES encrypts 64-bit blocks with a 56-bit key. In practice, data to be handled usually has an arbitrary length. The entire data may be available in advance, where block operations can be applied. The data may only be available a byte or even a bit at a time, where stream operations are more appropriate. In order to apply a block cipher in a variety of application, NIST FIPS PUB 81 defined four *modes of operation* [24] based on DES block cipher: the electronic codebook (ECB), cipher block chaining (CBC), cipher feedback (CFB), and output feedback (OFB) modes. In 2001, NIST defined another block operation mode in the special publication 800-38A [25]. The added mode is the Counter (CTR) mode. Although DES was the original chosen block cipher, the modes can be used with any symmetric block cipher, including 3-DES and AES. More recently, an additional block cipher mode of operations XTS-AES was standardized by IEEE and NIST [26].

5.2.1   Electronic Codebook (ECB) Mode

The ECB mode is the simplest mode. As shown in Figure 5.4, the input message is broken into independent blocks for encryption. For a given key K, ECB encryption handles a block at a time, analogous to the assignment of code words in a codebook. The encryption and decryption of the ECB mode are defined as follows:

$$C_i = E_K (P_i), \quad \text{for } i = 1, \ldots, N, \tag{5.4}$$

$$P_i = D_K (C_i), \quad \text{for } i = 1, \ldots, N. \tag{5.5}$$

The ECB mode is used when only a few blocks of information needs to be sent, for example, a session key encrypted using a master key. In both ECB encryption and ECB decryption, multiple forward cipher functions and inverse cipher functions can be computed in parallel.

   Advantages and limitations of ECB mode: The advantage of ECB mode is its computational efficiency and capability in parallel computing. However, ECB mode is not appropriate

Figure 5.4   The electronic codebook (ECB) mode. (a) Encryption. (b) Decryption.

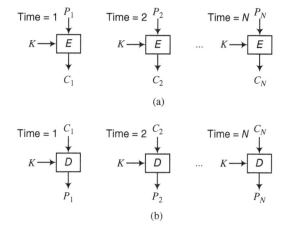

for a large amount of data since repetitions may show in ciphertext, due to its independent blocks. The main use of the ECB mode is to encrypt a small amount of data, e.g. just one or a few blocks.

5.2.2 Cipher Block Chaining (CBC) Mode

The CBC mode overcomes the weakness of repetitions and order independence in the ECB mode. In the CBC mode, the input message is also split into blocks, as shown in Figure 5.5. The encryption and decryption of a block in the CBC mode depends on current message block and all the previous ciphertext blocks. The CBC mode requires an initial value (i.e. IV) for the first block operation. The IV is usually a well-known value (e.g. all 0's), or otherwise is sent encrypted with the ECB mode. In CBC encryption, the first input block is formed by XORing the first input block message with the *IV*. Inputs to the other blocks are formed by XORing the current block message with the previous ciphertext. The encryption of the CBC mode is defined as follows:

$$\begin{cases} C_1 = E_K\,(P_1 \oplus IV), \\ C_i = E_K\,(P_i \oplus C_{i-1}), & \text{for } i = 2, \dots, N. \end{cases} \tag{5.6}$$

In CBC decryption, each ciphertext block is passed through the decryption algorithm. The result is XORed with the preceding ciphertext block to produce the plaintext block. The plaintext of the first block is revealed by first decrypting the ciphertext block, and then XORing with the *IV*. The decryption of the CBC mode is defined as follows:

$$\begin{cases} P_1 = D_K\,(C_1) \oplus IV, \\ P_i = D_K\,(C_i) \oplus C_{i-1}, & \text{for } i = 2, \dots, N. \end{cases} \tag{5.7}$$

Advantages and limitations of CBC mode: Since a ciphertext block depends on all blocks before it in the CBC mode, the encrypted message cannot be changed or rearranged without

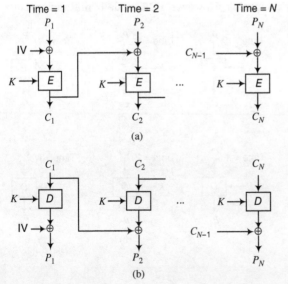

Figure 5.5 Cipher block chaining (CBC) mode. (a) Encryption. (b) Decryption.

totally destroying the subsequent data. However, the CBC mode needs an IV which must be known to both sender and receiver. If the IV is sent in clear text, then attack can be launched on the first block. Because any change of the first block can be compensated by changing the corresponding bits of the IV. Hence an IV must be a fixed value or be sent encrypted (e.g. using the ECB mode) before the rest of the message. The CBC mode is applicable whenever large amounts of data need to be sent securely, provided that all data is available in advance.

5.2.3 Cipher Feedback (CFB) Mode

In the cipher feedback (CFB) mode, message is treated as a stream of bits. In general, the CFB mode features the feedback of successive ciphertext segments into the input blocks of the forward cipher to generate output blocks that are XORed with the plaintext to produce the ciphertext. The standard of the CFB mode allows any number of bits (e.g. 1, 8, 64, 128, etc.) to be fed back, denoted by CFB-1, CFB-8, CFB-64, CFB-128, etc. It depends on the block size of the cipher used, for instance, DES (CFB-64) or AES (CFB-128).

As shown in Figure 5.6a, the encryption of the CFB mode starts with an IV. The block size of the cipher is b bits. The CFB mode also requires an integer parameter, denoted s, $1 \leq s \leq b$. It is the size of each plaintext block and the output of each ciphertext block. Let $MSB_s(X)$ be the left most significant s bits of X. Let $LSB_{(s-b)}(X)$ be the remaining $(b-s)$ bits

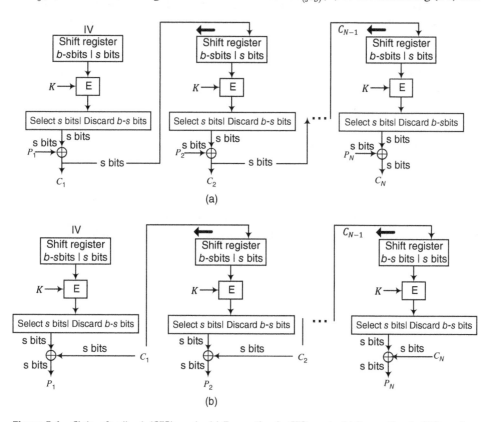

Figure 5.6 Cipher feedback (CFB) mode. (a) Encryption in CFB mode. (b) Decryption in CFB mode.

after the s-bit left shift register of the b-bit block X. The first ciphertext segment is produced by XORing the first plaintext segment with the *s* most significant bits of the first output block of the cipher. The remaining $(b - s)$ bits of the first output block of the cipher are discarded. The $(b - s)$ least significant bits of the IV are then concatenated with the *s* bits of the first ciphertext segment to form the second input block. The process is repeated with the successive input blocks until a ciphertext segment is produced from every plaintext segment. The encryption of the CFB mode is defined as follows:

$$\begin{cases} I_1 = IV, \\ C_1 = P_1 \oplus MSB_s(E_K(I_1)), \\ C_i = P_i \oplus MSB_s[E_K(LSB_{(b-s)}(I_{i-1})||C_{i-1})], & \text{for } i = 2, \dots, N. \end{cases} \quad (5.8)$$

In CFB decryption, simply replace the input of P_i to C_i. as follows:

$$\begin{cases} I_1 = IV, \\ P_1 = C_1 \oplus MSB_s(E_K(I_1)), \\ P_i = C_i \oplus MSB_s[E_K(LSB_{(b-s)}(I_{i-1})||C_{i-1})], & \text{for } i = 2, \dots, N. \end{cases} \quad (5.9)$$

Advantages and limitations of CFB mode: The CFB mode is usually applied as the stream mode, which is appropriate when data arrives in bits or bytes. The CFB mode is not appropriate for "noisy" transmission links. Because any corrupted data in the transmission will continue corrupting the values in the current and all future blocks. Moreover, the CFB model also requires an IV for initial operation. If the IV is revealed to an attacker, with a known plaintext block, e.g. the *ith* plaintext block, then the *ith* output of the forward cipher function can be determined easily from the *ith* ciphertext block of the message. With a properly secured IV, the CFB mode can be used for quantities of stream oriented data, and for authentication.

5.2.4 Output Feedback (OFB) Mode

The output feedback (OFB) mode is an alternative to the CFB mode. As shown in Figure 5.7, the input of the OFB mode is treated as a stream of bits. The output of cipher is XORed to the plaintext as the ciphertext. The output of the cipher is also fed back as the input to the cipher for the encryption of next block. Let I_i and O_i be the input and output encryption E of the *ith* block. Then the encryption of the OFB mode is defined as follows:

$$\begin{cases} I_1 = IV, \\ I_i = O_{i-1}, & \text{for } i = 2, \dots, N \\ O_i = E_K(I_i), & \text{for } i = 1, \dots, N \\ C_i = P_i \oplus O_i, & \text{for } i = 1, \dots, N. \end{cases} \quad (5.10)$$

For decryption, the OFB mode is defined as follows:

$$\begin{cases} I_1 = IV, \\ I_i = O_{i-1}, & \text{for } i = 2, \dots, N \\ O_i = E_K(I_i), & \text{for } i = 1, \dots, N \\ P_i = C_i \oplus O_i, & \text{for } i = 1, \dots, N. \end{cases} \quad (5.11)$$

Figure 5.7 Output feedback (OFB) mode. (a) Encryption in OFB mode. (b) Decryption in OFB mode.

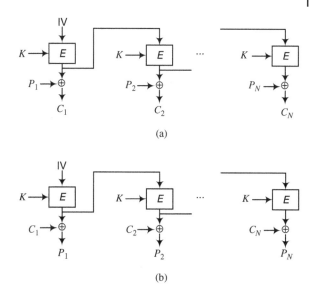

(a)

(b)

Advantages and limitations of OFB mode: The feedback in the OFB mode is independent of the input plaintext. The advantage is that bit errors in transmission do not propagate in the OFB mode. However, the OFB mode is more vulnerable to a message stream modification attack than the CFB mode. Since OFB is a Vernam cipher variant, the stream should never be reused (otherwise the 2 ciphertexts can be combined, cancelling these bits, and leaving a "book" cipher to solve). Moreover, sender and receiver need to remain in synchronization, or all data is lost.

5.2.5 The Counter (CTR) Mode

The counter (CTR) mode features the application of the forward cipher to a set of input blocks, i.e. counters, to produce a sequence of output blocks that are XORed with the plaintext to produce the ciphertext, as illustrated in Figure 5.8a. The size of a counter is equal to the size of a plaintext block. The only requirement stated in SP 800-38A is that the counter value must be different for each plaintext block that is encrypted. Typically the counter is initialized to some value and then incremented by 1 for each subsequent block. For encryption, a counter is encrypted first. The output is XORed with the plaintext to produce the ciphertext. The counter is incremented by 1 for the next block. The encryption for the CRT mode is defined as follows:

$$\begin{cases} O_i = E_K\,(i + Counter - 1), & \text{for } i = 1, \ldots, N \\ C_i = P_i \oplus O_i, & \text{for } i = 1, \ldots, N. \end{cases} \tag{5.12}$$

For decryption, swap the ciphertext and the plaintext in each block, it is operated as follows:

$$\begin{cases} O_i = E_K\,(i + Counter - 1), & \text{for } i = 1, \ldots, N \\ P_i = C_i \oplus O_i, & \text{for } i = 1, \ldots, N. \end{cases} \tag{5.13}$$

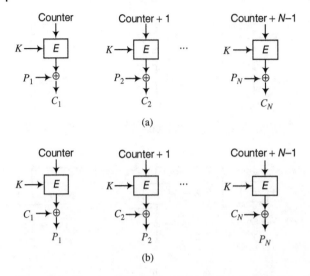

Figure 5.8 Counter (CTR) mode. (a) Encryption in CTR mode. (b) Decryption in CTR mode.

Advantages and limitations of CTR mode: An advantage of the CTR mode is its high efficiency, since the output of cipher can be calculated in advance. Moreover, since the encryption of each block is independent, parallel encryptions can be applied in the CTR mode. The CTR mode can also get random access to encrypted data blocks. However, the CTR mode must not reuse the same key nor counter value.

5.2.6 Last Block in Different Modes

One issue that arises with block cipher operation modes is how to handle the last block since a message may not be the multiple of a block size. In general, the last block needs to be padded (typically with 0's). At the receiver's side, the padding needs to be obvious. As an alternative, the last byte of the padding can be reserved as a count of how many bytes padding was used (including the count). For example, $[b_1, b_2, b_3, b_4, b_5, 0, 0, 3]$ means that there are 5 data bytes, then 3 bytes "padding+count." Note that if this is done, if the last block is an even multiple of 8 bytes or has exactly the same form as "padding+count," then an extra block is added. There are other, more esoteric, "ciphertext stealing" modes, which avoid the need for an extra block.

5.2.7 XTS-AES Mode

XTS-AES is an additional block cipher mode of operation. This mode is defined by IEEE Standard 1619-2007 and NIST Standard XTS-AES, which describes a method of encryption for data stored in sector-based devices [26]. The XTS-AES mode is based on the concept of an adjustable block cipher. The requirements for encrypting stored data, also referred to as "data at rest" differ from those for transmitted data. Figure 5.9 illustrates XTS-AES operation on a single block. The XTS-AES encryption of a single block involves two instances of the AES algorithm with two keys (i.e. K_1 and K_2), and a number of parameters. In essence,

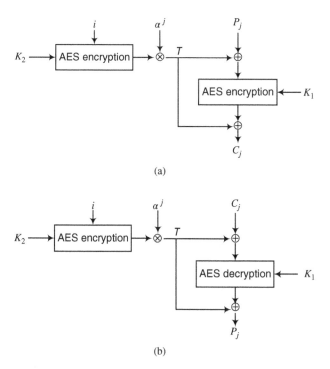

Figure 5.9 XTS-AES operation on a single block. (a) Encryption in XTS-AES. (b) Decryption in XTS-AES.

i is the 128-bit tweak for the sector, j is the block number within the sector. Each data unit (sector) is assigned a tweak value that is a non-negative integer. $\otimes$ is modular multiplication of two polynomials with binary coefficients module $x^{128} + x^7 + x + 1$, this is multiplication in $GF(2^{128})$; α is a primitive element of $GF(2^{128})$ that corresponds to polynomial x (i.e. $0000\ldots010_2$). The XTS-AES block encryption is defined as follows:

$$\begin{cases} T_j & = E_{K_2}(i) \otimes \alpha^j, \\ C_j & = T_j \oplus E_{K_1}(T_j \oplus P_j). \end{cases} \tag{5.14}$$

In the XTS-AES decryption, T is calculated in the same way. The other AES encryption function is replaced by AES decryption function, and swap C_j and P_j. Thus the XTS-AES block decryption is defined as follows:

$$\begin{cases} T_j & = E_{K_2}(i) \otimes \alpha^j, \\ P_j & = T_j \oplus D_{K_1}(T_j \oplus C_j). \end{cases} \tag{5.15}$$

An input message may have multiple blocks. Instead of using padding, XTS-AES uses a technique *ciphertext—stealing*. The ciphertext—stealing technique used in both XTS-AES encryption and decryption is shown in Figure 5.10. This technique ensures that the encrypted data remains the same size as the original plaintext, which it must fit on a block-oriented storage device.

Advantages and limitations of XTS-AES mode: The XTS-AES mode is suitable for parallel operation. Because there is no chaining, multiple blocks can be encrypted or decrypted

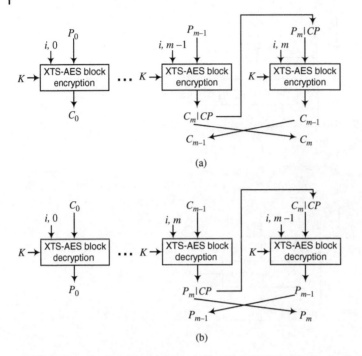

Figure 5.10 XTS-AES ciphertext−stealing mode. (a) The ciphertext-stealing in XTS-AES encryption (b) The ciphertext-stealing in XTS-AES decryption.

simultaneously. Therefore, XTS-AES has good efficiency. However, unlike CTR mode, XTS-AES mode includes a nonce (the parameter i) as well as a counter (parameter j). XTS-AES addresses security concerns related to stored data.

5.3 Public Key Infrastructure

All the cryptographic systems discussed so far are symmetric key systems, where one key is pre-shared with a sender and a receiver. If this key is disclosed, the entire communication would be compromised. Also, both parties are equal in symmetric cryptographic systems. All classical and modern block and stream ciphers are of this form, and still rely on the fundamental building blocks of substitution and permutation (transposition). There are two key issues considering symmetric systems:

- *Key distribution*: How to have secure communications in general without having to trust a Key Distribution Center (KDC) with your key?
- *Digital signatures*: How to verify a message comes intact from the claimed sender?

Public-Key Cryptography (PKC) was developed to address those issues.

5.3.1 Basics of Public Key Cryptography

PKC is asymmetric, which involves the use of two separate keys (i.e. public/private keys), in contrast to symmetric encryption which uses only one key. Anyone knowing the public key can encrypt messages or verify signatures, but cannot decrypt messages or create signatures, counter-intuitive though this may seem. It works by the clever use of number theory problems that are easy one way but hard the other. Note that public key schemes are neither more nor less secure than private key (security depends on the key size for both), nor do they replace private key schemes (they are too slow to do so), rather they complement them. Both also have issues with key distribution, requiring the use of some suitable protocol [27–29].

Public key algorithms must follow the important characteristic as follows. First, it is computationally infeasible to determine the decryption key given only knowledge of the cryptographic algorithm and the encryption key. Second, it is computationally easy to encrypt or to decrypt messages when the relevant key is known. Third, either of the two related keys can be used for encryption, with the other used for decryption (for some algorithms). Therefore, public key schemes utilize problems that are easy (P type) one way but hard (NP type) the other way, e.g. exponentiation in one way but logarithm in the other way; multiplication in one way but factoring in the other way. Public-key schemes can be used for either secrecy or authentication. PKC is asymmetric because those who encrypt messages or verify signatures cannot decrypt messages or create signatures. PKC involves the use of two keys:

- *Public-key*: which may be known by anyone and can be used to encrypt messages, and verify signatures;
- *Private-key*: which is known only to the recipient, used to decrypt messages, and sign (or create) signatures.

Figure 5.11 illustrates essential elements of a public-key cryptographic scheme. As shown in Figure 5.11a, the transmitter A encrypts the plaintext M using the receiver B's public key PU_b

$$C = E_{PU_b}(M). \tag{5.16}$$

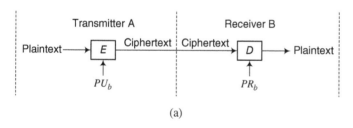

(a)

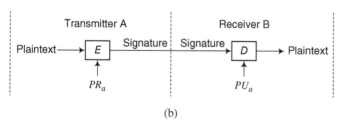

(b)

Figure 5.11 Public-key cryptography. (a) Encryption and decryption. (b) Digital signature.

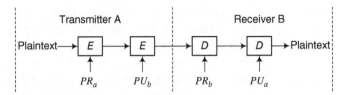

Figure 5.12 Public-key cryptographic systems.

The receiver B then decrypts the message using its own private key PR_b

$$M = D_{PR_b}(C). \tag{5.17}$$

Confidentiality is provided since B is the only one that can decrypt the ciphertext. As shown in Figure 5.11b, the transmitter A signs the plaintext using its own private key PR_a

$$S = E_{PR_a}(M). \tag{5.18}$$

The receiver B then verifies the signature using A's public key PU_a

$$M = D_{PU_a}(M). \tag{5.19}$$

Note that anyone who has A's public key can verify the signature. The two operations can be used together for both encryption and authentication, as illustrated in Figure 5.12. In this case, separate key pairs are used for each of these purposes. Sender A first signs the plaintext M to get $E_{PR_a}(M)$, then encrypts the result to get

$$C = E_{PU_b}(E_{PR_a}(M)). \tag{5.20}$$

Receiver B first decrypts the ciphertext to get $D_{PR_b}(C)$, then verifies the signature

$$M = D_{PU_a}(D_{PR_b}(C)). \tag{5.21}$$

It is not used typically in general practice but only in some specific applications, because public-key schemes have high computational cost.

5.3.2 Public-Key Applications

Depending on the application, the sender uses either the sender's private key or the receiver's public key, or both, to perform some type of cryptographic function. In broad terms, the use of public-key cryptosystems can be classified into the three categories:

- *Encryption and decryption*: The sender encrypts a message with the recipient's public key.
- *Digital signature*: The sender signs a message with its private key, either to the whole message or to a small block of data that is a function of the message.
- *Key exchange*: Two sides cooperate to exchange a session key, to be used for the symmetric key secure communications.

Some public key algorithms are capable for all three applications, whereas others can be used only for one or two of these applications.

5.3.3 Security of Public Key Schemes

Public key schemes are no more or less secure than private key schemes. It is the key size that determines the security in either type of schemes since brute-force exhaustive search attack is always theoretically possible. Nonetheless, the key sizes of public key schemes and private key schemes cannot be compared directly. Typically, a 64-bit private key scheme has roughly similar security to a 512-bit RSA (a public key scheme). Therefore, public key schemes use much larger keys than private key schemes. Besides key size, the security of public key schemes relies on the easy encryption/decryption and hard cryptanalysis problems. There is usually a firmer theoretical basis for determining the security of public key schemes. However, it requires the use of very large numbers to secure public key schemes. Hence public key schemes are relatively slow in computation compared to private key schemes.

5.4 The RSA Algorithm

The RSA algorithm is by far the most widely used public key encryption scheme. RSA was first published by Rivest et al. of MIT in 1977 [30]. Since that time RSA has reigned supreme as the most widely accepted and implemented general-purpose approach to public key scheme. It is based on exponentiation in a finite field over integers modulo a prime, using large integers (e.g. 1024 bits). Its security is due to the cost of factoring large numbers, which is a hard problem [31].

5.4.1 RSA Key Setup

RSA key setup is done once when a user establishes or replaces its public/private keys, using the steps as follows:

(1) Select two large prime numbers p and q at random.
(2) Compute system modulus $n = p \cdot q$, note that $\phi(n) = (p - 1)(q - 1)$.
(3) Select at random the encryption key e, such that $1 < e < \phi(n)$ and $GCD(e, \phi(n)) = 1$.
(4) Find decryption key d s.t., $d \equiv 1 \bmod \phi(n)$, for $0 \leq d < n$.
(5) The public key is $PU = \{e, n\}$.
(6) The private key is $PR = \{d, n\}$.

It is critically important that the factors p and q of the modulus n are kept secret, since if these become known, the system can be broken. Note that different users will have different moduli n.

RSA key setup example: Given two prime numbers $p = 13$ and $q = 19$, the public/private keys are derived as follows:

$$n = p \cdot q = 13 \times 19 = 247,$$

$$\phi(n) = (p - 1) \cdot (q - 1) = 12 \times 18 = 216.$$

The encryption key e is selected as $e = 11$, where $GCD(11, 216) = 1$. The decryption key d is found as $d = 59$ by calculating the multiplicative inverse of $e = 11$. It can be easily verified, i.e. $59 \times 11 = 649 = 3 \times 216 + 1$. Therefore, the public key of this example is $PU = \{11, 247\}$, and private key is $PR = \{59, 247\}$.

Note that in the real applications of RSA, the parameters p, q, n, e, d must all be very large integers. Primality testing such as Miller–Rabin algorithm can be used to test if a generated large integer is a prime number or not. Extended Euclidean Algorithm can be used to find d given e and $\phi(n)$.

5.4.2 RSA Encryption and Decryption

The actual RSA encryption and decryption computations are each a single exponentiation modulo n. Note that the message must be smaller than the modulus. The "magic" is in the choice of the exponents which makes the system work.

- To encrypt a message M, the sender:
  - Obtains public key of recipient $PU = \{e, n\}$.
  - Computes $C = M^e \bmod n, 0 \leq M < n$.
- To decrypt a ciphertext C, the recipient:
  - Uses the private key $PR = \{d, n\}$.
  - Computes $M = C^d \bmod n$.

Note that the message M must be smaller than the modulus n. If M is larger than n, it can be split into smaller blocks.

RSA works because of Euler's theorem, s.t., $a^{\phi(n)} \bmod n = 1$, where $GCD(a, n) = 1$. In RSA, we have $e \cdot d = 1 + k \cdot \phi(n)$ for some k. Hence, it can be verified that

$$C^d = M^{e \cdot d} = M^{1+k \cdot \phi(n)} = M^1 \cdot (M^{\phi(n)})^k \equiv M \bmod n.$$

RSA encryption and decryption example: Given message $M = 8$, $PU = \{11, 247\}$ and $PR = \{59, 247\}$, the encryption of RSA is process as follows:

$$C = M^e \bmod n = 8^{11} \bmod 247 = 31.$$

The decryption of this ciphertext is as follows:

$$M = C^d \bmod n = 31^{59} \bmod 247 = 8.$$

Note that in the real applications of RSA, the Chinese Remainder Theory and some techniques on efficient exponentiation can be used for increasing the computation efficiency.

5.4.3 RSA Security Analysis

There are possible approaches to attack the RSA algorithm. The defense against the brute-force attack is the same for RSA as for other cryptosystems, namely, using a large key space. Therefore, the larger the number of bits in d, the better security it is against brute-force key search. However, because the calculations involved both in key generation and in encryption/decryption are complex, the larger the size of the key, the slower the system will run. Besides brute-force attack, there are other possible approaches, including mathematical attacks (based on difficulty of computing $\phi(n)$, by factoring modulus n), timing attacks, and chosen ciphertext attacks (CCA).

5.4.3.1 Factoring Problem

One can identify three approaches to attacking RSA mathematically:

- Factor $n = p \cdot q$, hence compute $\phi(n)$ and then d according to p and q.
- Determine $\phi(n)$ directly and compute d.
- Find d directly.

Currently the best algorithm of factoring is the *Lattice Sieve*, which can factor a 200-decimal digits (663-bit) value. Therefore, it is assumed that 1024-bit or 2048-bit RSA is secure against factoring in the near future.

5.4.3.2 Timing attacks

Timing attack is a ciphertext only attack. It exploits timing variations in operations. For example, multiplying large number takes more time than multiplying small number. Or, it has different patterns for different instructions executed. In RSA, it exploits time taken in exponentiation. Although the timing attack is a serious threat, there are simple countermeasures that can be used, including using constant exponentiation time algorithms, adding random delays, or using blind values in calculations.

5.4.3.3 Chosen Ciphertext Attacks

In CCA, an attacker chooses ciphertext and gets decrypted plaintext back. Ciphertext is chosen to exploit properties of RSA to provide information to help cryptanalysis. RSA is in fact vulnerable to CCA. Nonetheless, we can counter CCA with random pad of plaintext, or use Optimal Asymmetric Encryption Padding (OAEP).

5.5 Diffie–Hellman (D–H) Key Exchange

One of the important service achieved by public key cryptographic system is more efficient key exchange. D–H key exchange is one of the first practical examples of public key exchange implementation [21]. D–H algorithm cannot be used to exchange an arbitrary message. It rather establishes a common key that is known only to the two participants. D–H algorithm is based on exponentiation in a finite field which is easy to compute. The security of D–H algorithm is based on the difficulty of computing discrete logarithms.

5.5.1 Finite-Field Diffie–Hellman

D–H algorithm starts by setting global parameters agreed by both participants. The global parameters of a finite-field D–H include: a large prime integer or polynomial q, and a being a primitive root mod q. Note that if a is a primitive root of the prime number q, then the numbers $(a \bmod q)$, $(a^2 \bmod q)$, ..., $(a^{q-1} \bmod q)$ are distinct and consist of the integers from 1 through $(q-1)$ in some permutation. Each participant generates their public and private keys. User A generates the key pairs as follows:

- First choose a secret key $x_A < q$;
- Then compute a public key $y_A = a^{x_A} \bmod q$.

User B generates the key pairs as follows:

- First choose a secret key $x_B < q$;
- Then compute a public key $y_B = a^{x_B} \bmod q$.

The two participants A and B first exchange their public keys, then create a common secret key K_{AB} as follows:

- At A: $K_{AB} = y_B^{x_A} \bmod q$.
- At B: $K_{AB} = y_A^{x_B} \bmod q$.

The key K_{AB} is used as a session key that can be applied to any symmetric-key encryption schemes.

D–H example: given the global parameters $q = 29$ and $a = 3$. The two participants select secret keys $x_A = 7$ and $x_B = 11$. The public keys are computed as follows:

$$y_A = 3^7 \bmod 29 = 12; \quad y_B = 3^{11} \bmod 29 = 15.$$

The shared session key the is as follows:

$$K_{AB} = 15^7 \bmod 29 = 12^{11} \bmod 29 = 17.$$

5.5.2 Elliptic-Curve Diffie–Hellman

The Elliptic-Curve Diffie–Hellman (ECDH) is based on Elliptic Curve Cryptography (ECC). An elliptic curve is a plane curve over a finite field $\mathbb{G}$:

$$y^2 = x^3 + ax + b, \tag{5.22}$$

and a distinguished point at infinity. The finite field $\mathbb{G}$ is assumed to have a prime order of p, or an order defined in binary 2^m. The domain parameters in ECDH are (p, a, b, g, n, h) or $(m, f(x), a, b, g, n, h)$ for prime $\mathbb{G}_p$ or $\mathbb{G}_{2^m}$, respectively, where g is a generator of the finite field. The basic idea of key distribution of ECDH is the same as finite field D–H. Two participants A and B have symmetric operations. Using A for illustration, ECDH generates and distribute a key as follows:

- A and B pick private keys $d_A \in [1, n-1]$ and $d_B \in [1, n-1]$;
- Each computes the corresponding public key $d_A \cdot g$ and $d_B \cdot g$;
- Exchange the public keys;
- Each computes the shared key locally $K_{AB} = d_A \cdot d_B \cdot g$.

Note that $(x \cdot g)$ indicates adding g to itself x times on the elliptic curve.

5.5.3 Diffie–Hellman Key Exchange Vulnerability

D–H key exchange is vulnerable to man-in-the-middle attack. Assume that A and B are legitimate users and C is the attacker. The man-in-the-middle attack is launched as follows:

- C generates two random private keys x_{C_1} and x_{C_2} and the corresponding public keys y_{C_1} and y_{C_2}.
- C sends y_{C_1} and y_{C_2} to B and A correspondingly.

- C intercepts y_A that is sent from A to B. C computes $k_2 = (y_A)^{x_{C_2}} \bmod q$.
- B computes $k_1 = (y_{C_1})^{x_B} \bmod q$.
- C intercepts y_B that is sent from B to A.
- C computes $k_1 = (y_B)^{x_{C_1}} \bmod q$.
- A computes $k_2 = (y_{C_2})^{x_A} \bmod q$.

After the attack, A and C share a key k_2 while B and C share a key k_1. In later communications, C can either impersonate B to communicate with A or vice versa. C may also be an eavesdropper by relaying communications.

5.6 Summary

In this chapter, several more cryptographic techniques are illustrated, including AES, block cipher modes of operations, and the introduction to public key cryptographic systems. RSA algorithm and D–H key exchange algorithm are described in further details for public key cryptographic systems. Those cryptographic techniques are used for security mechanisms designed for some wireless communications systems.

6

Message Authentication, Digital Signature, and Key Management

In Chapters 4 and 5, encryption has been illustrated to protect the secrecy of the message contents. Besides the secrecy, there are other security aspects for protection, for example, the integrity of the message, the identity of the sender, etc. In this chapter, *message authentication* and *digital signature* will be introduced to protect those aspects of message contents. Furthermore, key management will be discussed for different cryptographic systems.

6.1 Message Authentication

6.1.1 Message Authentication Functions

Message authentication ensures that a message is received from the alleged sender and in its original form. No alteration of the content shall be allowed during the transmission. Thus, message authentication is a method to verify that the integrity of a message is maintained during communications [32]. Message authentication can also be applied to validate identity of the originator, and provide non-repudiation of the origin (dispute resolution). There are three types of functions that may be used to produce an authenticator: message encryption, message authentication code, and hash function.

Message encryption is to provide confidentiality of the message during transmission. Nonetheless, message encryption by itself also provides a measure of authentication. If symmetric encryption is used with a suitable structure, redundancy or a checksum, then receiver is certain that the received content is not altered during transmission. Moreover, receiver can also authenticate the sender since only the two of them can create a ciphertext with meaningful content. If public-key encryption is used, encryption with the public key does not provide message authentication since anyone potentially knows public key could create the message. However, if sender signs a message using its private key and encrypts the message with the receiver's public key, then message authentication is provided the same as using symmetric ciphers. While message authentication can be provided in this way, it is not recommended because of the cost of two public key encryptions used on each message.

Message Authentication Code (MAC) is a cryptographic checksum with a fixed-size block of digits that is generated by an algorithm or a function. A MAC algorithm takes in both message and a secret key. It performs similarly to encryption but it is not required to be

Security in Wireless Communication Networks, First Edition. Yi Qian, Feng Ye, and Hsiao-Hwa Chen.
© 2022 John Wiley & Sons Ltd. Published 2022 by John Wiley & Sons Ltd.
Companion website: www.wiley.com/go/qian/sec51

reversible for an MAC algorithm. In practice, MAC algorithms are mostly irreversible due to the small fixed-sized output. MAC is appended to the original message before transmission. Receiver performs the same computation on message and checks if the result matches the received MAC. MAC provides assurance that a message is unaltered and comes from sender.

Hash Function creates a fixed-size block based on the original message. Different from an MAC algorithm, a hash function does not have a secret key. A hash value is created solely from the message. Before transmission, a hash value is created and appended to the original message. Receiver performs the same computation on the received message to create a hash value. If the hash value matches the received one, then the received message is considered unaltered. However, since there is no key used for a hash function, a hash value does not validate the identity of originator, nor does it provide non-repudiation.

6.1.2 Message Authentication Code

An overview of MAC usage is shown in Figure 6.1. At the sender side, the original message is passed through an MAC algorithm $F_K(\cdot)$ with a secret key K. The output of $F_K(\cdot)$ is the MAC, which is then appended to the original message for transmission. At the receiver side, the same algorithm is applied with the input of the received message and the secret key. The computed result is then compared with the received MAC for verification. If the two values match, then the integrity of the message is validated. It can be seen that MAC provides authentication only. The message is still transmitted in clear text. In order to provide confidentiality, encryption is required with a different key from the MAC algorithm. One can compute MAC either before or after encryption. However, it is generally regarded better to computer MAC before encryption. Note that a MAC is not a digital signature. A MAC does not provide non-repudiation since it lacks unique information of the originator.

There are several requirements for an MAC algorithm $F_K(\cdot)$ (with key K), listed as follows:

- With a known message M_1 and its MAC $F_K(M_1)$, it is infeasible (in practice) to find another message M_2 that generates the same MAC $F_K(M_2) = F_K(M_1)$.
- MACs generated from function F should be uniformly distributed.
- MAC generation should depend equally on all bits of the input message.

Note that a MAC algorithm $F_K(\cdot)$ is in general a many-to-one function, since the output is a fixed size block (which usually has a smaller size than the input message). Therefore, it is possible that different messages have the same MAC. However, it must be extremely

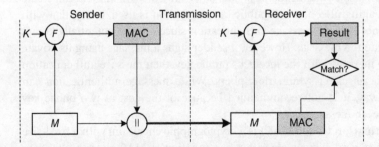

Figure 6.1 An illustration of message authentication code usage.

hard (e.g. only by exhaustive search) to find two messages that generate the same MAC in practice.

It is mentioned before that symmetric ciphers may be applied for message authentication. Symmetric block ciphers are good candidate for MAC algorithms since good block ciphers meet the three requirements very well. Any block cipher chaining mode can be used with the output of the final block being the MAC.

6.1.3 Hash Functions

A hash function maps arbitrary size of input message to a fixed-size block (i.e. a hash value) [33]. Using hash function for message authentication is similar to using MAC, as illustrated in Figure 6.2. The sender generates a hash value based on the original message. The hash value is appended to the message before transmission. At the receiver side, the same hash function is applied to the received message to generate a hash value, which is then compared with the received hash value for verification. If matches, then the integrity of the message is verified.

Hash functions for message authentication are considered cryptographic hash functions or secure hash functions. There are specific requirements for those cryptographic hash functions. For a cryptographic hash function $H(\cdot)$, the requirements are listed in the following:

- The function can be applied to message M with arbitrary size.
- The output hash value h has fixed length.
- It is easy to compute $h = H(M)$ for any message M.
- Given h, it is infeasible to find input x s.t. $H(x) = h$ (one-way property).
- Given x, it is infeasible to find y s.t. $H(y) = H(x)$ (weak collision resistance).
- It is infeasible to find any x, y s.t. $H(y) = H(x)$ (strong collision resistance).

Since hash value is a fixed size but input message can be any size, hash function is a many-to-one function. There exist multiple inputs that produce the same hash value. However, it should be no easier than brute-force attack to find such messages that generate the same hash value. There are several proposals for simple hash functions. One example is to simply split the input into blocks and then XOR all the blocks. Apparently, such a simple hash function cannot be used for message authentication because it does not satisfy the

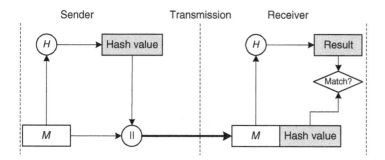

Figure 6.2 Hash function for message authentication.

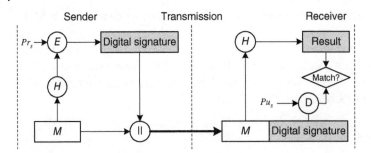

Figure 6.3 Hash function with digital signature.

requirements for cryptographic hash functions. It is easy to manipulate the original message without change the final hash value.

A hash function is generally public and not keyed. A MAC function on the other hand is keyed. Hash functions can be used in various ways to provide security protections. One such application is to use a digital signature to provide both message authentication and non-repudiation. An overview of this application is shown in Figure 6.3. The sender first create a hash value based on the input message, then signs the results with its private key to create a digital signature. Note that the encryption together with the hash function operates as a digital signature. The generated digital signature is appended to the original message. At the receiver side, a hash value is first created based on the received message. The received digital signature is verified using sender's public key. If the two values match, then the received message is unaltered, and non-repudiation is also verified besides message authentication.

6.1.4 Size of MAC and Hash Value

The security of a MAC or hash function depends on the size of the MAC or hash value. If the size of the MAC or hash value is too small, it will be susceptible to attacks especially *birthday* attack. Birthday attack is based on *Birthday Paradox* [34]. It is to compute the probability $P(n)$ that, in a set of n randomly chosen people, at least two of them have the same birthday. For simplicity, assume that there are 365 possible birthdays (excluding leap years, twins, etc.), then $P(n)$ is calculated as follows:

$$P(n) = \begin{cases} 0, & n = 1, \\ 1 - \overline{P}(n) = 1 - \dfrac{n! \binom{365}{n}}{365^n}, & 2 \leq n \leq 365, \\ 1, & n > 365, \end{cases} \tag{6.1}$$

where $\overline{P}(n)$ is the probability that no two people in the room have the same birthday. According to birthday paradox, $P(n) > 99.9\%$ is reached with $n \geq 70$ people, and $P(n) > 50\%$ can be reached with just $n \geq 23$ people. An illustration of the probability with regard to number of people is given in Figure 6.4. Birthday paradox shows that a collision can happen with high probability with a much small space than the original one. According to birthday paradox, a birthday attack works as follows: for a message with

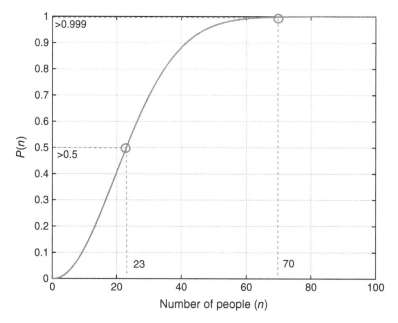

Figure 6.4 Illustration of birthday paradox.

length m, an attacker first generates $2^{m/2}$ variations of a valid message all with essentially the same meaning (e.g. "class at nine" and "class in the morning"). The attacker also generates $2^{m/2}$ variations of a desired fraudulent message (e.g. "class at noon"). According to birthday paradox, it is over 50% chance that a pair of messages can be found one in each set that generate the same hash value. If such a pair is found, then a forgery can substitute the original message and pass the hash value check. Therefore, MAC and hash values must be large to prevent birthday attack. For a message with length m, brute-force attack exploits strong collision resistance MAC/hash value at a cost of $2^{m/2}$. In general, 128-bit hash value is considered vulnerable, while 160 bits or larger is recommended. On the other hand, 128 bits or larger is recommended for MAC since a MAC is also protected by a secret key.

6.2 MAC and Hash Algorithms

There are many algorithms and functions to generate MAC and hash values. This section will describe some examples of both MAC and secure hash functions.

6.2.1 Data Authentication Algorithm

Data Authentication Algorithm (DAA) is a MAC algorithm that applies block cipher DES in a CBC mode. DAA is standardized in FIPS PUB 113 / ANSI X9.17 [35]. Figure 6.5 gives an overview of DAA. The input message M is divided into N blocks with each block D_i 64 bits and the last block D_N possibly padded zeros to be exactly 64 bits. The initial vector IV is set

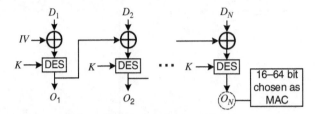

Figure 6.5 Data authentication algorithm.

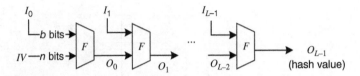

Figure 6.6 A basic structure of hash function.

to be zeros in DAA. Each block operation takes in current data block D_i XORed with the output of the previous block, then inputs to the DES using the secret key K. Mathematically, DAA is expressed as follows.

$$O_i = \begin{cases} IV, & i = 0, \\ DES_K(D_i \oplus O_{i-1}), & i = 1, 2, \dots, N. \end{cases} \tag{6.2}$$

The output of the final block operation O_N is the MAC value. The most significant 16 to 64 bits of O_N can be used as a MAC. However, as mentioned earlier, even using the entire 64 bits is considered too small for security.

6.2.2 A Basic Hash Function Structure

Many hash functions follow a basic structure as shown in Figure 6.6. It condenses an arbitrary size message M to a fixed-size block, by processing the input in blocks, through some compression function, either custom or block cipher based. An input is first padded to multiple of the block size b bits, and then split into L blocks, $I_0 \| I_1 \dots \| I_{L-1}$. Each block operation is a compression function F that takes in the input block message I_{i-1} of b bits and the result from the previous block operation of O_{i-1} of n bits, and outputs a result O_i that is n bits.

$$O_i = \begin{cases} F(I_i, IV), & i = 0, \\ F(I_i, O_{i-1}), & i = 1, 2, \dots, L - 1. \end{cases} \tag{6.3}$$

The final output O_{L-1} is the hash value. This has been proven a fundamentally sound structure. Newer designs simply refine the structure and add to the hash value length.

6.2.3 Secure Hash Algorithm (SHA)

Secure Hash Algorithm (SHA) was originally designed by US government agencies NIST and NSA in 1993 [36]. The corresponding standard is NIST FIPS PUB 180. SHA is based on design of Message-Digest Algorithm (MD4) with key differences. The original SHA was

Table 6.1 Comparison of SHA parameters (in *bits*).

	SHA-1	SHA-256	SHA-384	SHA-512
Message digest size	160	256	384	512
Maximum message size	2^{64}	2^{64}	2^{128}	2^{128}
Block size	512	512	1024	1024
Word size	32	32	64	64
Number of steps	80	64	80	80
Security	80	128	192	256

revised in 1995 as SHA-1, standardized as NIST FIPS PUB180-1. SHA-1 has 80 rounds with a block size of 512 bits and produces 160-bit hash values. In February 2005, an attack by X. Wang, Y. Yin, and H. Yu was announced [37]. The attack can find collisions in the full version of SHA-1, requiring fewer than 2^{69} operations, far fewer than the 2^{80} operations previously thought needed to find a collision with an SHA-1 hash. In 2017, the first collision for full SHA-1 was announced [38]. NIST issued revision FIPS 180-2 in 2002, commonly known as SHA-2, which adds three additional versions of SHA: SHA-256, SHA-384, and SHA-512. The standards are further updated to NIST FIPS 180-4 in 2012 [36]. A comparison of SHA parameters is shown in Table 6.1. The latest versions of SHA were designed for compatibility with increased security provided by the AES cipher. Although the structure of SHA-2 is similar to SHA-1, security is much improved in SHA-256, SHA-384, and SHA-512. Here, security refers to the fact that a birthday attack on a message digest of size n produces a collision with a work factor of approximately $2^{n/2}$.

6.2.4 SHA-512

As defined in NIST FIPS 180-4 [36], SHA-256, SHA-384, and SHA-512 share a similar structure. SHA-512 will be illustrated for details. The SHA-512 algorithm takes as input a message with a maximum length of less than 2^{128} bits and produces an output a 512-bit message digest. SHA-512 follows the structure as depicted in Figure 6.7. The algorithm consists of the following steps:

(1) Append padding string to the original message. Padding string is a sequence of bits starting with '1', followed by zeros to make the message length congruent to 896 modulo 1024. Note that the length of padding is between 1 and 1024 bits. That is to say, if the original message has the length congruent to 896 modulo 1024 (e.g. a length of 896 bits), padding will still be applied such that $1\|0\dots0$ (i.e. '1' followed by 1023 '0's).

(2) Append *Length* to the padded message. *Length* is a block of 128 bits that contains the length of the original message M (before padding). With padding and the length block, the message is divided into N blocks, with each block message M_i 1024 bits. For example, if the original message is 896 bits, the padded message is 1024 bits, *Length* records 896 using 128 bits. The total input has a length of 2048 bits.

(3) Initialize hash buffer value $H_0 = IV$, which consists of 512 bits or eight 64-bit words.

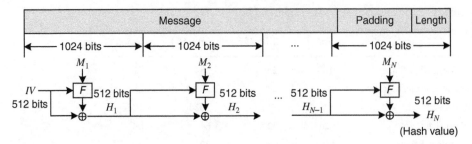

Figure 6.7 Overview of the SHA-512 algorithm.

(4) Process the message in 1024-bit (128-word) blocks in the compression function. The compression function is the core of SHA-512, labeled F in Figure 6.7.

(5) Output the 512-bit final state H_N as the resulting hash value.

6.2.4.1 SHA-512 Compression Function

The compression function F is the core of SHA-512. The overall structure of the compression function is shown in Figure 6.8. It processes the divided message blocks with 1024-bit or 128-word in each block ($M_1\|M_2\|\ldots\|M_N$) and outputs a 512-bit buffer value. For each compression function, the buffer value from the previous block operation is fed into the function besides the message block. The buffer value is updated by XORing the output of the compression function with the previous buffer value. Therefore, for the ith block operation, the updated buffer value H_i is computed as follows:

$$H_i = \begin{cases} F(M_1, IV) \oplus IV, & i = 1, \\ F(M_i, H_{i-1}) \oplus H_{i-1}, & i = 2, \ldots, N, \end{cases} \tag{6.4}$$

where IV is the initial buffer value H_0, which are the 64-bit binary expansion of the fractional part of the square root of the first eight prime numbers (i.e. 2, 3, 5, 7, 11, 13, 17, 19). In HEX, H_0 consists of the following eight 64-bit words:

$$H_0[A] = 6A09E667F3BCC908$$

$$H_0[B] = BB67AE8584CAA73B$$

$$H_0[C] = 3C6EF372FE94F82B$$

$$H_0[D] = A54FF53A5F1D36F1$$

$$H_0[E] = 510E527FADE682D1$$

$$H_0[F] = 9B05688C2B3E6C1F$$

$$H_0[G] = 1F83D9ABFB41BD6B$$

$$H_0[H] = 5BE0CD19137E2179$$

The compression function F uses a module that consists of 80 rounds. Each round (e.g. the tth round) takes as input the 512-bit buffer value and updates the contents of the buffer. In addition to the buffer value, each round also makes use of a 64-bit value W_t derived using a message schedule from the current 1024-bit message block being processed, and an

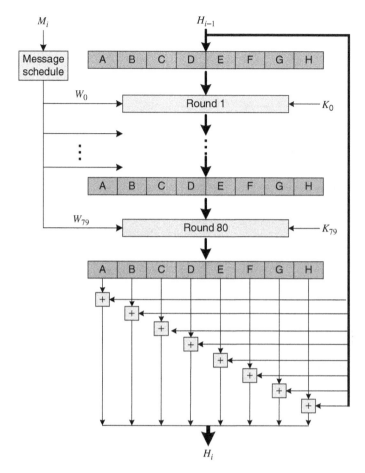

Figure 6.8 SHA-512 compression function F.

additive constant K_t that is based on the first 64 bits of the fractional parts of the cube roots of the first eighty prime numbers. The output of the 80th round is added to the input to the first round to produce the final buffer value for this message block, which forms the input to the next block operation of this compression function.

6.2.4.2 SHA-512 Round Function

The structure of each of the 80 rounds is shown in Figure 6.9. Each 64-bit word (i.e. A, B, C, D, E, F, G, or H) is shuffled along one place, and in some cases manipulated using a series of simple logical functions AND ($\wedge$), NOT ($\neg$), XOR ($\oplus$). In addition, $R^n(x)$ is circular right shift of the argument x by n bits in order to provide the avalanche and completeness properties of the hash function. The five logical functions are defined as follows:

- $Ma(A, B, C) = (A \wedge B) \oplus (A \wedge C) \oplus (B \wedge C)$,
- $\Sigma_0 = R^{28}(A) \oplus R^{34}(A) \oplus R^{39}(A)$,

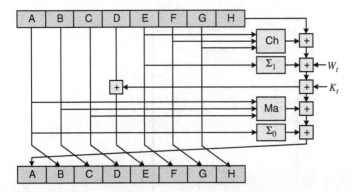

Figure 6.9 SHA-512 round function.

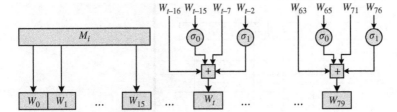

Figure 6.10 SHA-512 message schedule.

- $Ch(E, F, G) = (E \wedge F) \oplus (\neg E \wedge G)$,
- $\Sigma_1 = R^{14}(E) \oplus R^{18}(E) \oplus R^{41}(A)$,
- "+" is addition modulo 2^{64}.

The 1024 bit message block is derived into eighty 64-bit words $W_0, \ldots, W_{79}$ by following the message schedule shown in Figure 6.10. The first 16 values of W_t are taken directly from the 16 words of the current message block M_i. For the remaining 64 steps, the value of W_t can be calculated by

$$W_t = \sigma_1(W_{t-2}) + W_{t-7} + \sigma_0(W_{t-15}) + W_{t-16},$$

where + is addition modulo 2^{64}. The two logical functions σ_0 and σ_1 are illustrated as follows:

$$\sigma_0(x) = R^1(x) \oplus R^8(x) \oplus S^7(x),$$

$$\sigma_1(x) = R^{19}(x) \oplus R^{61}(x) \oplus S^6(x),$$

where $S^n(x)$ is the right shift of the argument x by n bits. Word derivation introduces a great deal of redundancy and interdependence into the message blocks that are compressed, which complicates the task of finding a different message block that maps to the same compression function output. The eighty additive constants $K_0, \ldots, K_{79}$ represent the first 64-bit of the fractional parts of the cube roots of the first eighty prime number. Table 6.2 lists the constants in hexadecimal values [36].

Table 6.2 Additive constants $K_0, K_1, \ldots, K_{79}$ for SHA-512.

428A2F98D728AE22	7137449123EF65CD	B5C0FBCFEC4D3B2F	E9B5DBA58189DBBC
3956C25BF348B538	59F111F1B605D019	923F82A4AF194F9B	AB1C5ED5DA6D8118
D807AA98A3030242	12835B0145706FBE	243185BE4EE4B28C	550C7DC3D5FFB4E2
72BE5D74F27B896F	80DEB1FE3B1696B1	9BDC06A725C71235	C19BF174CF692694
E49B69C19EF14AD2	EFBE4786384F25E3	0FC19DC68B8CD5B5	240CA1CC77AC9C65
2DE92C6F592B0275	4A7484AA6EA6E483	5CB0A9DCBD41FBD4	76F988DA831153B5
983E5152EE66DFAB	A831C66D2DB43210	B00327C898FB213F	BF597FC7BEEF0EE4
C6E00BF33DA88FC2	D5A79147930AA725	06CA6351E003826F	142929670A0E6E70
27B70A8546D22FFC	2E1B21385C26C926	4D2C6DFC5AC42AED	53380D139D95B3DF
650A73548BAF63DE	766A0ABB3C77B2A8	81C2C92E47EDAEE6	92722C851482353B
A2BFE8A14CF10364	A81A664BBC423001	C24B8B70D0F89791	C76C51A30654BE30
D192E819D6EF5218	D69906245565A910	F40E35855771202A	106AA07032BBD1B8
19A4C116B8D2D0C8	1E376C085141AB53	2748774CDF8EEB99	34B0BCB5E19B48A8
391C0CB3C5C95A63	4ED8AA4AE3418ACB	5B9CCA4F7763E373	682E6FF3D6B2B8A3
748F82EE5DEFB2FC	78A5636F43172F60	84C87814A1F0AB72	8CC702081A6439EC
90BEFFFA23631E28	A4506CEBDE82BDE9	BEF9A3F7B2C67915	C67178F2E372532B
CA273ECEEA26619C	D186B8C721C0C207	EADA7DD6CDE0EB1E	F57D4F7FEE6ED178
06F067AA72176FBA	0A637DC5A2C898A6	113F9804BEF90DAE	1B710B35131C471B
28DB77F523047D84	32CAAB7B40C72493	3C9EBE0A15C9BEBC	431D67C49C100D4C
4CC5D4BECB3E42B6	4597F299CFC657E2	5FCB6FAB3AD6FAEC	6C44198C4A475817

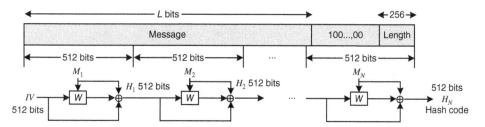

Figure 6.11 Overview of Whirlpool algorithm.

6.2.5 Whirlpool

Whirlpool is a hash function endorsed by the New European Schemes for Signatures, Integrity, and Encryption (NESSIE) project, a European Union-sponsored effort to put forward a portfolio of strong cryptographic primitives of various types [39]. The compression function of Whirlpool is based on a modified AES block cipher. It is intended to provide a comparable (if not better) security and performance to non-block-cipher based hash functions, e.g. SHA-2. The overall structure of Whirlpool is illustrated in Figure 6.11. It takes as input a message with a maximum length of less than 2^{256} bits and produces a 512-bit message digest. The input is processed in 512-bit blocks. The algorithm consists of the following steps:

(1) Append padding bits ($1\|0, \ldots, 0$) to make the message length congruent to 256 modulo 512. Append another 256 bits at the end of the padded message that is reserved for the "Length" of the original message M. The augmented message will be divided into N blocks, with each block message M_i 512 bits.

(2) Set IV to initialize hash matrix $H_0 = IV$.

(3) Process message in 512-bit (64-byte) blocks, using as its core, the block function W. In each block i, W takes the input of the current block message M_i and the previous output

Table 6.3 Comparison of Whirlpool block cipher W and AES.

	Whirlpool	AES
Block size	512	128
Key size	512	128, 192, 256
Matrix orientation	Row-wise	Column-wise
Number of rounds	10	10,12, 14
Key expansion	W found function	Dedicated algorithm
$GF(2^8)$ polynomial	$x^8 + x^4 + x^3 + x^2 + 1$	$x^8 + x^4 + x^3 + x + 1$
S-box	Recursive structure	Multiplicative inverse in $GF(2^8)$ plus affine transformation
Round constants	Successive entries of the S-box	Elements 2^i of $GF(2^8)$
Diffusion layer	Mix rows	Mix columns
Permutation	Shift columns	Shit rows

H_{i-1}, the output of W is XORed with the M_i and H_{i-1} again to get the H_i. The last block of the process H_N will be the 512 bits hash code.

The function W was specifically designed for use in the has function Whirlpool. It has the security and efficiency of AES but with a hash length that equals SHA-512. A comparison of Whirlpool block cipher W and AES is shown in Table 6.3.

Figure 6.12 shows the structure of the block function W. Each round in the algorithm involves the use of four different steps: add round key, substitute bytes, shift columns, and mix rows. The input is mapped by rows in W. The key schedule uses the same W round function, but with round constants $RC[I]$ (being S-box outputs) instead of subkeys.

6.2.6 Other MAC Functions

6.2.6.1 Keyed Hash Functions as MACs

Cryptographic hash functions are efficient and widely available; however, a hash function cannot be used directly for an MAC because the lack of a secret key. HMAC is an MAC created by existing hash functions, as standardized in the Internet standard RFC 2104 [40]. The overall structure of HMAC is shown in Figure 6.13. Given secret key K, it is first padded with zeros to the left end to create a b-bit string K^+ (e.g. if K is of length 160 bits and $b = 512$, then K will be appended with 44 zero bytes). *ipad* is a pad value of 00110110 (36 in HEX) repeated to fill block. *opad* is a pad value of 01011100 (5C in HEX) repeated to fill block. M is the message input to HMAC (including the padding specified in the embedded hash function). In summary, for a message M with a secret key K, HMAC creates an MAC as follows:

$$HMAC(K, M) = H[(K^+ \oplus opad) \| H[(K^+ \oplus ipad) \| M)]].$$

Any hash function H can be used in HMAC, e.g. MD5, SHA-2, Whirlpool, etc. the security of HMAC is based on the security of the hash function.

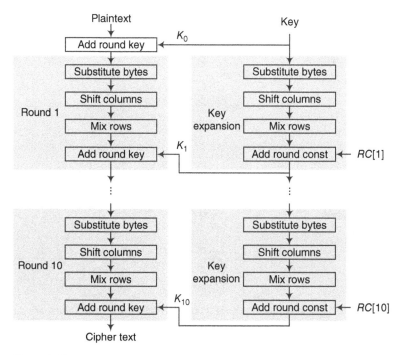

Figure 6.12 Whirlpool block function W.

Figure 6.13 HMAC overview.

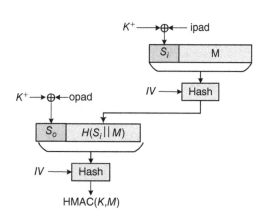

6.2.6.2 Cipher-Based MAC

Cipher-based MAC (CMAC) is a block cipher-based MAC algorithm that has been widely adopted in government and industry. The core of the CMAC algorithm is a refinement of the Cipher Block Chaining MAC (CBC-MAC) adopted by NIST. CMAC uses multiple keys to remove the restriction of CBC-MAC that only messages of one fixed length of $m \times n$ bits are processed, where n is the cipher block size and m is a fixed positive integer. CMAC is specified in NIST Special Publication 800-38B [41]. CMAC generates a l-bit MAC (i.e. CMAC tag T) using a b-bit block cipher E (e.g. AES or triple-DES) and a secret key K. Two subkeys K_1 and K_2 are first derived from K as follows:

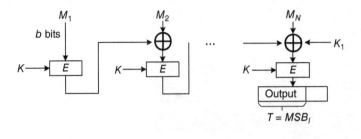

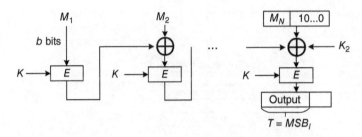

Figure 6.14 Cipher-based message authentication code (CMAC). (a) Message length is interger multiple of block size. (b) Message length is not interger multiple of block size.

- Calculates a temporary value $K_0 = E_K(0)$.
- If $MSB(K_0) = 0$, then $K_1 = SL^1(K_0)$, else $K_1 = SL^1(K_0) \oplus C$, where $SL^n(x)$ is left shift of the argument x by n bits and C is a certain constant that depends only on b.
- If $MSB(K_1) = 0$, then $K_2 = SL^1(K_1)$, else $K_2 = SL^1(K_1) \oplus C$.

For example, suppose $b = 4$, $C = (0011)_2$, and $K_0 = E_K(0) = (0101)_2$. Then $K_1 = (1010)_2$ and $K_2 = (0100)_2 \oplus (0011)_2 = (0111)_2$. Once the subkeys are generated, the CMAC tag T is generated as shown in Figure 6.14. The input message is split into b-bit blocks $M = M_1\|M_2\|, \dots, \|M_N$. If M_N is a complete block then apply K_1 as illustrated in Figure 6.14a. Else, pad M_N to a complete block with "$10, \dots, 0$," and then apply K_2 as illustrated in Figure 6.14b. For $i = 1$ to $N - 1$, calculate the output of each block operation as follows:

$$c_i = E_K(c_{i-1} \oplus M_i). \tag{6.5}$$

The output of the final block is $c_N = E_K(c_{N-1} \oplus M_N)$. The CMAC tag T is the l most significant bits of c_N, i.e. $T = MSB_l(c_N)$.

6.3 Digital Signature and Authentication

Message authentication is a procedure to verify that the received message comes from the alleged source and have not been altered. It does not protect sender and receiver against each other. Certain mechanisms must be provided to protect the two parties against each other. Digital signature provides the security capability to achieve such a goal. Thus, a digital signature function includes the authentication function with additional capabilities.

6.3.1 Digital Signature Properties

A digital signature needs to provide the ability to verify sender and time of signature, the ability to authenticate message contents, and the ability to be verified by third parties to resolve disputes. Therefore, a digital signature must have the following properties:

- Must depend on the message signed;
- Must use information unique to sender to prevent both forgery and denial;
- Must be relatively easy to produce;
- Must verify the sender and the time of the signature;
- Must authenticate the contents at the time of the signature;
- Must be verifiable by third parties, to resolve disputes.

A secure hash function is usually embedded in a digital signature scheme to meet the requirements. A variety of approaches have been proposed for the digital signature function. These approaches fall into two categories: *direct digital signatures* and *arbitrated digital signature.*

Direct digital signatures involve only the communicating parties with direct application of public-key algorithms. For sender X and receiver Y, a digital signature may be formed by encrypting the entire message M with the sender's private key Pr_x, or by encrypting a hash value of the message $H(M)$ with the sender's private key. Confidentiality can be provided by further encrypting the entire message plus signature using either public or symmetric key schemes. A message with message authentication, digital signature, and confidentiality can be created as follows:

$$X \rightarrow Y : C = E(Pu_y, [M \| E(Pr_x, H(M))]).$$

The digital signature $E(Pr_x, H(M))$ is applied to the hash value only. It is important to perform the digital signature function first and then an outer confidentiality function. In case of dispute, a third party must view the message and its signature. A direct digital signature depends on the security of the sender's private key. It may have issues if a private key is compromised thus signatures can be forged. As a counter measurement, it is necessary to add time stamps and timely key revocation schemes.

Arbitrated digital signatures can address the problems that are associated with direct digital signatures. An arbiter is a third party that is independent of both sender and receiver. An arbiter must also be a trusted party so that the mechanism can work properly. A variety of arrangements are available to apply arbitrated digital signatures, with either private or public key algorithms. Let X, Y, and A be sender, receiver, and arbiter, respectively. The following arrangements apply private key algorithm (with pre-shared secret keys K_{xa} between X and A, and K_{ya} between Y and A):

$$X \rightarrow A : M \| E(K_{xa}, [ID_X \| H(M)]),$$
$$A \rightarrow Y : E(K_{ya}, [ID_X \| M \| E(K_{ya}, [ID_X \| H(M)]) \| T]),$$

where T is a time stamp. Note that message M is revealed to the arbiter in this arrangement. Another private key based arrangement without revealing the message is as follows:

$$X \rightarrow A : ID_X \| E(K_{xy}, M) \| E(K_{xa}, [ID_X \| H(E(K_{xy}, M))]),$$
$$A \rightarrow Y : E(K_{ya}, [ID_X \| E(K_{xy}, M)] \| E(K_{ya}, [ID_X \| H(E(K_{xy}, M)) \| T]).$$

Note that this arrangement requires another secret key K_{xy} shared between sender and receiver. Public key encryption can also be applied to arbitrated digital signatures. The following arrangement is based on public key encryption and message is not revealed to the arbiter.

$$X \to A : ID_X \| E(Pr_X, \| [ID_X \| E(Pu_y, E(Pr_X, M))]),$$
$$A \to Y : E(Pr_a, [ID_X \| E(Pu_y, E(Pr_X, M)) \| T]),$$

where Pu_n and Pr_n are public and private keys of entity n, respectively.

6.3.2 Digital Signature Standard and Algorithm

The Digital Signature Standard (DSS) was originally proposed in 1991 and revised in 1993 in response to public feedback concerning the security of the scheme. There was a further minor revision in 1996. In 2000, an expanded version of the standard was issued as NIST FIPS 186-2, which incorporates digital signature algorithms based on RSA and on elliptic curve cryptography. The DSS makes use of the SHA, and presents a new digital signature technique, the Digital Signature Algorithm (DSA). DSA is the US government approved signature scheme, which is designed to provide strong signatures without allowing easy use for encryption. As discussed in the last chapter, RSA can be applied as a digital signature approach, as shown in Figure 6.15. The RSA creates digital signature by encrypting the hash value with the sender's private key. The verification is done by decrypting the signature with the sender's public key. The DSA digital signature approach is different from the RSA approach in how the message signature is generated and verified.

The DSA signature scheme is superior to the RSA approach in two ways. First, the DSA signature creates a 320-bit signature, which is much smaller than the 1024-bit result of RSA. Second, the DSA approach is faster due to its fast computation (mostly done by modular function over a 160-bit number). Nevertheless, unlike RSA, the DSA approach cannot be used for encryption or key exchange. Although it is a public key technique. The DSA is based on the difficulty of computing discrete logarithms, and is based on a scheme originally presented by ElGamal and Schnorr. As shown in Figure 6.16, DSA signature uses the message hash value, global public values, private key and random value k to create a 2 part signature (s, r). Verification is done by computing a function of the hash value, public key,

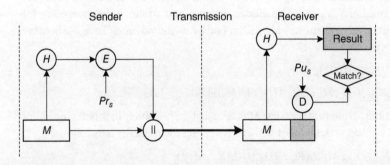

Figure 6.15 RSA approach for digital signature.

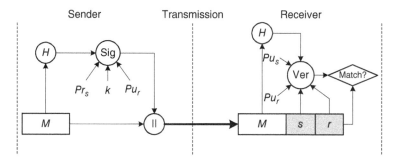

Figure 6.16 DSS approach for digital signature.

(s, r), to generate a result that is compared with r. The details are given as follows. A set of global parameters (p, q, g) is chosen, where p is a large prime number that is $2^{L-1} < p < 2^L$. L is between 512 and 1024 bits and is a multiple of 64. And q is a 160-bit prime number that is a prime factor of $(p - 1)$. g is chosen as $g = h^{(p-1)}/q$, where $h < p - 1$, $h^{(p-1)/q}(\text{mod } p) > 1$. With the global parameters distributed to all DSA users, each user chooses a random private key $x, x < q$ and computes their public key $y = g^x(\text{mod } p)$. To create a digital signature, the sender first generates a random signature key $k, k < q$. This value will be destroyed after use, and never be reused. The sender then computes the signature pair (s, r) as follows:

$$r = (g^k(\text{mod } p))(\text{mod } q), \tag{6.6}$$

and

$$s = (k^{-1} \cdot H(M) + x \cdot r)(\text{mod } q). \tag{6.7}$$

On the receiving side, the receiver verifies the signature as follows:

$$w = s^{-1}(\text{mod } q),$$
$$u_1 = (H(M) \cdot w)(\text{mod } q),$$
$$u_2 = (r \cdot w)(\text{mod } q),$$
$$v = (g^{u_1} \cdot y^{u_2}(\text{mod } p))(\text{mod } q).$$

If v matches r then the signature is verified.

6.3.3 The Elliptic Curve Digital Signature Algorithm

The Elliptic Curve Digital Signature Standard (ECDSA) was a public key cryptography developed for the American National Standards Institute by the Accredited Standards Committee on Financial Services, X9 (ANSI X9.62) [42]. The Elliptic Curve Digital Signature Standard (ECDSA) is the elliptic curve analog of DSA [43].

6.3.3.1 ECDSA Domain Parameters
ECDSA domain parameters are common to a group of users and may be public. The domain parameters are the foundation to generate public/private key pairs required for digital signature generation and verification. Specifically, domain parameters for ECDSA are of the form (q, FR, a, b, G, n, h), where q is the field size; FR is an indication of the basis used; a and b

Table 6.4   ECDSA security parameters.

Bit length of n	Maximum cofactor h
160–223	2^{10}
224–255	2^{14}
256–383	2^{16}
384–511	2^{24}
≥ 512	2^{32}

are two field elements that define the equation of the curve, G is a base point of prime order on the curve (i.e. $G = (x_G, y_G)$), n is the order of the point G, and h is the cofactor (which is equal to the order of the curve divided by n). A domain parameter seed is optional if the elliptic curve is randomly generated in a verifiable fashion. ECDSA specifies five ranges for n and maximum sizes for the corresponding cofactor h, as listed in Table 6.4. Two arithmetic fields are defined for curves in ECDSA. One is the prime finite field GF_p where p is an odd prime number, and the other is the binary finite field GF_{2^m}.

Recommended Curves over Binary Fields:

$$E : \quad y^2 + xy = x^3 + x^2 + b, \tag{6.8}$$

for each field degree m, where the cofactor is $h = 2$. Alternatively, the recommended Koblitz curves have the form:

$$E_a : \quad y^2 + xy = x^3 + ax^2 + 1, \tag{6.9}$$

where $a = 0$ or 1, the cofactor is $h = 2$ if $a = 1$, and $h = 4$ if $a = 0$.

Recommended Curves over Prime Fields:

$$E : \quad y^2 \equiv x^3 - 3x + b \quad (\text{mod} \quad p), \tag{6.10}$$

for each prime p of prime order n. The selection $a \equiv -3$ for the coefficient of x was made for reasons of efficiency, as explained in IEEE Std 1363-2000 [44].

6.3.3.2   ECDSA Private/Public Keys

An ECDSA key pair consists of a private key d and a public key Q that is associated with a specific set of ECDSA domain parameters; d, Q and the domain parameters are mathematically related to each other. Two methods (i.e. Algorithms 6.1 and 6.2) can be used to generate a digital signature key pair (d, Q) and can be generated for a set of domain parameters (q, FR, a, b, G, n, h).

In comparison, Algorithm 6.1 requires 64 more bits for d so that bias produced by the mod function is negligible. Note that if N is invalid, or the bit string b cannot satisfy the required security strength, the algorithms cannot generate the key pair.

6.3.3.3 ECDSA Digital Signature Generation

Given a message m, an ECDSA digital signature (r, s) is generated as follows:

(1) Select a random integer k, $1 \leq k \leq n - 1$.
(2) Compute $kG = (x_1, y_1)$ and convert x_1 to an integer c_{x_1} through Algorithm 6.2.
(3) Compute r as follows:

$$r = x_1 \mod n. \tag{6.11}$$

(If $r = 0$, select another k and start again.)
(4) Compute the multiplicative inverse $k^{-1} \mod n$.
(5) Compute the hash value of m through a hash function specified by NIST FIPS 180 [45] and convert the result to an integer e. The security strength of the chosen hash function shall meet or exceed the security strength required for the digital signature process.
(6) Compute s as follows:

$$s = k^{-1}(e + dr) \mod n. \tag{6.12}$$

(If $s = 0$, start again.)
(7) Return the ECDSA digital signature (r, s).

Algorithm 6.1 ECDSA key pair generation using extra random bits

---

Input: Domain parameters (q, FR, a, b, G, n, h);
Output: ECDSA key pair (d, Q);
    $N \leftarrow$ length of (n);
    $b \leftarrow$ a random string of $(N + 64)$ bits
    Convert bit string b to the non-negative integer c according to Algorithm 6.3;
    $d \leftarrow (c \mod (n - 1))) + 1$;
    $Q \leftarrow dG$;
    Return (d, Q);

Algorithm 6.2 ECDSA key pair generation by testing candidates

---

Input: Domain parameters (q, FR, a, b, G, n, h);
Output: ECDSA key pair (d, Q);
    $N \leftarrow$ length of (n);
    while $c > n - 2$ **do**
        $b \leftarrow$ a random string of N bits
        Convert bit string b to the non-negative integer c according to Algorithm 6.3;
    end while
    $d \leftarrow c + 1$;
    $Q \leftarrow dG$;
    Return (d, Q);

Algorithm 6.3 Conversion of a bit string to an integer

Input: An n-bit sequence $\{b_1, b_2, \ldots, b_n\}$
Output: Integer C;
    Let $\{b_1, b_2, \ldots, b_n\}$ be the bits of b from leftmost to rightmost;
    $C = \sum_{i=1}^{n} 2^{n-i} b_i$;
    Return C;

6.3.3.4 ECDSA Digital Signature Verification

An ECDSA digital signature (r, s) of message m is verified as follows:

(1) Verify that both r and s are positive integers smaller than n.
(2) Compute the hash value of m and convert the result to an integer e.
(3) Compute $w = s^{-1} \bmod n$;
(4) Compute $u_1 = ew \bmod n$ and $u_2 = rw \bmod n$.
(5) Compute $X = u_1 G + u_2 Q$. (Reject the signature if $X = \mathcal{O}$.)
(6) Convert the x-coordinate x_1 of X to an integer c_{x_1}, and compute v as follows:

$$v = c_{x_1} \bmod n. \tag{6.13}$$

(7) If $v = r$, accept the signature.

Proof that signature verification works: Given that $s = k^{-1}(e + dr) \bmod n$, it can be seen that,

$$k \equiv s^{-1}(e + dr) \equiv s^{-1}e + s^{-1}rd \equiv we + wrd \equiv u_1 + u_2 d \quad \bmod \quad n.$$

Thus

$$X = u_1 G + u_2 Q = (u_1 + u_2 d)G = kG,$$

and $v = r$ is required.

6.3.4 Authentication Protocols

Authentication is a security feature that convince parties of each other's identity. Session key establishment and exchange often occur with an authentication process. An authentication may be one-way or mutual depending on the application scenarios. Whether the authentication is one-way or mutual, the key issues of authentication include confidentiality and timeliness. That is to say, it is critical for authentication to protect session keys during the process (if applied), and be strong against replay attacks. Replay attacks are where a valid signed message is copied and later resent. Such replays, at worst, could allow an opponent to compromise a session key or successfully impersonate another party. At minimum, a successful replay can disrupt operations by presenting parties with messages that appear genuine but are not. Different techniques can be applied for authentication protocols to counter replay attacks. Possible choices are:

- using sequence numbers: generally impractical since must remember last number used with every communicating party;

- using timestamps: needs synchronized clocks among all parties involved, which can be problematic;
- using challenge/response: using unique, random, unpredictable nonce, but not suitable for connectionless applications because of handshake overhead.

One kind of authentication protocols uses symmetric encryption. In general, a two-level hierarchy of symmetric encryption keys can be used to provide confidentiality for communication in a distributed environment. One level involves master keys that are for each pair of users and a trusted Key Distribution Center (KDC). The KDC is responsible for generating session keys, and for distributing those keys to the parties involved. The other level involves the session keys that enable direct communications among authenticated pair of parties.

Needham–Schroeder Protocol is one example of two-level hierarchy based authentication protocol. It is used by two parties A and B, who both trust a common KDC. But A and B do not trust each other, thus require authentication first. A and B share a secret key K_a and K_b, respectively, with the KDC. The KDC gives one part the information needed to establish a session key with the other. After authentication, A and B share a session key K_s. The protocol operates as follows:

$$A \rightarrow KDC : ID_A\|ID_B\|N_1,$$
$$KDC \rightarrow A : E(K_a, [K_s\|ID_B\|N_1\|E(K_b, [K_s\|ID_A])]),$$
$$A \rightarrow B : E(K_b, [K_s\|ID_A]),$$
$$B \rightarrow A : E(K_s, N_2),$$
$$A \rightarrow B : E(K_s, f(N_2)).$$

Nonces N_1 and N_2 are added to protect the authentication against replay attacks.

Public key encryption can also be applied to authentication protocols. In this case, it is generally assumed that each of the two parties is in possession of the current public key of the other. A third party is involved, known as an Authentication Server (AS). The AS is also assumed to be a trusted party.

Denning–Sacco authentication protocol is one example of public key based authentication protocol. In this protocol, the AS only provides public-key certificates. The session key is chosen and encrypted by A; hence, there is no risk of exposure by the AS. The protocol operates as follows:

$$A \rightarrow AS : ID_A\|ID_B,$$
$$AS \rightarrow A : E(PR_{as}, [ID_A\|PU_a\|T]\|E(PR_{as}, [ID_B\|PU_b\|T])),$$
$$A \rightarrow B : E(PR_{as}, [ID_A\|PU_a\|T]\|E(PR_{as}, [ID_B\|PU_b\|T$$
$$\|E(PU_b, E(PR_a, [K_s\|T]))]).$$

The timestamps T protect against replays of compromised keys. This protocol is compact but requires synchronization of clocks.

6.4 Key Management

A slight touch of session key distribution has been given in the previous section of authentication protocols. Key management and distribution are complex topics. In this section, a general idea and some examples of key management schemes will be given.

6.4.1 Key Distribution with Symmetric Key Encryptions

A secret key must be shared between the two parties for symmetric encryption to work. Other features such as message authentication and digital signature may also depend on pre-shared keys. Those keys must be protected from an unauthorized access. Moreover, keys are more likely to be comprised after being used for a period of time. Therefore, frequent key changes are usually desirable to enhance the security. Once a key is generated or updated, it needs to be distributed to the two parties (e.g. A and B). Key distribution can be achieved in a number of ways:

(1) Physical delivery from A to B: it is the simplest key distribution mechanism. However, it is only applicable when there is personal contact between recipient and key issuer.
(2) A third party C selects and delivers a key to A and B: key is known to C. It is only applicable when C is trusted by both A and B.
(3) A and B can use a previous key to encrypt a new key: it is only applicable when a secret key has been shared between A and B, usually through one of the two ways mentioned earlier. This method suffers that if an attacker gains access to one key, then all subsequent keys will be revealed.
(4) A third party C relays a key between A and B: a new key is usually chosen by A or B. It is only applicable when A and B both have secure communications with a trusted third party C.

If end-to-end encryption is provided at the network level, then a secret key is needed for every pair of hosts on the network that wish to communicate. If encryption is done at the application level, then a key is needed for every pair of users that require communications. For N hosts/users, the number of required symmetric keys is $[N(N-1)]/2$, as illustrated in Figure 6.17. With a network that supports 10 000 application users, as many as 50 million keys may be required for application-level encryption. As number of communication parties grow, only option (4) (or variants of it) is a practical solution to the huge growth in number of keys potentially needed. One advantage of option (4) is that *key hierarchy* is applied to reduce the number of keys. In this scheme, a Key Distribution Center (KDC) is responsible for distributing keys to pairs of users that need communications. Each user must share a unique secret key with the KDC so that secure communications can be established between each user and the KDC. The unique secret keys are the first level of key hierarchy. A second level includes session keys that are used for the duration of a logical connection. More levels of keys may be applied based on different requirements.

A typical key distribution scenario with key hierarchy works as follows:

(1) A requests from the KDC a session key to protect a logical connection to B. The message includes the identities of A and B and a unique nonce N_1.

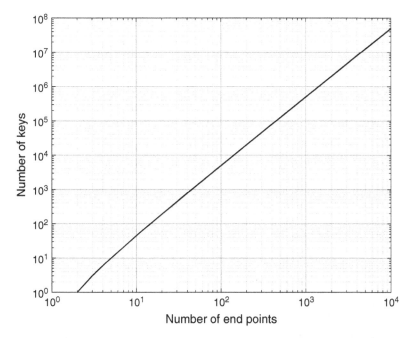

Figure 6.17 Number of keys required for different number of end points.

(2) The KDC responds with a message encrypted using K_a that includes a one-time session key K_s to be used for the session, the original request message to enable A to match response with appropriate request, and info for B.

(3) A stores the session key for use in the upcoming session and forwards to B the information from the KDC for B, namely, $E(K_b, [K_s\|ID_A])$. Because this information is encrypted with K_b, it is protected from eavesdropping. At this point, a session key has been securely delivered to A and B, and they may begin their protected communications. Two additional steps are desirable in the following.

(4) Using the new session key K_s for encryption, B sends a nonce N_2 to A.

(5) Also using K_s, A responds with $f(N2)$, where f is a function that performs some transformation on N_2 (e.g. adding one). These steps assure B that the original message it received (step 3) was not a replay. Note that the actual key distribution involves only steps 1 through 3 but that steps 4 and 5, as well as 3, perform an authentication function.

For very large networks, a single KDC may not be sufficient. One may set up a hierarchy of KDCs where each KDC handles its own local domain. For communications among users within the same local domain, the local KDC is responsible for key distribution. If two entities in different domains desire a shared key, then the corresponding local KDCs can communicate through a (hierarchy of) global KDC(s).

6.4.2 Symmetric Key Distribution Using Public Key Cryptosystems

A public key cryptosystem can be applied to encrypt secret keys for distribution. Although public key cryptosystems are also based on some pre-shared parameters (including public

Initiator A   Responder B

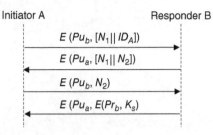

Figure 6.18 Key distribution with confidentiality and authentication.

keys), revealing those parameters is more controllable than revealing secret keys that are used in symmetric encryptions. A direct application of a public key cryptosystem is as follows: assuming A and B are communicating parties that have exchanged public keys Pu_a and Pu_b. ID_A and ID_B are the identities of A and B, respectively. Let A initiate the key distribution process, and let B choose the session key K_s. Detailed processes are illustrated in Figure 6.18. The first three messages are for mutual authentication. Nonce N_1 is applied in the first message for A to authenticate B. B replies an encrypted $N_1 \| N_2$ to identify itself as well asking authentication of A. A replies an encrypted N_2 in the third message and completes the authentication. In the last message, B chooses a session key K_s, encrypts it and also signs it before transmitting to A. The result is that this scheme ensures both confidentiality and authentication in the exchange of a secret key.

A public-key cryptosystem can also be applied in a hybrid way. Session keys are chosen and distributed by a KDC. The KDC shares a secret master key that is used for session key distribution with each user. A public key scheme is used to distribute the master keys in the beginning of the distribution process. The advantage of the hybrid scheme is that, for a configuration with a single KDC serving a widely distributed set of users, the master keys can be more efficiently and securely distributed.

6.4.3 Distribution of Public Keys

In public key cryptosystems, one problem is that the public key is public, hence any participant can send his or her public key to any other participant or broadcast the key to the community. Its major weakness is forgery, anyone can create a key claiming to be someone else and broadcast it. Several techniques have been proposed for the distribution of public keys, which can mostly be grouped into the categories as follows.

- *Public announcement*: Users announce public keys to recipients or broadcast to community at large. The major weakness for this category of method is forgery. Anyone can create a key claiming to be someone else and broadcast it. Until the forgery is discovered they can masquerade as the claimed user.
- *Using publicly available directory*: A greater degree of security can be achieved by maintaining a publicly available dynamic directory of public keys. Maintenance and distribution of the public directory would have to be the responsibility of some trusted entity or organization. Participants register securely with a directory, they can replace keys at any time. The directory is periodically published. This scheme is clearly more secure than individual public announcements but still has vulnerabilities to tampering or forgery.

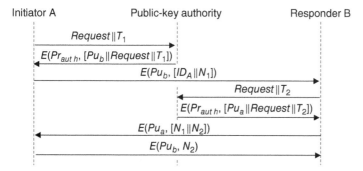

Figure 6.19 Public key authority.

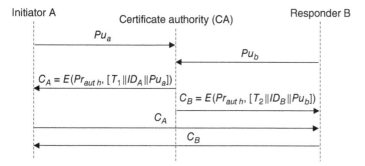

Figure 6.20 Public-key certificates.

- *Using public key authority*: a public key authority is applied to provide tighter control over the distribution of public keys from the directory for stronger security. As shown in Figure 6.19, participants (e.g. A and B) must know the public key Pu_{auth} for the public key authority. To request a public key, the initiator A contacts directly to the public key authority. No direct contact with B is needed to obtain its public key. The final two steps are for the mutual authentication between A and B. Nonetheless, some drawbacks exist in this scenario. First, the public key authority becomes a bottleneck in the system, for a user must appeal to the authority for a public key for every other user that it wishes to contact. Second, the directory of names and public keys is still vulnerable to tampering.
- *Using public key certificates*: a public key certificate binds an identity to public key, with all contents signed by a trusted Certificate Authority (CA). By using certificates, participants are able to exchange keys without contacting a public-key authority. As shown in Figure 6.20, participants (e.g. A and B) can present their public key to the CA, and obtain a certificate. When A needs to fetch the public key of B, B will present its certificate C_B to A. A verifies the certificate by way of the attached trusted signature from the CA. A participant can also convey its key information to another by transmitting its certificate. Other participant can verify that the certificate was created by the authority, provided that they know its public key.

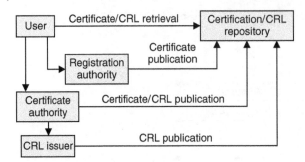

Figure 6.21 Public key infrastructure.

6.4.4 Public Key Infrastructure

RFC 2822 (Internet Message Format) defines public-key infrastructure (PKI) as the set of hardware, software, people, policies, and procedures needed to create, manage, store, distribute, and revoke digital certificates based on asymmetric cryptography (public key cryptography) [46]. The principal of PKI is to enable secure, convenient, and efficient acquisition of public keys. Figure 6.21 shows interrelationships among the elements of public key infrastructure. **User** is a generic term used to denote entities that are in the subject field of a public key certificate, such as end users, servers, routers, etc. **Certificate authority** is the issuer of certificates and certificate revocation lists in most cases. A CA may also support a variety of administrative functions. For example, a user needs to get registered with CA to create its certificate. **Registration authority** is an optional component that performs a number of administrative functions. A user may also register itself with the RA to get its certificate. **CRL issuer** is an optional component that publishes CRLs. **Repository** is the "directory" for storing certificates and CRLs so that they can be retrieved by users.

6.4.5 X.509 Authentication Service

X.509 is a standard defined for public key infrastructure certificate and certificate revocation list profile [47]. The standard was subsequently revised to address some of the security concerns. The second version was issued in 1993. A third version was issued in 1995 and revised in 2000. The X.509 standard has been universally accepted for formatting public-key certificates. X.509 certificates are widely used in network security applications, including IP security, secure sockets layer (SSL), secure electronic transactions (SET), and S/MIME. X.509 defines both a framework for the authentication services by the X.500 directory, and alternative authentication protocols based on the use of public key certificates. Public key cryptography and digital signatures are the foundation of X.509. There is no dictated algorithm in X.509 standard, nonetheless RSA is recommended for public key encryption.

The format of X.509 certificate and revocation list are shown in Figure 6.22. The heart of the X.509 scheme is the public-key certificate associated with each user. The CA signs the certificate for user A with its private key as follows:

$$CA \ll A \gg = CA\,(V, SN, AI, CA, TA, UCA, A, UA, Ap),$$

where A is the participant, V is the version number, SN is the serial number that is unique within CA to identify a certificate, AI is the signature algorithm identifier, CA is the issuer's

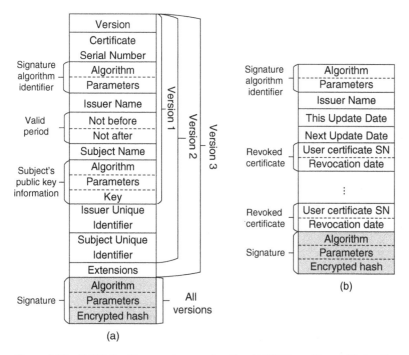

Figure 6.22 X.509 Certificate and revocation list. (a) X.509 certificate. (b) Certificate revocation list.

name, *TA* is the valid period, *UCA* is the optional unique identifier of the CA, *A* is the subject name, *UA* is the optional unique identifier of the user A, and *Ap* includes the subject's public key information. *UCA* and *UA* are only applied in the second and third versions of X.509. Extensions only exist for the third version of X.509.

User certificates generated by a CA have the characteristics that any user with access to the public key of the CA can verify the user public key that was certified, and no party other than the CA can modify the certificate without this being detected. Because certificates are unforgeable, they can be placed in a directory without the need for the directory to make special efforts to protect them. X.509 version 3 includes a number of optional extensions. Each extension consists of an extension identifier, a criticality indicator, and an extension value. The certificate extensions fall into three main categories:

- *Key and policy information*: convey additional information about the subject and issuer keys, plus indicators of certificate policy. A certificate policy is a named set of rules that indicates the applicability of a certificate to a particular community and/or class of application with common security requirements.
- *Subject and issuer attributes*: support alternative names, in alternative formats, for a certificate subject or certificate issuer and can convey additional information about the certificate subject; e.g. postal address, email address, or picture image.
- *Certification path constraints*: allow constraint specifications to be included in certificates issued for CA's by other CA's that may restrict the types of certificates that can be issued by the subject CA or that may occur subsequently in a certification chain.

As mentioned earlier, a certificate includes a period of validity, i.e. *TA*. Typically a new certificate is issued just before the expiration of the old one. In addition, it may be desirable on occasion to revoke a certificate before it expires. To support this, each CA must maintain a list consisting of all revoked but not expired certificates issued by that CA, known as the Certificate Revocation List (CRL). Each CRL posted to the directory is signed by the issuer and includes the issuer's name, the date the list was created, the date the next CRL is scheduled to be issued, and an entry for each revoked certificate. Each entry consists of the serial number of a certificate and revocation date for that certificate. Because serial numbers are unique within a CA, the serial number is sufficient to identify the certificate. When a user receives a certificate in a message, the user must determine whether the certificate has been revoked, by checking the directory CRL each time a certificate is received, this often does not happen in practice.

6.5 Summary

In this chapter, message authentication has been discussed at first. MAC and hash functions are two widely used mechanisms to provide message authentication. The difference is that an MAC is keyed and a hash function has no secret key. Hash functions can be used as the core of an MAC, e.g. HMAC. The security of HMAC depends on the hash function that is applied. Besides message authentication, digital signature has been discussed in this chapter. A digital signature is an algorithm or scheme for verifying the authenticity of digital messages or information. In the last, key management has been briefly discussed in this chapter. Both symmetric and asymmetric key distributions have been illustrated. Hierarchical key distribution mechanisms are needed when a communication system has large number of users. Both symmetric and public key mechanisms are applied in key distribution mechanisms, serving different purposes.

Part III

Security for Wireless Local Area Networks

Part II

7

WLAN Security

Wireless Local Area Network (WLAN) is a wireless computer network within a limited area. Typical deployment of a WLAN can be found in a hotel, a school building, and many other places. Because of the wireless communication method, WLAN enables flexible access and mobility within its transmission coverage. Meanwhile, WLAN is more vulnerable to attacks than wired connections due to the lack of physical connections. In this chapter, the security solutions of WLAN technologies are introduced.

7.1 Introduction to WLAN

The most popular modern WLAN technology is Wi-Fi, which is a trademark of the *Wi-Fi Alliance* [48]. The core technology of Wi-Fi is based on IEEE 802.11 standards [49]. Wi-Fi is almost the de facto feature for all devices and appliances that has communication capability, such as smart phones, laptop computers, tablets, etc. Even legacy devices have been upgraded with Wi-Fi capability, e.g. smart TVs, smart watches, smart refrigerators, etc. Without loss of generality, WLAN and Wi-Fi will be used interchangeably in the rest of this chapter.

7.1.1 Wi-Fi Operating Modes

Wi-Fi supports infrastructure based mode and ad-hoc mode. An infrastructure based Wi-Fi typically comprises of an Access Point (AP) and several wireless stations (STAs). An AP is normally a wireless router, which is the base station of a WLAN. An STA is a Wi-Fi enabled device, such as a smart phone or a laptop computer. The process of joining a Wi-Fi network is illustrated in Figure 7.1. The STA initiates the process by sending an association request to the AP. The AP then runs an authentication process. An association response is sent back to the STA if the authentication is successful. In an infrastructure based Wi-Fi network, all STAs only communicate with the AP. The AP constantly broadcasts beacon frames to manage the network. Beacon frames include MAC header, timestamp, beacon interval, capability information, service set identifier (SSID), supported data rates, radio parameters, etc.

There are two types of communications in an infrastructure based Wi-Fi network: one is to connect an STA to the external network (e.g. Internet), and the other is to connect an STA to another STA within the same Wi-Fi network. As shown in Figure 7.2, in both types

Security in Wireless Communication Networks, First Edition. Yi Qian, Feng Ye, and Hsiao-Hwa Chen.
© 2022 John Wiley & Sons Ltd. Published 2022 by John Wiley & Sons Ltd.
Companion website: www.wiley.com/go/qian/sec51

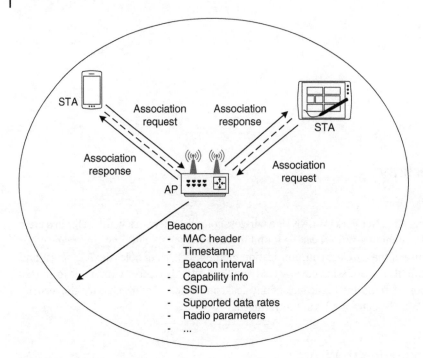

Figure 7.1 Overview of WLAN access.

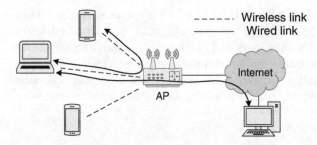

Figure 7.2 Communication flows and Internet connections in a WLAN.

of communications, the STA transmits data to the AP first. Then the AP relays the data to the external network or to another STA depending on request. Wi-Fi ad-hoc mode supports direct communications between the two Wi-Fi devices without a pre-deployed infrastructure, for example, using a laptop or a smart phone as a Wi-Fi hotspot to create an ad-hoc Wi-Fi network. Wi-Fi Direct is a certification mark by the Wi-Fi Alliance for devices supporting a technology that enables ad-hoc Wi-Fi [50]. Wi-Fi Direct can be used to share content, synchronize data, play audio and video, and many other things that may or may not require external network connection.

7.1.2 Challenges in WLAN Security

WLAN defined by the IEEE 802.11 standards focuses on PHY layer and MAC layer protocols. Unfortunately, security was not well considered by the original IEEE standards.

The security discussion of WLAN will be focused on MAC layer issues in the rest of this chapter. Readers are recommended to consult IPSec and TLS/SSL for security design at network layer and transport layer. In the lower layers, providing WLAN security is harder than providing security in wired network for several reasons as indicated in the following.

- Users share the open medium in a WLAN while a physical break is required to access the medium for a wired network.
- Users are roaming everywhere in a WLAN or between WLANs. It is harder to manage WLAN in a timely and secure manner.
- Some APs may be rogue ones that can hardly be identified by users.
- Some of the existing security solutions for WLAN have flaws.

7.1.3 Tricks that Fail to Protect WLAN

Many users are aware of the security issues in Wi-Fi. A few techniques have been adopted to secure a Wi-Fi network. However, some of them are virtually useless despite the convincing description. For example, *MAC authentication*, where a MAC address filter is applied to the AP. On one hand, MAC addresses are transmitted in clear text over the air, thus can be easily captured. MAC addresses can also be easily cloned and defeated. On the other hand, it is extremely difficult to manage MAC filtering. Someone may disable *Dynamic Host Configuration Protocol (DHCP)* and force static IP addresses. However, IP schemes are easy to figure out since the IP addresses are sent over the air in clear text, similar to MAC addresses. Therefore, it is easy for an intruder to figure out an IP scheme and enter a static IP address. Another trick is *SSID hiding*. Unfortunately, an SSID is only hidden once from AP beacon suppression. There are four other SSIDs that are broadcast from an AP, i.e. probe request, probe response, association request, and re-association request. Moreover, an SSID must be transmitted in clear text as required by IEEE 802.11 standards, otherwise a Wi-Fi network cannot function properly. As a result, there is no such thing as hiding an SSID. Another misleading security method is *antenna placement and signal suppression*. While legitimate devices have power and antenna limitations set by authorities, hackers' equipment can be much more powerful and versatile. For example, directional high-gain antennas can pick up a weak Wi-Fi signal from several kilometers away. Moreover, lowering the signal strength will cause more harm to legitimate users. We can see that all the trivial ways of protecting WLAN are useless. Therefore, serious cryptography schemes are needed to protect WLAN.

7.2 Evolution of WLAN Security

The evolution of WLAN security is summarized in Figure 7.3. The Wired Equivalent Privacy (WEP) was the first one proposed to secure WLAN. However, it was a total failure. IEEE 802.11 with authentication based on IEEE 802.1X standards was the next WLAN security. Because of the introduction of IEEE 802.1X standard, it provides a better authentication using RADIUS and EAP. However, the weakness of encryption and key management was not addressed in this security generation. The next stage of WLAN security is IEEE 802.11 with Wi-Fi Protected Access (WPA). WPA was intended by the

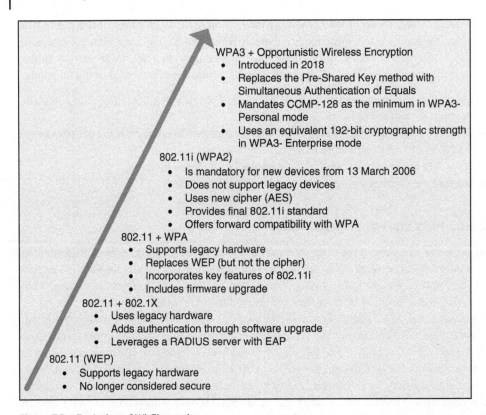

WPA3 + Opportunistic Wireless Encryption
- Introduced in 2018
- Replaces the Pre-Shared Key method with Simultaneous Authentication of Equals
- Mandates CCMP-128 as the minimum in WPA3-Personal mode
- Uses an equivalent 192-bit cryptographic strength in WPA3- Enterprise mode

802.11i (WPA2)
- Is mandatory for new devices from 13 March 2006
- Does not support legacy devices
- Uses new cipher (AES)
- Provides final 802.11i standard
- Offers forward compatibility with WPA

802.11 + WPA
- Supports legacy hardware
- Replaces WEP (but not the cipher)
- Incorporates key features of 802.11i
- Includes firmware upgrade

802.11 + 802.1X
- Uses legacy hardware
- Adds authentication through software upgrade
- Leverages a RADIUS server with EAP

802.11 (WEP)
- Supports legacy hardware
- No longer considered secure

Figure 7.3 Evolution of Wi-Fi security.

Wi-Fi Appliance as an intermediate measure to take the place of WEP while developing the full IEEE 802.11i standard. While not all, WPA implements much of the IEEE 802.11i standard, including authentication with 802.1X, and some of the key management schemes. WPA can be implemented through firmware upgrade on legacy wireless network interface cards designed for WEP. However, the legacy APs before 2003 may not be upgraded to support WPA. TKIP is adopted for WPA for encryption. Although WPA still uses RC4 for encryption, it employs a per-packet key to prevent the types of attacks that compromised WEP. A better message integrity check mechanism called Michael is also applied in WPA to mitigate the weakness of CRC-32.

IEEE 802.11i standard defines a later Wi-Fi security protocol. Although WPA is considered secure temporarily, a better solution is still needed. WPA2 has replaced WPA starting from September 2004, and it is mandatory for all new Wi-Fi devices from 13 March 2006 [51]. WPA2, while not being equivalent to IEEE 802.11i, implements all the mandatory elements. A major difference from WPA is the mandatory support of AES-Counter Mode with Cipher Block Chaining Message Authentication Code Protocol (AES-CCMP) in WPA2 [52]. Wi-Fi Alliance announced the release of WPA3 in January 2018 with several security improvements over WPA2 [53]. WPA3 introduces three new capabilities for personal and enterprise Wi-Fi networks. Two of the features provide robust protection when a weak password is chosen. The other feature is a 192-bit security suite for higher security requirements. WPA3 is currently optional and retains interoperability with WPA2. In addition to WPA3, a new opportunistic wireless encryption (OWE) is also introduced in 2018 to enhance user privacy in open Wi-Fi networks.

7.3 Wired Equivalent Privacy

WEP is a set of security schemes introduced as part of the original IEEE 802.11 wireless networking standard in September 1998. WEP was intended to protect wireless confidentiality comparable with that of a traditional wired network. It is also intended to achieve some other security goals such as access control and data integrity. Although WEP has serious flaws and thus all the security goals are failed, it still shares some of the major design of more advanced security schemes, especially its successor, the WPA. Detailed design and flaws of WEP will be discussed in the rest of Section 7.3.

7.3.1 WEP Access Control

WEP provides access control or authentication to WLAN. As shown in Figure 7.4, it requires an STA to be authenticated by the AP before association. The authentication process is based on a simple challenge–response protocol. WEP authentication process includes four handshakes as follows:

(1) An STA sends authentication request to the AP.
(2) The AP generates a random value r and sends it back to the STA as a challenge.
(3) The STA encrypts r with the shared secret key k between STAs and the AP as

$$E_k(r) = r \oplus \text{RC4}(IV \| k), \tag{7.1}$$

where $\oplus$ is the XOR function; and IV is the pre-defined initial vector (IV). The encrypted message is the response to the AP.
(4) The AP decrypts the response and verifies the value. If the decrypted value matches the challenge r, access is granted to the STA. Otherwise access is denied.

The security of WEP authentication depends on the secret key k that is known to legitimate STAs and the AP, and the cryptographic encryption function. The core function of the encryption is RC4.

Figure 7.4 WEP authentication process.

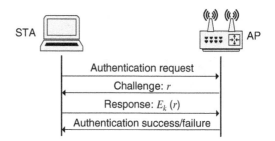

7.3.2 WEP Integrity and Confidentiality

WEP utilizes Cyclic Redundancy Check (CRC) checksum to provide data integrity to WLAN transmissions. An overview of the message encryption process is shown in Figure 7.5. CRC value of a plaintext is computed and appended to the plaintext. Particularly, CRC-32 is chosen for WEP. The CRC checksum is named as Integrity Check Value (ICV) in WEP security. Sender encrypts the entire frame of plaintext and ICV before transmission. At the receiver side, the ciphertext is decrypted first. The receiver then computes the ICV with the decrypted plaintext and compares it with the received ICV. If the two values match, the message is assumed to remain unchanged during transmission. WEP confidentiality is provided by encrypting message before transmission. The cryptographic encryption algorithm is RC4 algorithm, the same as the one for WEP authentication process.

The overview of WEP security is shown in Figure 7.6. Each seed input into RC4 consists of an IV and pre-shared secret key k. The IV is a 24-bit value generated by a pseudo-random number generating function. Note that the IV is changed for each message. Thus the seed is updated for each input message. With a given seed, RC4 produces a pseudo-random keystream. This keystream is XORed with the message and ICV to get the ciphertext. The same IV is also attached to the ciphertext being transmitted to the receiver. At the receiver side, pre-shared key is concatenated with IV to get the seed. The seed is input into RC4 to generate keystream. The functions of WEP security are almost symmetric at both the STA and the AP sides. Except that IV is generated at the AP side if the message is transmitted from the AP.

7.3.3 WEP Key Management

The IEEE WEP security standard defines two types of keys: *default key* and *key mapping keys*.

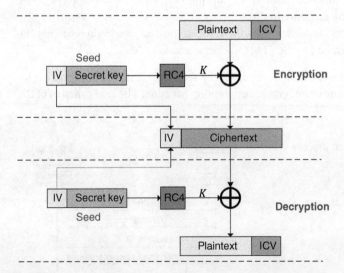

Figure 7.5 Overview of WEP encryption and message integrity.

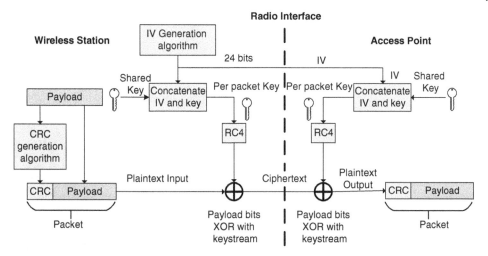

Figure 7.6 The overview of WEP security.

Figure 7.7 WEP key management—
default key.

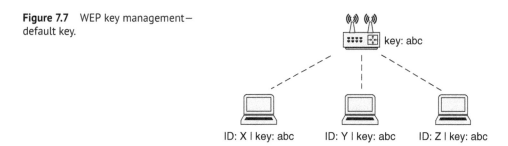

- **Default key** is also called shared key, group key, multicast key, or broadcast key. The default key is 40-bit or 104-bit long. It is static, thus it is manually installed in every STA and the AP. Each STA uses the same shared secret key, as shown in Figure 7.7. If a member leaves the group and should not have access to the network, then the default key needs to be changed for all devices simultaneously. In practice, WEP supports multiple default keys (i.e. four default keys) to help smooth the change of keys. When multiple default keys are applied, one of the keys currently used is called the active key. The active key is used to encrypt messages. A message header contains a key ID that allows the receiver to find out which key should be used to decrypt the message.
- **Key mapping key** is also called individual key, per-station key, or unique key. As shown in Figure 7.8, each STA is assigned with a different key. The AP keeps a table of all the keys. Key mapping key is not generally implemented in WEP enabled devices.

7.3.4 WEP Security Problems

As mentioned earlier, WEP security has flaws in almost all of its security features and thus is not recommended for users.

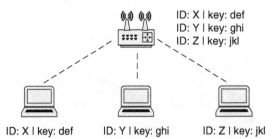

ID: X I key: def
ID: Y I key: ghi
ID: Z I key: jkl

Figure 7.8 WEP key management–key mapping key.

ID: X I key: def   ID: Y I key: ghi   ID: Z I key: jkl

7.3.4.1 Problems in WEP Access Control

The four-way handshake access control used by WEP has three major drawbacks. First, authentication process in WEP is one-way only. While the STA is authenticated by the AP, the AP is trusted without being authenticated. Therefore, a rogue AP can be easily deployed if WEP security is applied. Second, the same shared secret key is used for both authentication and encryption. More frequent usage increases the wear out of the shared key. Thirdly, if the STA is an attacker, then the keystream generated by RC4 can be revealed by simply calculating $E_k(r) \oplus r$ without knowing the pre-shared key. Once the keystream is recovered, the attacker can encrypt any subsequent challenges. In other words, a rogue AP can be set up by the attacker.

7.3.4.2 Problems in WEP Integrity

The problems of WEP integrity come from the weak cryptographic checksum computed by CRC-32. Due to the simplicity of CRC-32, it is highly possible that different messages result in the same ICV value for WEP integrity. In fact, a CRC checksum is generated by dividing message polynomial by a given CRC polynomial. The remainder is the CRC value. Assuming the CRC polynomial is

$$p = p_n x^n + p_{n-1} x^{n-1} + \cdots + p_0 x^0,$$

it is easy to verify that

$$(p_1 + p_2) \bmod c = (p_1 \bmod c) + (p_2 \bmod c). \tag{7.2}$$

Therefore, calculating ICV is a linear function, where

$$\text{ICV}(p + q) = \text{ICV}(p) + \text{ICV}(q). \tag{7.3}$$

If a CRC-32 valid plaintext is XORed with a ciphertext, the modified message will pass the ICV check after decryption. However, the original message has been altered, thus the integrity is not guaranteed in WEP security.

7.3.4.3 Problems in WEP Confidentiality

WEP relies on RC4 for confidentiality. Although RC4 is easy for software implementation, it is harder to guarantee security compared with block cipher. Particularly, implementation of RC4 in WEP has two major problems: *IV reuse* and *weak key*. IV reuse causes security issues in seed generation for RC4. A seed of RC4 in WEP consists of IV and the shared key. Because the shared key is not changed, the same seed is generated if the IV is reused. Since IVs are only 24 bits, there are only 2^{24} unique values. In other words, after around

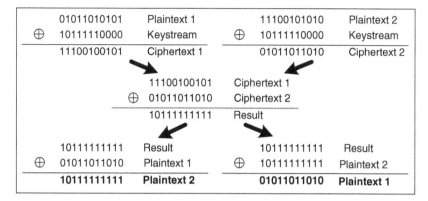

Figure 7.9 An example of security issue from IV reuse.

17 million messages, IVs are reused. For example, on a 11 Mbps transmission link, reuse of IVs will occur after

$$2^{24} \text{ packets} \times \frac{1500 \text{ bytes}}{\text{packet}} \times \frac{8 \text{ bits}}{1 \text{ byte}} /11 \text{ Mbps} = 18,300 \text{ seconds} \approx 5 \text{ hours.}$$

A reused seed in RC4 results in the same keystream, which is used to encrypt and decrypt a message. If the same keystream is reused to encrypt two messages, then the message may be revealed without knowing the shared key or the IV. An example is illustrated in Figure 7.9. Assuming that either Plaintext 1 (01011010101) or Plaintext 2 (11100101010) is known to the attacker. The same IV is used to encrypt both Plaintext 1 and Plaintext 2. Without loss of generality, assuming the produced key stream is 10111110000, which is unknown to the attacker. Then Ciphertext 1 and Ciphertext 2 are 11100100101 and 01011011010, respectively. Ciphertext can be captured by the attacker. After computing an intermediate result Ciphertext 1 ⊕ Ciphertext 2, the attacker can reveal Plaintext 1 with given Plaintext 2 or vice versa.

Moreover, keys must be carefully chosen in WEP security. If a weak key is chosen, then the seed values input into RC4 algorithm result in a sequence of non-random output. In this case, the first few bytes of the output reveals too much information about the key. With some known plaintext, such as protocol preambles or fixed frames in a protocol, attackers can further exploit the weak IVs and to get the shared key. A weak key will further shorten the breaking time of the shared key because the number of sufficient messages to be captured can be much less than waiting for an IV reuse, only about one million messages due to the short length of IV.

7.3.4.4 Problems in WEP Key Management

WEP key management lacks a centralized controller, where all keys are manually distributed. For a small scale WLAN with a few STAs, such as a home Wi-Fi, it is possible to manage keys. However, for a large scale WLAN that has a large number of STAs, such as an enterprise Wi-Fi, it is difficult to change security keys. Nonetheless, since a single set of keys is shared by all the STAs and the AP, frequent change of keys is necessary. For example, if an employee quits the job, then the security key needs update to prevent disclosure. Moreover, if default key mode is used in WEP, STAs in the same WLAN can decrypt each other's messages since the key is also the master key.

7.3.5 Possible WEP Security Enhancement

Security of WEP can be enhanced with some minor improvements of the scheme. To avoid IV reuse, longer ICV space shall be used. If 48-bit IV is applied, then an IV reuse is almost impossible. Moreover, instead of transmitting IV in plaintext, it would be a better protection if a hashed value of IV is applied. To avoid a weak key, one can filter out such weak keys and use other keys instead, or one may discard the first 256 outputs of RC4 algorithm so that it is more difficult for an attacker to observe the correlation between the input and the output of RC4 algorithm. Other kinds of protection may also be added to WEP, e.g. firewall and Virtual Private Network (VPN). However, those minor improvements can only mitigate the attacks to WEP. Even with those improvements, WEP is by no means a robust security scheme for WLAN. Although WEP is still available in most APs, it is recommended to avoid using WEP to protect WLANs. A more secure protection (i.e. the IEEE 802.11i standard) is recommended. The IEEE 802.11i standard depends on the IEEE 802.1X standard. An overview of IEEE 802.1X is presented in Section 7.4.

7.4 IEEE 802.1X Authentication Model

IEEE 802.1X standard (interchangeable with 802.1X hereafter for simplicity) is part of the IEEE 802.1 group of networking protocols [54]. It is defined for Port-based Network Access Control (PNAC). IEEE 802.1X provides an authentication mechanism to devices that are to attach to an LAN or a WLAN.

7.4.1 An Overview of IEEE 802.1X

IEEE 802.1X standard was originally defined for IEEE 802.3 Ethernet in 2001. It defines the encapsulation of the Extensible Authentication Protocol (EAP) over IEEE 802 standard, which is also known as EAPoL (EAP over LAN). In 2004, EAPoL was clarified to suit IEEE 802.11 wireless and fiber distributed data interface in IEEE 802.1X. IEEE 802.1X authentication model involves three parties: a *supplicant*, an *authenticator*, and an *authentication server*, as shown in Figure 7.10.

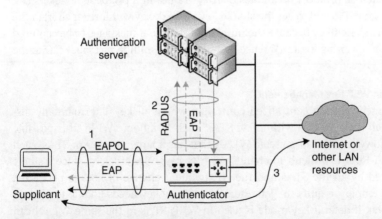

Figure 7.10 IEEE 802.1X authentication model.

- The supplicant is a client device that wishes to get access to the network (i.e. an LAN or a WLAN). For instance, a supplicant can be a desktop computer in an LAN or a smart phone in a WLAN. The authenticator is a network device, such as an Ethernet switch in an LAN or a wireless AP in a WLAN that has direct communication links with the supplicants.
- The authenticator controls the security access to a protected network. A supplicant is not allowed to access the protected side of the network (e.g. Internet or enterprise database) through the authenticator before the identity of the supplicant is validated and authorized.
- The authentication server is a host that supports Remote Authentication Dial In User Service (RADIUS) and EAP protocols.

In IEEE 802.1X standard, a supplicant requests access by providing *credentials*, such as user name/password or a digital certificate to the authenticator. The credentials are forwarded to the authentication server by the authenticator for verification. The supplicant will be allowed to access the protected side of the network if the credentials are verified by the authenticator server. Once the credentials are verified, the supplicant can access the protected resources through the authenticator without involving the authentication server.

7.4.2 Protocols in IEEE 802.1X

A summary of the IEEE 802.1x protocol architecture for WLAN is shown in Figure 7.11. Major protocols in 802.1X include EAP, EAPoL, and RADIUS.

EAP (RFC 3748) is an authentication framework without a specific authentication mechanism [55]. EAP is designed to provide some common functions and negotiation of authentication methods transport the messages of actual authentication protocols (e.g. TLS' Transport Layer Security). There are several EAP methods defined for IEEE 802.11 WLAN, including EAP-TLS (RFC 2716) [56], EAP-TTLS, PEAP, and EAP-SIM.

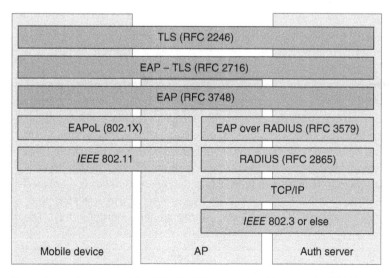

Figure 7.11 Summary of the IEEE 802.1X protocol architecture for WLAN.

Table 7.1 EAP messages.

Message	Description
EAP request	It carries messages from the supplicant to the authentication server.
EAP response	It carries messages from the authentication server to the supplicant.
EAP success	It signals successful authentication.
EAP failure	It signals authentication failure.

Octets: 12	2	1	1	2	Variable	4
MAC header	Protocol type	Protocol version	Packet type	Packet body length	Packet body	FCS

Figure 7.12 EAPoL frame format.

EAP includes four types of messages, as described in Table 7.1. In IEEE 802.1X, although transmitting all the EAP messages, an authenticator does not understand what is inside the EAP messages, it recognizes only EAP success and failure.

EAPoL is designed for IEEE 802.1X to give a generic network sign-on to access network resources. EAPoL is used to encapsulate EAP messages into LAN protocols (e.g. Ethernet), and to carry EAP messages between supplicants and the authenticator. The EAPoL frame format is shown in Figure 7.12. Each EAPoL frame includes 12-byte MAC header, 2-byte Ethernet type, 1-byte version, 1-byte packet type, 2-byte packet body length, variable length depending on packet body, and 4-byte frame check sequence.

RADIUS (RFC 2865-2869, RFC 2548) is a networking protocol that provides centralized authentication, authorization, and accounting management for users who connect and use a network service. RADIUS is used to carry EAP messages between the authenticator and the authentication server in IEEE 802.1X standard. RADIUS utilizes MS-MPPE-Recv-Key attribute to transport the session key from the authentication server to the authenticator. The format of the MS-MPPE-Recv-Key attribute is shown in Figure 7.13. In IEEE 802.11i, RADIUS is mandated by WPA and is optional for Robust Security Network (RSN).

The MS-MPPE-Recv-Key attribute contains a session key for encrypting packets received by the NAS from the remote host by the Microsoft Point-to-Point Encryption Protocol (MPPE) [57]. An MS-MPPE-Recv-Key attribute frame contains a 1-byte Vendor type, 1-byte Vendor length, 2-byte Salt and a variable size (in the even multiple of 16) of String. The Salt field is used to ensure the uniqueness of the keys used to encrypt each of the encrypted attributes. The String field consists of the Key-length, Key sub-fields, and the optional Padding sub-field. The actual encryption key is contained in the Key sub-field. The Padding sub-field is presented when the combined length of the unencrypted Key-length and Key sub-fields are not an even multiple of 16. The String field must be encrypted prior to transmission.

Octets: 1	1	2	Variable
Vendor type	Vendor length	Salt	String

Figure 7.13 MS-MPPE-Recv-Key attribute format.

7.4.3 Mapping the IEEE 802.1X model to WLAN

IEEE 802.1X is not defined specifically for WLAN. Nonetheless, it is easy to map IEEE 802.1X model to WLAN.

- Supplicant → Mobile device (STA)
- Authenticator → Access point (AP)
- Authentication server → Server application running on the AP
- Port → Logical state implemented in software in the AP

In order to apply IEEE 802.1X in IEEE 802.11i, some more features need to be added. First, a successful authentication not only switches on the port for the supplicant, but also exchanges a session key between the supplicant and the authentication server. Second, the session key is sent to the AP in a secure way. In other words, there is a shared key between the AP and the authentication server. In practice, this shared key is usually set up manually.

7.5 IEEE 802.11i Standard

Despite of being the first security protocol for IEEE 802.11, WEP is a total failure. After the collapse of WEP, the IEEE started to develop a new security architecture, which is standardized as IEEE 802.11i [58]. The IEEE 802.11i standard defines two security protocols for WLAN, widely known as WPA and Wi-Fi Protected Access II (WPA2).

7.5.1 Overview of IEEE 802.11i

Comparing with WEP, the access control model of IEEE 802.11i is based on IEEE 802.1X. It has a more flexible authentication framework based on EAP. Thus the authentication in IEEE 802.11i can be based on strong protocols (e.g. TLS). The authentication process also results in a shared session key, which can prevent session hijacking. Different keys are derived from the session key using a one-way function for encryption and integrity in IEEE 802.11i. Besides keys, the functions of integrity protection and encryption are improved in IEEE 802.11i compared with WEP.

In order to allow backward compatibility, IEEE 802.11i defines an optional protocol called Temporal Key Integrity Protocol (TKIP). TKIP has the same structure as WEP. The core encryption algorithm of TKIP is still based on RC4, while the security problems in WEP are avoided. Function *Michael* replaces CRC-32 for integrity protection in TKIP. Applying TKIP is widely known as WPA, or IEEE 802.11i draft. More importantly, the IEEE 802.11i standard defines the concept of Robust Security Network (RSN). An overview of RSN association is depicted in Figure 7.14. AES is the chosen cryptographic algorithm for both confidentiality and data integrity in RSN. Although RSN is a good security solution, it requires hardware upgrade to implement. Applying RSN/AES is widely known as WPA2, or just IEEE 802.11i.

7.5.2 IEEE 802.11i Access Control

User authentication and key management of IEEE 802.11i are generally based on IEEE 802.1X. WPA supports two authenticated key management protocols: the *IEEE 802.1X and EAP authentication*, and the *pre-shared key (PSK) authentication*. The IEEE 802.1X

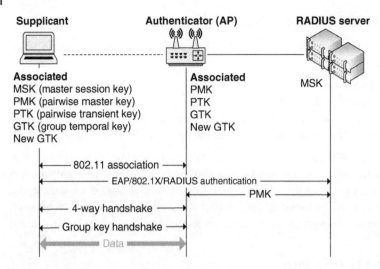

Figure 7.14 Robust security network association.

and EAP authentication based WPA is also referred to as WPA-Enterprise or WPA-802.1X mode because it is preferred by enterprise environments. WPA-802.1X is available with both WPA and WPA2 for confidentiality and integrity protection. To apply access control, WPA-802.1X requires a centralized RADIUS authentication server. Mutual authentication is required by WPA-802.1X to prevent user from joining rogue networks. The PSK based authentication is designed for home and small office network. It is also referred as WPA-Personal or WPA-PSK. WPA-PSK requires neither authentication server nor EAP framework. WPA-PSK requires users to manually enter password (i.e. the master key) in the AP or the wireless gateway, and enter the same password in each STA. During networking operation, all wireless network device within the WLAN network (i.e. STAs and APs) encrypt network traffic using a 256-bit key.

The WLAN authentication process based on IEEE 802.1X is shown in Figure 7.15. The supplicant first associates its connections with an AP. The AP then sends the supplicant an EAP *identity request*. The supplicant responds by sending its EAP *identity* to the AP. The identity usually comprises of the supplicant's username and domain, e.g. a company email address. Then, the AP forwards this request to the RADIUS authentication server over its uncontrolled port. After receiving the supplicant's EAP *identity*, the RADIUS authentication server first requests a domain user certificate that is associated with the identity it just received. The AS then sends the server certificate to the supplicant through the AP. Upon receiving the server certificate, the supplicant validates it and sends its own domain user certificate back to the RADIUS authentication server through the AP. Note that the supplicant needs to obtain the certificate over an Ethernet connection to the network or some other means of connection that does not use IEEE 802.1X. The RADIUS authentication server checks with the active directory domain controller and certificate authority to ensure that the domain user account information in the *identity* packet is indeed associated with the domain user certificate received from the supplicant. Once confirmed, the RADIUS authentication server sends an authentication success message to the AP. If the verification fails, an authentication failure message is sent to the AP instead. After the authentication process, the AS and the supplicant have established a session for

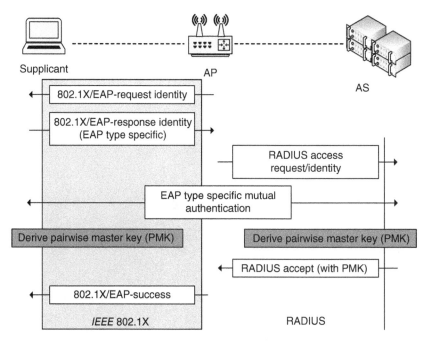

Figure 7.15   WLAN authentication based on IEEE 802.1X.

secure communications. The authentication server and the supplicant process a mutually authenticated master key. The master key represents decision to grant access based on authentication. The supplicant and the authentication server have derived the Master Session Key (MSK) and Pairwise Master Key (PMK). The PMK is also distributed to the AP from the AS.

7.5.3   IEEE 802.1i Key Management

The key hierarchies of WPA-802.1X is shown in Figure 7.16. PMK is a 256-bit master key to derive Pairwise Transient Keys (PTK) at both the supplicant and the AP. PTK includes a Key Encryption Key (KEK), a Key Integrity Key (KIK), a Data Encryption Key (DEK), and a Data Integrity Key (DIK). The KEK and the KIK are derived to protect key handshakes. DEK and DIK are derived to protect unicast message transmission between the STA and the AP. Group Master Key (GMK) is a 128-bit key randomly generated at the AP. From GMK, a Group Transient Key (GTK) is derived at the AP. GTK includes a Group Encryption Key (GEK) and a Group Integrity Key (GIK). GEK and GIK are used to protect broadcast message transmission between the STA and the AP.

In both TKIP and AES-CCMP modes, the same four-way handshake process is applied to achieve PTK distributions. As shown in Figure 7.17, the four-way handshake process is as follows:

(1) The AP sends a nonce-value to the STA (ANonce). The client now has all the attributes to construct the PTK.

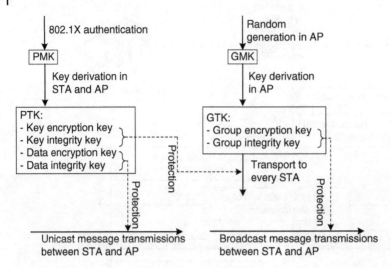

Figure 7.16 Key hierarchies in WPA-802.1X.

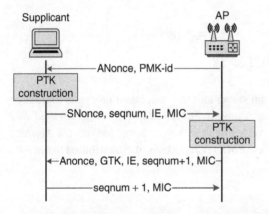

Figure 7.17 Four-way handshake in IEEE 802.11i.

(2) The STA sends its own nonce-value (SNonce) to the AP together with a sequence number (seqnum), information element (IE), and a message integrity code (MIC).

(3) The AP constructs and sends the GTK and seqnum+1 together with another MIC. This sequence number will be used in the next multicast or broadcast frame, so that the receiving STA can perform basic replay detection.

(4) The STA sends a confirmation to the AP.

The GTK used in the network may need to be updated due to the expiration of a preset timer. When a device leaves the network, the GTK also needs to be updated. This is to prevent the device from receiving any more multicast or broadcast messages from the AP. To handle the updating, IEEE 802.11i defines a Group Key Handshake that consists of a two-way handshake:

(1) The AP sends the new GTK to each STA in the network. The GTK is encrypted using the KEK assigned to that STA, and protects the data from tampering, by use of a MIC.

(2) The STA acknowledges the new GTK and replies to the AP.

Keys for TKIP mode: PTK is derived by first concatenating PMK, AP's MAC address (MAC1), STA's MAC address (MAC2), AP nonce (Nonce1), and STA nonce (Nonce2). The product is then put through a 512 − *bit* pseudo random function (PRF). A hash function is usually applied as the PRF in this case.

$$PTK = \mathrm{PRF}_{512}(PMK\|MAC1\|MAC2\|Nonce1\|Nonce2)$$
$$= KEK\|KIK\|DEK\|DIK$$

GTK is derived by first concatenating GMK, AP's MAC address (MAC), and Group nonce (GNonce). The product is then put through a 256-bit PRF.

$$GTK = \mathrm{PRF}_{256}(GMK\|MAC\|GNonce)$$
$$= GEK\|GIK$$

Keys for AES-CCMP mode: PTK is derived in a similar way, where the PRF is 384-bit, since DEK and DIK are the same.

$$PTK = \mathrm{PRF}_{384}(PMK\|MAC1\|MAC2\|Nonce1\|Nonce2)$$
$$= KEK\|KIK\|DEK\|DIK$$

GTK is also derived in a similar way, where the PRF is 128-bit, since GEK and GIK are the same.

$$GTK = \mathrm{PRF}_{128}(GMK\|MAC\|GNonce)$$
$$= GEK\|GIK$$

7.5.4 IEEE 802.11i Integrity and Confidentiality

Both the TKIP and the AES-CCMP modes defined by IEEE 802.11i standards provide integrity and confidentiality.

7.5.4.1 TKIP Mode

The TKIP mode is defined and mandatory for WPA, which still uses RC4 and the core function so that it can be backward compatible to legacy devices. TKIP corrects the flaws in WEP so that the security is much enhanced in two aspects. First, while CRC-32 is still applied as the ICV to the message, an 8-byte MIC is also computed and appended to the message. The MIC is computed by function *Michael*. The purpose of countermeasures with Michael is to reliably detect an attack and close down communications to the attacked STA for a period time. Specifically, if two incorrect Michael MIC codes are received within one minute, the AP will reset the TKIP session with a different key and change all the future keystreams. Michael limits an attacker to one try per minute for the entire network and thus protect the integrity of transmission. Second, the IV space is increased to 48 bits in TKIP in order to prevent IV reuse. The IV is also used as replay counter to prevent replay attacks. Moreover, TKIP applies per-packet keys instead of using the pre-shared

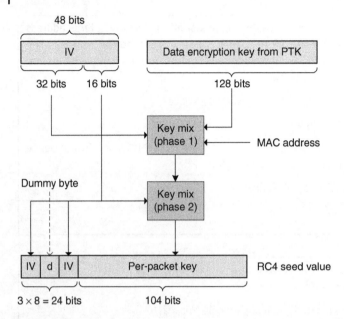

Figure 7.18 TKIP—Generating RC4 keys.

key to prevent attacks based on weak keys. As shown in Figure 7.18, the per-packet key is derived from DEK, MAC address, and part of the IV.

7.5.4.2 AES-CCMP Mode

The AES-CCMP mode is the standard encryption protocol for use with the WPA2 standard. It is more secure than TKIP protocol. CCMP means CTR mode and CBC-MAC. CCMP uses Counter Mode with CBC-MAC (CCM) for data confidentiality and CBC-MAC for authentication and integrity. CCMP encryption is based on AES processing and uses a 128-bit key and a 128-bit block size.

CBC-MAC protects the integrity of the MAC header, CCMP header, and the Medium Access Control Protocol Data Unit (MPDU). The mutable fields, including the retry bit in the Frame Control words and the Duration field, are set to zero. The input is padded with zeros if the length is not multiple of 128 bits. The initial block of CBC-MAC comprises of 5 fields, 8-bit flag, 8-bit priority, 48-bit source address, 48-bit packet number, and 16-bit data length. The final 128-bit block of CBC encryption is truncated to upper 64 bits to get CBC-MAC value.

The encryption is using AES in CTR mode. Both MPDU and CBC-MAC are encrypted while MAC and CCMP headers are not. The format of the counter is similar to the CTC-MAC initial block, including 8-bit flag, 8-bit priority, 48-bit source address, 48-bit packet number, and 16-bit counter. The counter is initialized with 1 and is incremented after each encrypted block.

7.5.5 Function *Michael*

Michael is a keyed hash function with a 64-bit key, and the output is a 64-bit value, as illustrated in Algorithm 7.1 [59]. The input key is converted to two 32-bit key words, and

---

Algorithm 7.1 Michael

Input: key (k_0, k_1), message $m_0 m_1, \ldots m_{n-1}$;
Output: ICV;
    $(L, R) \leftarrow (k_0, k_1)$
    for $i = 0$ to $n - 1$ **do**
      $L \leftarrow L \oplus m_i$;
      $(L, R) \leftarrow F(L, R)$;
    end for
    $ICV \leftarrow (L, R)$;

---

the message is divided into blocks with 32 bits each. The input message is padded at the end with a single byte with the hexadecimal value 5A followed by between 4 and 7 zero bytes as illustrated in Figure 7.19. The number of 0's is chosen so that the overall length of the message with padding is a multiple of 4 bytes.

The round function $F(\cdot)$ is illustrated in Algorithm 7.2. In this function, $\lll$ is left rotation by bytes, i.e. $(x \lll n)$ indicates n-byte left rotation of x. And $\ggg$ is right rotation by bytes, i.e. $(x \ggg n)$ indicates n-byte right rotation of x. Function $swap(\cdot)$ is a byte level swapping function, i.e. given 4 bytes $ABCD$, $swap(ABCD) = BADC$. Moreover, $\oplus$ is XOR and $\boxplus$ is the addition modulo 2^{32}.

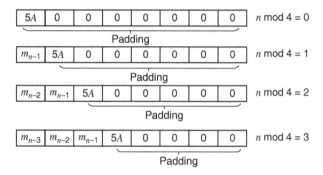

Figure 7.19 Padding in Michael.

Algorithm 7.2 Round function F of Michael

Input: (L, R);
Output: (L, R);
    $R \leftarrow R \oplus (L \lll 17)$;
    $L \leftarrow (L \boxplus R) \bmod 2^{32}$;
    $R \leftarrow R \oplus swap(L)$;
    $L \leftarrow (L \boxplus R) \bmod 2^{32}$;
    $R \leftarrow R \oplus (L \lll 3)$;
    $L \leftarrow (L \boxplus R) \bmod 2^{32}$;
    $R \leftarrow R \oplus (L \ggg 2)$;
    $L \leftarrow (L \boxplus R) \bmod 2^{32}$;

7.5.6 Weakness in 802.11i

The known weakness in WPA is in the implementation of RC4 [60]. Specifically, the 2-byte WPA frame counter, i.e. TSC biases in the RC4 keystream, thus permits mounting effective statistical, plaintext-recovering attack when the same plaintext is encrypted in many different frames. Nonetheless, WPA is not intended to be a long-term secure solution and users are recommended to implement WPA2 whenever available. The only major weakness found in WPA2 is the Key Reinstallation Attacks (KRACK) targeting the four-way handshake of the authentication protocol [61]. In normal operation, when a client joins a network and triggers the four-way handshake, a key is installed after receiving message 3 for data confidentiality. Message 3 may be retransmitted multiple times due to unreliable network conditions. Each time the client receives a copy of message 3, it reinstalls the same session key with incremental nonce (packet number) and replay counter used by the encryption algorithm. An attacker can force nonce resets by collecting and replaying message 3. With nonce resets, an attacker can decrypt packets, and/or launch attacks such as replay attack and forgery. KRACK may also be launched against group key in WPA2.

7.6 Wi-Fi Protected Access 3 and Opportunistic Wireless Encryption

The latest Wi-Fi security introduces WPA3 and opportunistic wireless encryption (OWE) with new features to simplify and enhance Wi-Fi security for open/public network, WPA3-Personal and WPA3-Enterprise.

7.6.1 WPA3-Personal

One of the major new features introduced in WPA3 is Simultaneous Authentication of Equals (SAE), which provides more robust password-based authentication, even when the chosen password falls short of typical complexity recommendations [53]. SAE replaces the four-way handshake method (PSK) used in WPA2 [58], thus being robust against KRACK attacks. SAE considers devices equally rather than requester and authenticator. As shown in Figure 7.20, both entities send their authentication information independently in the SAE handshake process. SAE is based on elliptic-curve Diffie–Hellman where the description is simplified in the figure. Because the PSK generated in the SAE process is for authentication only, SAE offers *forward secrecy*. In traditional PSK based systems, an attacker can hold on to encrypted data and later decrypt it once the password is hacked. In comparison, SAE requires a password change each time a connection is established. In this way, even if an attacker gets access to the network, passwords used for data prior to the connection would not be available.

7.6.2 WPA3-Enterprise

WPA3-Enterprise builds upon WPA2 to ensure the consistent application of security protocols across the network. The major enhancement of WPA3-Enterprise is its application of equivalent 192-bit cryptographic strength security protocols and cryptographic tools. Therefore, enterprise, government, and other institutions can implement

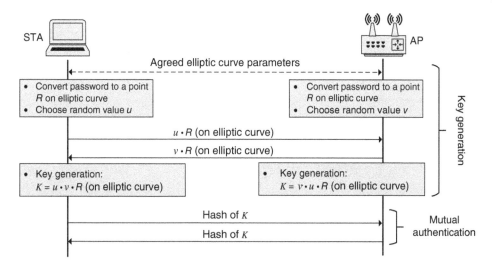

Figure 7.20 Overview of SAE process for identity authentication.

Table 7.2 Cryptographic tools used in WPA3-Enterprise.

Security feature	Tools
Authenticated encryption	GCMP-256
Key derivation and confirmation	HMAC-SHA384
Key establishment and authentication	ECDH and ECDSA (384-bit elliptic curve)
Robust management frame protection	BIP-GMAC-256

WPA3-Enterprise for stronger security. The cryptographic tools used in WPA3-Enterprise are listed in Table 7.2.

7.6.3 Opportunistic Wireless Encryption

Open/public Wi-Fi networks in locations like coffee shops, airports, etc. do not apply Authentication and Key Agreement (AKA). As a result, wireless traffic is not encrypted. OWE is defined to encrypt wireless traffic in open/public networks [63]. As shown in Figure 7.21, OWE is based on Diffie–Hellman algorithm (i.e. $F(DH(\cdot))$) and HMAC-based Extract-and-Expand Key Derivation Function (HKDF) to establish the PMK that can be used for traffic encryption. Both elliptic curve cryptography and finite field cryptography are supported by OWE. Once the secret z is established through Diffie–Hellman algorithm, the HKDF first outputs prk as follows:

$$prk = \text{HKDF-extract}(C|A|group, z), \tag{7.4}$$

where C and A are the public keys of the STA and AP, *group* is from the Diffie–Hellman parameter element. Finally, the PMK is generated as follows:

$$PMK = \text{HKDF-expand}(prk, \text{``OWE Key Generation''}, n), \tag{7.5}$$

where n is the number of bits of the digest produced by that hash algorithm.

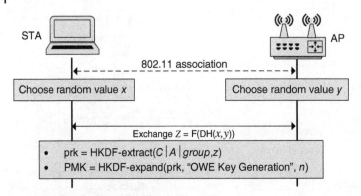

Figure 7.21 Overview of OWE process.

7.7 Summary

WLAN security is discussed in this chapter. The original WLAN security defined by the IEEE 802.11 standard failed to protect the network. Although WEP is still supported by current Wi-Fi devices, it is not recommended to apply for security. IEEE 802.11i defines the TKIP mode and the AES-CCMP, also known as WPA and WPA2, respectively, for better WLAN security. The TKIP mode has addressed some of the security issues in WEP and has backward compatibility with legacy devices. The AES-CCMP is mandatory for current WLAN devices due to better security. IEEE 802.11i also defines two types of authentications: WPA-Enterprise and WPA-Personal. WPA-Enterprise is based on IEEE 802.1X and EAP authentication, while WPA-Personal is based on the pre-shared key authentication. WPA3 is recently introduced as a successor to WPA2. In particular, WPA3 introduces traffic encryption to open/public networks, SAE to WPA3-Personal, and 192-bit security strength to WPA3-Enterprise.

8

Bluetooth Security

Bluetooth is an open standard for short-range radio communications. It is designed for wireless personal area networks (WPAN). Many wireless devices, such as smart watch, wireless headphone, wireless keyboard, wireless mouse, vehicular on-board audio system, etc., are based on Bluetooth technology. In this chapter, the security of current Bluetooth technology are introduced.

8.1 Introduction to Bluetooth

8.1.1 Overview of Bluetooth Technology

Bluetooth devices form an ad-hoc network for data transmission. As shown in Figure 8.1, a Bluetooth ad-hoc network consists of two or more devices within a small area. Among all the devices, one is a *master* device and the rest are active *slave* devices. Such an ad-hoc network is called a *piconet*. Within a piconet, the slave devices have direct connection with the master device. However, there are no direct connections between any slave devices. A Bluetooth piconet (interchangeable with piconet in the rest of the chapter) supports eight active devices, i.e. one master device and seven active slave devices. Up to 255 further slave devices can be supported if they are inactive. A Bluetooth device may join several piconets simultaneously. For example, a slave device can be interconnected to both master devices in two piconets; the master device of a piconet can be a slave device in another piconet. It is clear that a Bluetooth device cannot be the master device for two or more piconets simultaneously. If the devices of several piconets are interconnected (by forming extra piconets), then the interconnected piconets form a scatternet. The purpose of a scatternet is to connect more Bluetooth devices in a larger area. A Bluetooth scatternet can support up to 80 active devices [64].

Bluetooth uses a technique called *spread spectrum frequency hopping* to transmit data among 79 different frequencies. Each Piconet follows a different hopping sequence that defines up to 1600 hops per second by the master device. Frequency hopping helps to reduce jamming of Bluetooth transmissions in a fixed frequency. Moreover, frequency hopping increases the difficulty of eavesdropping since an eavesdropper may need to monitor all 79 frequencies to collect all the information. However, frequency hopping itself is not enough to secure Bluetooth communications. More secure features are needed for Bluetooth.

Security in Wireless Communication Networks, First Edition. Yi Qian, Feng Ye, and Hsiao-Hwa Chen.
© 2022 John Wiley & Sons Ltd. Published 2022 by John Wiley & Sons Ltd.
Companion website: www.wiley.com/go/qian/sec51

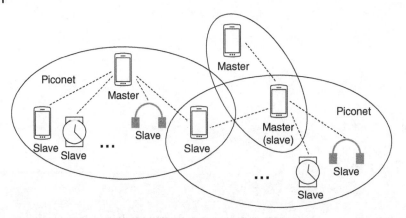

Figure 8.1 Overview of Bluetooth networks.

Table 8.1 Major evolution of Bluetooth versions.

Version	v1.1	v1.2	v2.0/v.2.1	v3.0	v4.0	v5.0
Date	2001/02	2003/11	2004/11 2007/07	2009/04	2009/12	2016/12
Data rate	732.2 kbps	1 Mbps	2.1 Mbps	24 Mbps	2.1/24 Mbps	2 Mbps
Major features	Signal strength indication	Adaptive frequency hopping	Enhanced data rate (EDR)	High speed channel	Low energy protocol	High spectral efficiency

There have been several versions of Bluetooth standards since 2001. Table 8.1 lists the major evolution of Bluetooth versions. The major improvements include both data transmission performance and security. In particular, while Bluetooth v2.0 + Enhanced Data Rate (EDR) published in November 2004 provided faster transmission speed, Bluetooth v2.1 + EDR published in July 2007 provided a significant security improvement on link key generation and management in the form of Secure Simple Pairing (SSP). The research of Bluetooth technology is still undergoing. Newer version with better performance and security will be developed. In the rest of this chapter, details of current Bluetooth security solutions are discussed.

8.1.2 Bluetooth Vulnerabilities and Threats

Bluetooth security is provided because of security threats that may put Bluetooth transmissions at risk. Being a wireless transmission technology, Bluetooth is susceptible to general wireless network threats, for example, eavesdropping, man-in-the-middle attacks, denial of service attacks, etc. Moreover, some attacks can threat Bluetooth security more directly. Some Bluetooth specific threats are Bluesnarfing, Bluejacking, Bluebugging, car whisperer, and fuzzing attacks [65–67].

8.1.2.1 Bluesnarfing

Bluesnarf is the most famous Bluetooth attack, first identified by Marcel Holtmann in September 2003 [67]. A Bluesnarf attack mostly happens to a cell phone with Bluetooth capability. The attack is launched by connecting to the OBEX Push Profile (OPP). The OPP is an easy way to exchange business cards and the objects between devices. In this way, an authentication may not be necessary. It is not a problem for OBEX Push. However, a Bluesnarf attacker can perform an OBEX GET request for known filenames. If the file system is not secure to protect the path and file names, a Bluesnarf attacker can easily get user data such as contact list and calendar file. In the worst case, the attacker can fetch the entire system if all file names are disclosed.

8.1.2.2 Bluejacking

Bluejacking is a possible attack to all Bluetooth-enable mobile devices, such as wearable devices, cell phones, etc. A Bluejacking attacker initiates the attack by sending unsolicited messages to a user. At this point, there is no harm to the user other than receiving an unexpected message. However, if the message tricks the user to respond, which usually causes authentication verification, then serious damage may follow. For example, a Bluejacking attack could be launched to a wearable device by sending a message that asks the user to confirm synchronizing daily tracking data. It is highly possible to trick the user to confirm such a request. Especially for those wearable devices only use vibration as message alerts.

8.1.2.3 Bluebugging

Bluebugging is an Bluetooth attack mostly targets cell phones. Bluebug is a Bluetooth security loophole on some Bluetooth-enabled cell phones. Using this loophole, it may only take a few seconds of a Bluebugging attacker to issue authentication commands via a covert channel without alerting the user. Bluebugging allows an attacker to gain control of the compromised cell phone. For example, an attacker may initiate phone calls from the compromised device, send text messages, read user's contact list, set call forwards, establish Internet connection, and several others that can cause privacy leakage and possible financial loss to a user.

8.1.2.4 Car Whisperer

Developed by European security researchers, Car Whisperer is a software tool to exploit a key implementation issue in hands-free Bluetooth car kits in vehicles. Through Car Whisperer, an attacker can send audio to the on-board speakers or eavesdrop from the microphone in car. Despite being a threat to Bluetooth security, Car Whisperer was intended to alert car manufacturers of their hands-free car kits and other Bluetooth devices that use standard passkeys (e.g. 0000 or 1234) so that such security threat could be mitigated. Recommendations include using random passkeys, or using some direct interaction with the device (e.g. near-field communication initiated Bluetooth connection).

8.1.2.5 Fuzzing Attacks

Bluetooth fuzzing attacks are lunched to detect vulnerability that potentially exists in the protocol stack. Therefore, fuzzing attacks may not harm a Bluetooth user instantly. To launch a fuzzing attack, an attacker sends malformed or non-standard data to a Bluetooth device and observes the reactions from the device. Potential vulnerability may be observed if a device's response is slowed or stopped. Fuzzing attacks are by no means harmless, because serious damages can be made based on the vulnerability found by a fuzzing attack. Developers may launch fuzzing attacks when developing security protocol stack to target possible vulnerability.

8.1.3 Bluetooth Security Services and Security Modes

8.1.3.1 Bluetooth Security Services
Bluetooth standard specifies three basic security services: *authentication*, *authorization*, and *confidentiality*. Since Bluetooth networks do not have a physical infrastructure, all Bluetooth security services are provided only between Bluetooth devices over the wireless transmission links. The security services in Bluetooth include:

- *Authentication*: Entity authentication is to verify the identity of communicating devices. User authentication is not provided natively by Bluetooth.
- *Authorization*: Bluetooth security allows the control of resources. A Bluetooth device must be authorized to use a service.
- *Confidentiality*: User data is encrypted before transmission, thus confidentiality is provided in addition to the frequency hopping mentioned earlier.

Besides the three security services, other security services are not provided in the Bluetooth standard. Some commonly recognized security services such as identity, non-repudiation, etc., must be provided with additional mechanisms if necessary.

8.1.3.2 Bluetooth Security Modes
Although all versions of Bluetooth specifications support the three basic security services, the implementation of each security service depends on one of the four security modes defined by various Bluetooth specifications [66, 68].

Security mode 1 (non-secure): No security functionality is provided in this mode. Mode 1 is only supported in v2.0 + EDR and earlier versions of Bluetooth standard. It is not supported by devices based on Bluetooth v2.1 + EDR and later versions of Bluetooth standard. Those devices must implement one of the other security modes.

Security mode 2 (service level enforced security): Security procedures are initiated after Link Manager Protocol (LMP) link establishment but before Logical Link Control and Adaptation layer Protocol (L2CAP) channel establishment. Policies for access control and interfaces with other device users are controlled by a centralized security manager [69]. Authorization may be applied in this mode to decide if a Bluetooth device should be granted access to a specific service. Both authentication and confidentiality are applied in mode 2 at the LMP layer. All Bluetooth devices can support security mode 2. However, Bluetooth v.2.1 + EDR versions can only support security mode 2 for backward compatibility.

Security mode 3 (link level enforced security): Security procedures are initiated before the physical link is fully established. Authentication and confidentiality are mandated in mode 3 for all connections from and to the Bluetooth device. Both one-way and mutual authentications are supported in mode 3. Only Bluetooth v2.0 + EDR or earlier versions can support security mode 3.

Security mode 4 (service level enforced security): In mode 4, security procedures are initiated after a link setup, which is similar to mode 2. There are three security requirements for services protected by security mode 4: *authenticated link key required*, *unauthenticated link key required*, or *no security required*. If a link key is required, security mode 4 applies the same authentication and encryption algorithms that are based on the link key as in modes 2 and 3. The difference is that security mode 4 applies a scheme called SSP to generate the link key. In particular, SSP uses Elliptic Curve Diffie Hellman (ECDH) techniques for key exchange and link key generation. Security mode 4 is mandatory for communications between Bluetooth v2.1 + EDR and later versions.

8.2 Link Key Generation

In all Bluetooth security modes 2, 3, and 4, a link key is needed for authentication and encryption algorithms. Depending on different security modes, there are two methods to generate the link key. Modes 2 and 3 use one of the methods, and security mode 4 uses the other.

8.2.1 Link Key Generation for Security Modes 2 and 3

For security modes 2 and 3, two associated Bluetooth devices (a master and a slave) generate the link key simultaneously in the process as shown in Figure 8.2. The slave device (device A) initiates the process by generating a random value IN_RAND and sends it to the master device (device B). Based on IN_RAND and a pre-shared secret value PIN, both devices derive the initialization key K_{init} as follows:

$$K_{init} = E_{22}(PIN, IN\_RAND),$$

where $E_{22}(\cdot)$ is the encryption algorithm applied in this process. If the secret value PIN is less than 16 bytes, then the physical address of the device BD_ADDR will be used to supplement the PIN value to generate the initialization key.

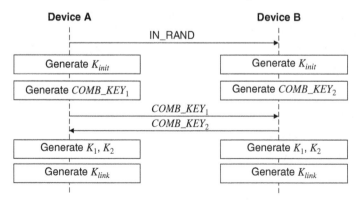

Figure 8.2 Link key generation in modes 2 and 3.

Once the initialization key is generated at both sides, the two devices will then generate a random value LK_RAND_1 and LK_RAND_2 at each side. Based on random values and the previously generated initialization key K_{init}, the two devices will generate the combination keys $COMB\_KEY$ as follows:

$$COMB\_KEY_1 = LK\_RAND_1 \oplus K_{init},$$

and

$$COMB\_KEY_2 = LK\_RAND_2 \oplus K_{init}.$$

The combination keys are exchanged between the two devices after generation. In the next step, device A calculates K_1 based on LK_RAND_1 and BD_ADDR_1 as follows:

$$K_1 = E_{21}(LK\_RAND_1, BD\_ADDR_1),$$

where $E_{21}(\cdot)$ is the encryption algorithm applied in this process. Moreover, device A also calculates K_2 based on $COMB\_KEY_2$, K_{init}, and BD_ADDR_2 as follows:

$$K_2 = E_{21}(COMB\_KEY_2 \oplus K_{init}, BD\_ADDR_2).$$

On the other side, device B calculates K_2 and K_1 in a similar way as follows:

$$K_2 = E_{21}(LK\_RAND_2, BD\_ADDR_2),$$

and

$$K_1 = E_{21}(COMB\_KEY_1 \oplus K_{init}, BD\_ADDR_1).$$

Finally, the same link key is generated by both devices as follows:

$$K_{link} = K_1 \oplus K_2.$$

The link key K_{link} will be applied in other security algorithms in both security modes 2 and 3. Security mode 4 has a different approach to generate the link key as illustrated in Section 8.2.2.

8.2.2 Link Key Generation for Security Mode 4

For security mode 4, two associated Bluetooth devices, a master (device B) and a slave (device A), generate the link key through the process as shown in Figure 8.3. Link key generation in mode 4 is based on ECDH, which involves public key cryptography in the process.

In the first step, the two devices exchange their public keys, i.e. Pu_1 for device A and Pu_2 for device B. Then a DH key K_{DH} is generated based on the two public keys using ECDH algorithm ($E_{192}(\cdot)$) as follows:

$$K_{DH} = \begin{cases} P_{192}(Pr_1, Pu_2), \text{ by device A,} \\ P_{192}(Pr_2, Pu_1), \text{ by device B.} \end{cases}$$

After generating K_{DH}, devices A and B need to establish shared parameters (N_1, N_2, r_1, r_2) through an *association model*, which has three different protocols: *Numeric Comparison*, *Out-of-Band*, and *Passkey Entry*. Assuming that the association model is completed and the

Figure 8.3 Link key generation in mode 4.

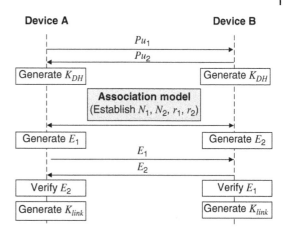

parameters are established (the association model will be illustrated later). Then devices A and B will generate values E_1 and E_2 as follows:

$$E_1 = f_3(K_{DH}, N_1, N_2, r_1, IOCAP_1, BD\_ADDR_1, BD\_ADDR_2),$$
$$E_2 = f_3(K_{DH}, N_2, N_1, r_1, IOCAP_2, BD\_ADDR_2, BD\_ADDR_1),$$

where $f_3(\cdot)$ is a function that uses the HMAC based SHA-256. Devices A and B exchange E_1 and E_2 after computing the values. The two devices then verify the received E_1 or E_2 values. If verifications are succeeded, both devices continue to derive the link key as follows:

$$K_{link} = f_2(K_{DH}, N_1, N_2, btlk, BD\_ADDR_1, BD\_ADDR_2),$$

where $f_2(\cdot)$ is also a function that uses the HMAC based SHA-256, and $btlk$ is a predefined bit string. The three protocols of the association model are illustrated in Section 8.2.3.

8.2.3 Association Model in Mode 4

The association model in security mode 4 has three protocols to establish parameters (N_1, N_2, r_1, r_2). Each protocol is designed to fit a specific type of Bluetooth devices. The three protocols are *Numeric Comparison*, *Out-of-Band*, and *Passkey Entry*, as detailed in the following.

8.2.3.1 Numeric comparison
This protocol was designed for the type of Bluetooth devices that are capable of displaying a number (typically six digits) and allowing input from a user for response. For pairing, a six-digit number is displayed on both devices. If the two match, a (human) user shall confirm it and continue pairing process. The difference between this process and the use of PINs is that this six-digit number is not applied as an input in the later pairing process. The final link key is not determined by the six-digit number. Thus, even if the six-digit value is revealed to an attacker, it cannot be used to generate the link key or the encryption key. The process of the numeric comparison protocol is illustrated in Figure 8.4. Each device generates a 128-bit random value, i.e. N_1 and N_2, respectively. These two values are used

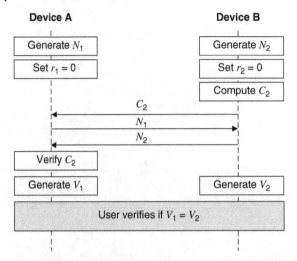

Figure 8.4 Association model–numeric comparison.

to prevent replay attacks. Values r_1 and r_2 are set to 0. Then, device B (master) computes a commitment C_2 as follows:

$$C_2 = f_1(Pu_2, Pr_1, N_2, 0),$$

where $f_1(\cdot)$ is a function that uses the HMAC based SHA-256. Note that Pu/Pr are the public/private key pairs for the corresponding devices. Commitment C_2 is to prevent an attacker from changing the values at a later time. Device B shall transfer C_2 to device A. The random values N_1 and N_2 are exchanged at this stage as well. Upon receiving C_2, device A verifies it by computing the commitment value with the same inputs. If the two values match, then device A computes a six-digit value V_1 as follows:

$$V_1 = g(Pu_1, Pu_2, N_1, N_2),$$

where $g(\cdot)$ is the hash function SHA-256. Similarly, device B computes a six-digit value V_2 as follows:

$$V_2 = g(Pu_1, Pu_2, N_1, N_2).$$

Note that V_1 and V_2 are the two six-digit values displayed on each device. User shall compare the two values and confirm if they match. If they do not match then the protocol aborts. If they match, the association model succeeds and the link key generation process continues with the parameters.

8.2.3.2 Out-of-Band (OOB)

This protocol was designed for a pair of Bluetooth devices that one has input capability (e.g. a Bluetooth keyboard) and the other has a display (e.g. a tablet, input capability is not required). Moreover, none-Bluetooth wireless technologies need to be involved in OOB protocol, thus the name out-of-band. One of the most popular none-Bluetooth technology for OOB protocol is the Near Field Communication (NFC). The NFC can be used to establish a Bluetooth connection by simply touch the two NFC-enabled devices and confirm the process. Let OOB-IO be a device that has a mechanism allowing it to communicate with the

Table 8.2 Scenarios for two devices using OOB protocol.

Scenario	Device 1	Device 2
1	OOB-IO	OOB-O
2	OOB-O	OOB-IO
3	OOB-IO	OOB-IO

Bluetooth controller and another device. Let OOB-O be a device that cannot communicate with the Bluetooth controller and can only transmit data. There are three scenarios that are practical for OOB model, as shown in Table 8.2. Because OOB model requires at least a device to be a reader, there must be a device with OOB-IO capability. The other can be either OOB-IO or OOB-O. The scenario where both devices are OOB-O does not exist.

To establish a connection using OOB protocol, the device with display shows a six-digit number, and the user enters on the device with input capability to confirm the process. Although the six-digit number is input to another device by the user, it is not applied in the link key generation process. The OOB protocol is illustrated in Figure 8.5. Either device A or device B can be the master because the OOB protocol is symmetric. Without loss of generality, device B is assumed to be the master for illustration. To begin the process, device A sets r_1 to a random value *RAND* and r_2 to 0. Device B sets r_2 to a random value *RAND* and r_1 to 0. Then, both devices A and B compute commitments C_1 and C_2 of their public keys. The random values and commitments need to be exchanged between the two devices. Depending on communication capability of the devices, there are two scenarios as shown in the shaded area of Figure 8.5. In the first scenario, if OOB communication is possible only in one direction, i.e. OOB-IO and OOB-O devices, then authentication will be based on the random value r. If the random value r is not sent to the other device, it is assumed to be zero. In this example, assuming device B is not capable of OOB communication (i.e. OOB-O), then r_1,

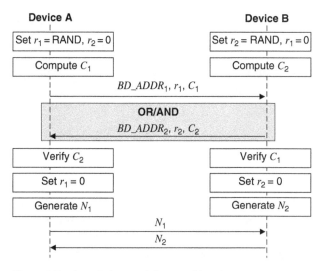

Figure 8.5 Association model—out-of-band.

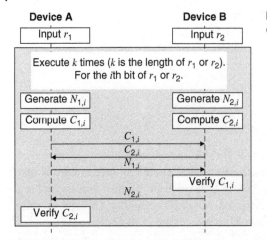

Device A   **Device B**

Input r_1 Input r_2

Execute k times (k is the length of r_1 or r_2). For the ith bit of r_1 or r_2.

Generate $N_{1,i}$ Generate $N_{2,i}$

Compute $C_{1,i}$ Compute $C_{2,i}$

$C_{1,i}$

$C_{2,i}$

$N_{1,i}$

Verify $C_{1,i}$

$N_{2,i}$

Verify $C_{2,i}$

Figure 8.6 Association model—passkey entry.

BD_ADDR_1 and C_1 are sent from device A to device B. Device B does not send any information to device A. Random value r_2 is set to 0 at both sides. In the second scenario, if both devices A and B are capable of OOB communication (i.e. OOB-IO), then they exchange their random values, Bluetooth addresses, and commitments. After exchanging information, both devices shall verify the received commitment values. Then, device A sets r_1 to 0 and device B sets r_2 to 0. Random values N_1 and N_2 are generated at each side. Finally, N_1 and N_2 are exchanged for further process.

8.2.3.3 Passkey entry

In this protocol, a *passkey* is needed to generate the parameters. Unlike the other models, passkey entry model runs iteratively based on the *length* of the passkey. A passkey is agreed by both devices before initiating the process. In practice, this passkey may be generated and displayed on one device. The process of passkey entry protocol is illustrated in Figure 8.6. The user inputs it into the other device. The random values r_1 and r_2 function as the passkey in this protocol. Therefore, $r_1 = r_2$. Let the passkey (r_1 or r_2) have a key length of k bits, then the rest operations shall be repeated for k times to generate N_1 and N_2. In the ith iteration, for example, device A chooses a 128-bit nonce $N_{1,i}$ and calculates a commitment value $C_{1,i}$ as follows:

$$C_{1,i} = f_1(Pu_1, Pu_2, N_{1,i}, r_{1,i}),$$

where $r_{1,i}$ is the ith bit of r_1. Device B also chooses a 128-bit nonce $N_{2,i}$ and calculates a commitment value $C_{2,i}$ as follows:

$$C_{2,i} = f_1(Pu_2, Pu_1, N_{2,i}, r_{2,i}).$$

The commitment values are exchanged after that. Device A also sends its nonce $N_{1,i}$ to device B. Device B then computes $C_{1,i}$ and compares it with the received one for verification. If the two values match, then device B sends its nonce $N_{2,i}$ to device A. After all k iterations, the final $N_{1,k}$ and $N_{2,k}$ shall be used as N_1 and N_2 for further process to generate the link key.

8.3 Authentication, Confidentiality, and Trust and Service Levels

8.3.1 Authentication

Bluetooth supports one-way authentication [68]. Figure 8.7 shows an one-way authentication process between device A (slave) and device B (master), where device A is authenticated by device B. As the supplicant, device A sends authentication request and its Bluetooth MAC address BD_ADDR_1 to device B. After receiving the request, device B computes a 128-bit random challenge AU_RAND and replies to device A. With the Bluetooth address of the supplicant and the random challenge available at both sides, the two devices compute two values ACO and $SRES$ as follows:

$$SRES\|ACO = E_1(BD\_ADDR_1, AU\_RAND, K_{link}),$$

where $E_1(\cdot)$ is the SAFER+ algorithm, $SRES$ is the 32-bit authentication response, and ACO is the authenticated ciphering offset that to be used for encryption process. The supplicant (device A) sends its $SRES$ to device B for verification. If the received $SRES$ matches the one computed by device B, then the verification succeeds. Device B confirms the success of authentication with device A. If an authentication fails, then a Bluetooth device shall wait for a short period of time before it can initiate another authentication process again. *Security of Bluetooth authentication depends solely on the link key.* Because the addresses of both devices are considered public parameters, they are exchanged in clear text during the process.

Mutual authentication: Bluetooth security does not specify a separate protocol for mutual authentication. Mutual authentication is achieved by executing the one-way authentication process twice. Once device A is authenticated by device B, the same authentication process is performed again with device B being the supplicant. An authentication supplicant is not associated with being the slave or the master device. In the example described earlier, device A is defined as the slave and device B is defined as the master. When device B is switched to the supplicant for mutual authentication, it is still the master device.

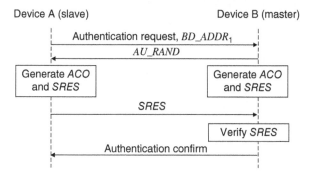

Figure 8.7 One-way Bluetooth authentication.

8.3.2 Confidentiality

Confidentiality is supported by all Bluetooth versions. It is achieved by encrypting data transmission. In total, Bluetooth standard defines three encryption modes, as follows:

- *Encryption mode 1*: No encryption.
- *Encryption mode 2*: Encrypt unicast data traffic only and leave broadcast data traffic unencrypted.
- *Encryption mode 3*: Encrypt all data traffic.

For those applications that do not need confidentiality, encryption mode 1 provides data transmission in cleartext. Encryption mode 2 provides confidentiality to partial traffic, excluding broadcast traffic. If confidentiality is needed for all data traffic, then encryption mode 3 should be applied. The encryption mechanism is the same for both encryption modes 2 and 3. A stream cipher E_0 is the core of the encryption mechanism. As all stream ciphers do, a KEYSTREAM is generated by E_0 for each incoming data block. Both encryption and decryption operations are performed by XORing the input data block with the KEYSTREAM.

The overall encryption mechanism is illustrated in Figure 8.8. The stream cipher E_0 is based on a cipher key K_c, which is generated using an internal key generator (KG) E_3 as follows:

$$K_c = E_3(K_{link}, EN\_RAND, COF),$$

where K_{link} is the 128-bit link key, EN_RAND is a 128-bit random number generated by the master device, and ciphering offset (COF) is a 96-bit value. If the current link key is not a master key, then the COF is the same as ACO produced in authentication. If the current link key is a master key, then the COF is derived from BD_ADDR of the master device as follows:

$$COF = BD\_ADDR \| BD\_ADDR.$$

The stream cipher E_0 does not take K_c as a direct input if its effective key length is less than 128 bits. Instead, K_c is input to a *Constraining Mechanism*, which generates a 128-bit

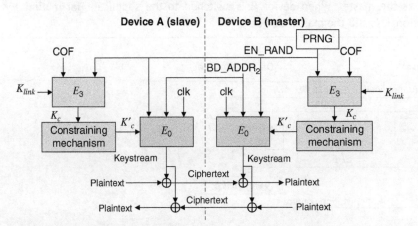

Figure 8.8 Encryption mechanism for Bluetooth security modes 2 and 3.

key constraint key K'_C. The constraining mechanism is applied to meet the restrictions on some encryption hardware. Besides K'_C, the inputs to E_0 also include the address of Device B BD_ADDR_2, the random value EN_RAND, and the current clock value clk. Finally, the 132-bit KEYSTREAM is computed as follows:

$$\text{KEYSTREAM} = E_0(K'_c, BD\_ADDR_2, EN\_RAND, clk).$$

Once KEYSTREAM is generated, encryption is performed by XORing the plaintext data block with the KEYSTREAM as follows:

$$\text{CIPHERTEXT} = \text{PLAINTEXT} \oplus \text{KEYSTREAM},$$

and decryption is performed by XORing the ciphertext data block with the KEYSTREAM as follows:

$$\text{PLAINTEXT} = \text{CIPHERTEXT} \oplus \text{KEYSTREAM}.$$

All three security services have been illustrated so far. Readers should have a clear view of the four security modes defined in Bluetooth standard. The detailed E_0 and E_3 algorithms will be illustrated in Section 8.4.

8.3.3 Trust and Security Service Levels

Besides four security modes, Bluetooth standard also defines two levels of trust and three levels of security services. As listed in Table 8.3, the two Bluetooth trust levels are *trusted* and *untrusted*. A trusted device has a fixed relationship with another device. A trusted device shall also have full access to all services of another device. An untrusted device does not have a fixed relationship. Thus restricted access is granted to an untrusted device.

The three levels of security services are listed in Table 8.4. These levels enable flexible configurations of security services depending on application requirements. Security level 1 requires both authorization and authentication. Security level 2 requires

Table 8.3 Bluetooth trust levels.

	Policies
Trusted	Fixed relationship and full access
Untrusted	Restricted access to services

Table 8.4 Bluetooth service levels.

	Security services	Policies
Level 1	Authorization	Automatic access is granted only to trusted devices;
	Authentication	untrusted devices need manual authorization
Level 2	Authentication	Access to an application is allowed only after
		an authentication procedure
Level 3	Open	Access is granted automatically

authentication only. Security level 3 has no security service required. None of the security level requires an encryption. In fact, since encryption has a mode that applies no encryption, forcing encryption does not necessarily achieve confidentiality.

Bluetooth Security Architecture

The Bluetooth standards define security services, security modes, as well as trust and service levels. The overall Bluetooth security architecture allows applications configure trust relationships. For example, trusted devices still have restricted access. Because Bluetooth security controls operate at the link layer, and the link layer is transparent to the security controls imposed by the application layers [68]. Therefore, more security policies can be defined by the security manager, and enforce user-based authentication through the application layers.

8.4 Cryptographic Functions for Security Modes 1, 2, and 3

Bluetooth security mechanisms are based on a few cryptographic functions [70]. In this section, the functions for security modes 1, 2, and 3 are illustrated, including SAFER+, E_1, E_{21}, E_{22}, E_3, and E_0.

8.4.1 SAFER+

Functions E_{21}, E_{22}, and E_3 used to generate keys are developed around the core function SAFER+. SAFER+ was originally a submission by Cylink Corporation for the advanced encryption standard [71]. SAFER+ is based on the SAFER (Secure And Fast Encryption Routine) family of ciphers. Similar to the finally selected AES, SAFER+ is a block cipher that operates on 128-bit data blocks. The key size of SAFER+ can be 128-bit, 192-bit, and 256-bit. Bluetooth security algorithms are based on the SAFER+ configuration with 128-bit key [70]. The following illustration of SAFER+ will be given based on the setting of 128-bit key.

8.4.1.1 Overview of the SAFER+ Structure
Figure 8.9 shows the overview of the SAFER+ structure. The SAFER+ is a block cipher that has eight rounds. In the encryption structure, each round function takes in the output of the previous round and two round subkeys. The first encryption round takes in the 128-bit plaintext. The output of the final encryption round (i.e. round 8) is input to an output transformation function, together with one round subkey. The decryption structure follows the reverse structure of encryption. The ciphertext is input to an input transformation function, together with one round subkey. The output is then input to eight decryption round functions. The output of the 8th decryption round is the plaintext.

8.4.1.2 SAFER+ Round Function
A SAFER+ round function consists of a *Key-Controlled Substitution* function and an *Invertible Linear Transformation* function. The key-controlled substitution function is shown in

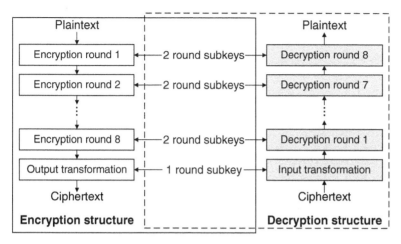

Figure 8.9 Overview of the SAFER+ structure.

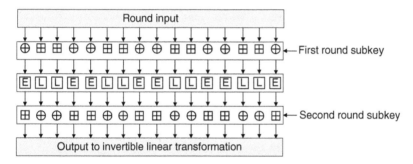

Figure 8.10 Key-controlled substitution.

Figure 8.10. Let $\oplus$ denote bitwise modulo-two addition of bytes, and $\boxplus$ denote modulo-256 addition of bytes. $E(\cdot)$ is a function that computes a value as follows:

$$E(x) = (45^x \bmod 257) \bmod 256.$$

$L(\cdot)$ is a function that computes a value as follows:

$$L(x) = \log_{45}(x) \bmod 256.$$

The output of the key-controlled substitution is input to the invertible linear transformation function, which is show in Figure 8.11. The invertible linear transformation function is based on the *Pseudo-Hadamard Transform* (PHT). As shown in Figure 8.12, the 2-PHT follows a "butterfly" structure. the 2-PHT function takes input a and b, and output $2a + b$ and $a + b$. Mathematically, the 2-PHT operates on the matrix H_2 in encryption, and the inverse matrix H_2^{-1} in decryption as follows:

$$H_2 = \begin{bmatrix} 2 & 1 \\ 1 & 1 \end{bmatrix}, \quad H_2^{-1} = \begin{bmatrix} 1 & -1 \\ -1 & 2 \end{bmatrix}.$$

Finally, the byte values go through the *Armenian shuffle* function, which is the coordinate permutation as follows: 9, 12, 13, 16, 3, 2, 7, 6, 11, 10, 15, 14, 1, 8, 5, 4.

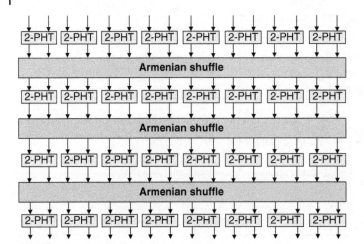

Figure 8.11 Invertible linear transformation.

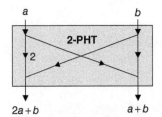

Figure 8.12 2-PHT function.

8.4.1.3 SAFER+ Key Schedule for 128-Bit Key

Besides the output of the previous encryption round (except for the first round, which takes input the plaintext), each SAFER+ encryption round function also takes input a round key. There are nine encryption round keys to be generated. Each of the first eight round keys consists of two 128-bit round subkeys. The last round key consists of one round subkey. In total, the SAFER+ key schedule generates 17 round subkeys. The SAFER+ key schedule is illustrated in Figure 8.13. Let K_i be the ith round subkey. The input key is appended with a parity byte. Then K_1 is the first 16 bytes of the expanded key, which is the input key. From K_2 to K_{17}, the expanded key first has a 3-bit left circular shift for each byte. Each round subkey K_i is generated by adding the first 16 bytes of the expanded key to a *bias*

$$B_i = (b_i[0], b_i[1], \ldots, b_i[15]), \quad i = 2, 3, \ldots, 17,$$

using modulo 256 addition 'Σ'. For each subkey K_i, the bias vector is defined as follows:

$$b_i[j] = \left[\left(45^{45^{17i+j+1} \bmod 257} \bmod 257 \right) \bmod 256 \right], \quad \text{for } j = 0, \ldots, 15.$$

8.4.2 Function $E_1(\cdot)$

In Bluetooth authentication, function $E_1(\cdot)$ is applied to generate SRES and ACO. As shown in Figure 8.14, the cores of function $E_1(\cdot)$ are SAFER+ and a slightly modified SAFER+. For better illustration, SAFER+ will be denoted by S_+, and the modified SAFER+

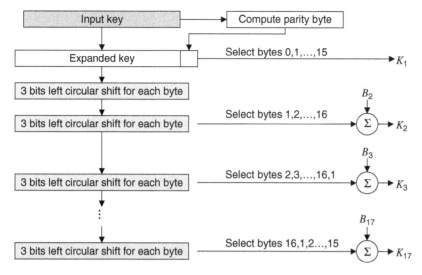

Figure 8.13 SAFER+ key schedule.

Figure 8.14 Overview of E_1.

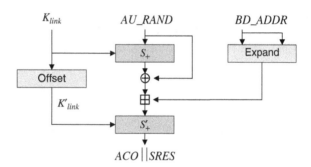

will be denoted by S'_+. The inputs to $E_1(\cdot)$ are link key K_{link}, a random value AU_RAND and BD_ADDR. The 8-byte input key K_{link} passes through an *offset* function and generates a 16-byte K'_{link} as follows:

$$K'_{link}[0] = (K_{link}[0] + 233) \bmod 256; \quad K'_{link}[1] = (K_{link}[1] \oplus 229);$$

$$K'_{link}[2] = (K_{link}[2] + 233) \bmod 256; \quad K'_{link}[3] = (K_{link}[3] \oplus 167);$$

$$K'_{link}[4] = (K_{link}[4] + 179) \bmod 256; \quad K'_{link}[5] = (K_{link}[5] \oplus 299);$$

$$K'_{link}[6] = (K_{link}[6] + 149) \bmod 256; \quad K'_{link}[7] = (K_{link}[7] \oplus 131);$$

$$K'_{link}[8] = (K_{link}[8] \oplus 233); \quad K'_{link}[9] = (K_{link}[9] + 229) \bmod 256;$$

$$K'_{link}[10] = (K_{link}[10] \oplus 223); \quad K'_{link}[11] = (K_{link}[11] + 193) \bmod 256;$$

$$K'_{link}[12] = (K_{link}[12] \oplus 179); \quad K'_{link}[13] = (K_{link}[13] + 167) \bmod 256;$$

$$K'_{link}[14] = (K_{link}[14] \oplus 149); \quad K'_{link}[15] = (K_{link}[15] + 131) \bmod 256.$$

Note that L bytes of *BD_ADDR* are expanded to 16 bytes by a function *Expand*:

$$\text{Expand}(X[0, \ldots, L-1], L) = (X[i \bmod L; i = 0, \ldots, 15]),$$

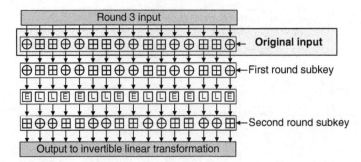

Figure 8.15 Modified key-controlled substitution for the third round in S'_+.

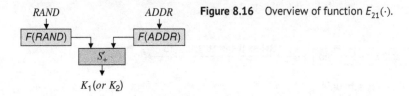

Figure 8.16 Overview of function $E_{21}(\cdot)$.

where $L = 6$ in function E_1. The final output of E_1 is the combination of a 4-byte SRES and a 12-byte ACO. The modification applied in S'_+ is to add the original input to the third round input, as shown in Figure 8.15.

8.4.3 Function $E_{21}(\cdot)$

Function $E_{21}(\cdot)$ is to generate the unit key K_1 or K_2 with inputs RAND and ADDR. As shown in Figure 8.16, the core function of $E_{21}(\cdot)$ is S'_+. Taken inputs *RAND* and *ADDR*, K_1 or K_2 are computed as follows:

$$K_1(\text{or } K_2) = E_{21}(RAND, ADDR) = S'_+(F(RAND), F(ADDR)),$$

where the 16-byte $F(RAND)$ is computed as follows:

$$F(RAND) = RAND[0, \ldots, 14] \cup (RAND[15] \oplus 6),$$

where $\cup$ defines the union operator. The $F(ADDR)$ is computed as follows:

$$F(ADDR) = \cup_{i=1}^{15} ADDR[i \bmod 6].$$

8.4.4 Function $E_{22}(\cdot)$

As shown in Figure 8.17, function $E_{22}(\cdot)$ is to generate the initial key K_{init} with Inputs *PIN* and *RAND* as follows:

$$K_{init} = E_{22}(PIN, RAND) = S'_+(F(PIN', N), F(RAND, N)),$$

where $F(PIN, N)$ is generated based on PIN and the length of PIN N. First, a 16-byte *PIN'* is generated by passing *PIN* through a constraining function as follows:

$$PIN' = \begin{cases} PIN[0, \ldots, N-1] \cup ADDR[0, \ldots, (5, 15-N)^-], & N < 16 \\ PIN[0, \ldots, N-1], & N = 16 \end{cases}$$

Figure 8.17 Overview of function $E_{22}(\cdot)$.

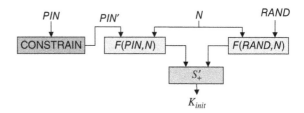

Figure 8.18 Overview of function $E_3(\cdot)$.

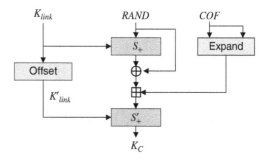

where $(5, 15 - N)^-$ outputs the smaller value between 5 and $(15 - N)$. It can be seen that if N is smaller than 16 bytes, then ADDR is used. If $N = 16$, then $PIN' = PIN$. Once PIN' is generated, $F(PIN, N)$ is computed as follows:

$$F(PIN, N) = \cup_{i=0}^{15} PIN'[i \bmod N].$$

The other input to S'_+ is $F(RAND, N)$, which is generated based on RAND and N as follows:

$$F(RAND, N) = RAND[0, \dots, 14] \cup (RAND[15] \oplus N).$$

The S'_+ function computes final output of E_{22} that is K_{init}.

8.4.5 Function $E_3(\cdot)$

As shown in Figure 8.18, function $E_3(\cdot)$ is to generate the encryption key K_C such that

$$K_C = E_3(K_{link}, RAND, COF).$$

Note that function $E_3(\cdot)$ has the same structure as function E_1. The difference is that the inputs are K_{link}, RAND and COF. And L in function Expand is 12 instead of 6.

8.4.6 Function $E_0(\cdot)$

Function $E_0(\cdot)$ is the stream cipher for Bluetooth data encryption. The core of $E_0(\cdot)$ is built around four independent linear feedback registers (LFSRs) and a finite state machine (FSM). An overview of function $E_0(\cdot)$ is shown in Figure 8.19. The four linear feedback registers $LFSR_i(t)$ compute x_{it} for $i = 1, 2, 3, 4$. The four LFSRs are each fully characterized by the following four feedback polynomials:

$$LFSR_1(t) = t^{25} + t^{20} + t^{12} + t^8 + 1;$$
$$LFSR_2(t) = t^{31} + t^{24} + t^{16} + t^{12} + 1;$$

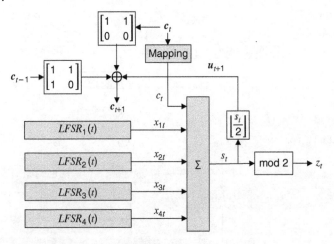

Figure 8.19 Overview of function E_0.

$$LFSR_3(t) = t^{33} + t^{28} + t^{24} + t^4 + 1;$$
$$LFSR_4(t) = t^{39} + t^{36} + t^{28} + t^4 + 1.$$

The outputs of the LFSRs are passed through an FSM, which is introduced to add nonlinearity so that finding the initial state is difficult from observing key stream data. At time t, the FSM functions as follows:

$$s_t = x_{1t} + x_{2t} + x_{3t} + x_{4t} + c_t,$$

where c_t takes on only the values 0, 1, 2, and 3. The c_t is updated as t increases from a vector $\mathbf{c}_t = (c_{0t}, c_{1t})^T$. The mapping function is to compute

$$c_t = 2c_{1t} + c_{0t}.$$

To compute the next c_{t+1}, first find the result of u_{t+1}, where $u_{t+1} = \lfloor s_t/2 \rfloor$ is a binary vector $\mathbf{u}_{t+1}$. Then a vector $\mathbf{c}_{t+1}$ is computed as follows:

$$\mathbf{c}_{t+1} = \begin{pmatrix} c_{0t+1} \\ c_{1t+1} \end{pmatrix} = \mathbf{u}_{t+1} \oplus \begin{pmatrix} 1 & 1 \\ 0 & 0 \end{pmatrix} \mathbf{c}_t \oplus \begin{pmatrix} 1 & 1 \\ 1 & 0 \end{pmatrix} \mathbf{c}_{t-1}.$$

Once s_t is generated from the FSM, the final output symbol z_t is the binary result obtained as follows:

$$z_t = s_t \bmod 2 = x_{1t} \oplus x_{2t} \oplus x_{3t} \oplus x_{4t} \oplus (c_t \bmod 2).$$

The output of E_0 is to generate the KEYSTREAM for data encryption and decryption. The final issue remains is how to deal with the inputs to $E_0(\cdot)$ in the initialization process.

Initialization of E_0: To generate the output for KEYSTREAM, $E_0(\cdot)$ takes inputs K'_C, EN_RAND, BD_ADDR and clk. The four inputs are used to derive 132-bit initial values, where 128 bits are for the LFSRs, and 2 bits each for $\mathbf{c}_{-1}$ and $\mathbf{c}_0$. The 132-bit initial values are derived by generating 200 stream cipher bits using the generator as follows. First shift in the three inputs K'_C, BD_ADDR, clk bits and the 6-bit constant 113 (208 bits total) by initializing the LFSRs:

- Open all LFSR switches.
- Initial the LFSRs as follows:

$$LFSR_1 : A[2]C[1]K[12]K[8]K[4]K[0]C_{24};$$

$$LFSR_2 : A[3]A[0]K[13]K[9]K[5]K[1]C_L[0]001;$$

$$LFSR_3 : A[4]C[2]K[14]K[10]K[6]K[2]C_{25};$$

$$LFSR_4 : A[5]A[1]K[15]K[11]K[7]K[3]C_U[0]111,$$

where $C_L[0] = C_3, C_2, C_1, C_0$, and $C_U[0] = C_7, C_6, C_5, C_4$.
- Set the initial states of the LFSRs to zero.
- Start to shift the input bits.
- Close feedback switch of the four LFSRs after 25, 31, 33, and 39 clock instants, respectively.
- At $t = 39$, set bits $\mathbf{c}_{39} = 0$ and $\mathbf{c}_{38} = 0$.
- Continue to shift in remaining inputs bits.

The process will continue until 200 outputs are produced. Finally, keep $\mathbf{c}_t$ and $\mathbf{c}_{t-1}$ and load the last 128 generated bits into the four LFSRs. *The last 128 bits are the initial values of the LFSRs.* They are fed back into the LFSRs to generate KEYSTREAM bits. The final state of $\mathbf{c}_t$ and $\mathbf{c}_{t-1}$ are initial values for $\mathbf{c}_{-1}$ and $\mathbf{c}_0$. Once initialized, E_0 is ready to generate KEYSTREAM bits for data block encryption/decryption.

8.5 Cryptographic Functions in Security Mode 4 (SSP)

Bluetooth security mode 4 applies SSP to generate link key for authentication. SSP includes functions that are different from those for security modes 1, 2, and 3. The functions used in SSP includes $P_{192}(\cdot), f_1(\cdot), g(\cdot), f_2(\cdot)$, and $f_3(\cdot)$. Each will be discussed in this section.

8.5.1 Function $P_{192}(\cdot)$

Function $P_{192}(\cdot)$ is the elliptic curve used to generate K_{DH} [72]. The elliptic curve E specified in $P_{192}(\cdot)$ is as follows:

$$E : y^2 + ax + b \bmod p,$$

where p is a prime modulus with order r, base point x-coordinate G_x and base point y-coordinate G_y; $a = \bmod (-3, p)$; and b is a defined number. The parameters set for the elliptic curve used in P_{192} are as follows:

$$p = 6277101735386680763835789423207666416083908700390324961279;$$

$$r = 6277101735386680763835789423176059013767194773182842284081;$$

$$b = 64210519 \; E59C80E7 \; 0FA7E9AB \; 72243049 \; FEB8DEEC \; C146B9B1;$$

$$G_x = 188DA80E \; B03090F6 \; 7CBF20EB \; 43A18800 \; F4FF0AFD \; 82FF1012;$$

$$G_y = 07192B95 \; FFC8DA78 \; 631011ED \; 6B24CDD5 \; 73F977A1 \; 1E794811.$$

With a given curve E, $P_{192}(u, V)$ is computed as the x-coordinate of the uth multiple uV of the point V.

8.5.2 Function $f_1(\cdot)$

In SSP, function $f_1(\cdot)$ is to compute the commitments. The function makes use of the HMAC based SHA-256 with 128-bit key X. For better illustration of the function, let the inputs to $f_1(\cdot)$ be 192-bit U, 192-bit V, 128-bit X, and 8-bit Z, then the output of $f_1(\cdot)$ is:

$$f_1(U, V, X, Z) = \text{HMAC-SHA-256}_X(U \| V \| Z)/2^{128},$$

which returns the most significant 128 bits of the hash value. Readers shall map the actual inputs that are used in SSP. Note that for Numeric Comparison and OOP protocols, $Z = 0$. For the Passkey protocol, the most significant bit of Z is set to one and the least significant bit is from one bit of the passkey.

8.5.3 Function $g(\cdot)$

Function $g(\cdot)$ uses the hash function SHA-256 to compute the output and applies the least significant 32 bits as the output of $g(\cdot)$. Let the four inputs be 192-bit U, 192-bit V, 128-bit X, and 8-bit Y, then the output of g is:

$$g(U, V, X, Y) = \text{SHA-256}(U \| V \| X \| Y) \bmod 2^{32}.$$

The checksum used for the numeric comparison protocol is the least significant 6 digits (in decimal), i.e. $g(U, V, X, Y) \bmod 10^6$.

8.5.3.1 Function $f_2(\cdot)$

Function f_2 is the final step to generate the link key in SSP. Similar to $f_1(\cdot)$, function f_2 also uses the HMAC based SHA-256, with 192-bit key W. Let the inputs to f_2 be 192-bit W, 128-bit U, 128-bit V, 32-bit X, 48-bit Y, and 48-bit Z, then the output of f_2 is:

$$f_2(W, U, V, X, Y, Z) = \text{HMAC-SHA-256}_W(U \| V \| X \| Y \| Z)/2^{128}$$

which returns the most significant 128 bits of the hash value as the link key.

8.5.3.2 Function $f_3(\cdot)$

Function $f_3(\cdot)$ also uses the HMAC based SHA-256, with a 192-bit key W. Let the inputs to f_3 be 192-bit W, 128-bit N, 128-bit U, 128-bit V, 16-bit X, 48-bit Y, and 48-bit Z, then the output of f_3 is:

$$f_3(W, N, U, V, X, Y, Z) = \text{HMAC-SHA-256}_W(N \| U \| V \| X \| Y \| Z)/2^{128}$$

where the most significant 128 bits of the hash value are the output of $f_3(\cdot)$, i.e. E_1 or E_2.

8.6 Summary

Bluetooth security is discussed in this chapter. Bluetooth standard specifies authentication, authorization, and confidentiality for securing data transmission. Four different security modes are provided to fit various Bluetooth transmission requirements. Besides security modes, Bluetooth standard also defines two trust levels and three service levels so that flexible configurations of security services can be designed in security policies. Bluetooth specifications have provided security features; however they do not guarantee secure Bluetooth connections from all adversary penetrations. An adequate level of knowledge and understanding must be provided for Bluetooth users. For example, a user shall not respond to suspicious Bluetooth messages or connection requests. The standard passkey must be avoided in Bluetooth connection. If using Bluetooth technology in an organization, security policies must be established to address the use of Bluetooth-enabled devices and users' responsibilities.

9

Zigbee Security

Zigbee is a wireless personal area network protocol designed to supply wireless communications at low cost and low power requirements. It is one of the promising communication technologies to support wireless sensor networks and Internet of things (IoTs). In this chapter, the security of Zigbee will be studied.

9.1 Introduction to Zigbee

9.1.1 Overview of Zigbee

Zigbee is a wireless personal area network (WPAN) protocol based on the IEEE 802.15.4 standard [73]. Zigbee specification defines a simpler and less expensive technology than other WPANs, such as Bluetooth or ad-hoc Wi-Fi, thus making Zigbee a perfect candidate for short-range low-rate wireless data transmission, such as wireless light switches, in-home electrical meters, traffic management systems, and other equipment [74, 75]. The IEEE 802.15.4 standard defines Zigbee to operate at the industrial, scientific, and medical (ISM) frequency bands (e.g. 2.4 GHz in most jurisdictions worldwide). An overview of the IEEE 802.15.4 protocol stack is shown in Figure 9.1. The IEEE standard defines the lower level device layers, i.e. the physical layer (PHY) and the media access control layer (MAC).

The PHY manages physical radio frequency (RF) transceiver and resource allocation, including channel selection, energy and signal management, etc. IEEE 802.15.4 standard defines Zigbee PHY to operate on one of the three possible unlicensed frequency bands, as shown in Table 9.1. The standard defines channels with 5 MHz each. The low power consumption of Zigbee limits transmission distances from 10 to 100 m line-of-sight, depending on power output and environmental characteristics [76]. The basic framework supports a 10-m communications range with a maximum theoretical data rate of 250 *kbps*. The MAC manages access to the physical channel and network beaconing. The IEEE standard defines collision avoidance through carrier sense multiple access-collision avoidance (CSMA/CA) and beacon-enabled networking. The IEEE 802.15.4 standard does not use standard Ethernet frames defined in IEEE 802.1D or IEEE 802.1Q. IEEE 802.15.4 PHYs only support frames of up to 127 bytes. An end device usually carries limited 802.15.4 functionality to reduce cost and complexity. Zigbee devices can transmit data over long distances by passing data through a mesh network of intermediate devices to reach more distant ones.

Security in Wireless Communication Networks, First Edition. Yi Qian, Feng Ye, and Hsiao-Hwa Chen.
© 2022 John Wiley & Sons Ltd. Published 2022 by John Wiley & Sons Ltd.
Companion website: www.wiley.com/go/qian/sec51

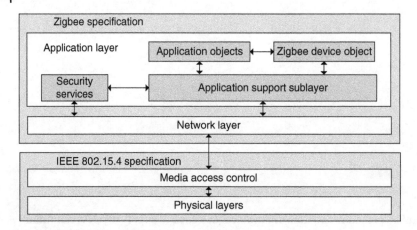

Figure 9.1 Overview of the Zigbee protocol stack.

Table 9.1 Zigbee operating frequency bands.

Frequency bands	Number of channels	Supporting areas
868.0–868.6 MHz	One communication channel	Europe
902–928 MHz	10 channels (2003), extended to 30 (2006)	North America
2400–2483.5 MHz	Up to 16 channels	Worldwide

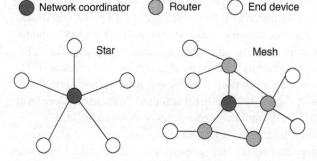

Figure 9.2 Zigbee network topologies.

Other higher-level layers and interoperability sublayers are not defined in IEEE 802.15.4 specification. The upper layers of Zigbee are standardized through the use the Zigbee standard as defined by the Zigbee Alliance. Zigbee network layer supports the formation of both *star* and *mesh* (or *peer-to-peer*) topologies, as shown in Figure 9.2. The network coordinator maintains overall network knowledge; a router carries full 802.15.4 functionality.

9.1.2 Security Threats Against Zigbee

IEEE 802.15.4 specification was designed with security supported in the MAC layer. Zigbee specification further designed security in upper layers in addition to the IEEE specification.

However, security is occasionally not well implemented by developers. The possible attacks against Zigbee applications can generally be classified into three categories: physical, key, and replay/injection.

- *Physical attacks*: In practice, many radios residing on Zigbee network employ a hard-coded encryption key. The encryption key is loaded in the memory once the device is powered on. With physical access to a Zigbee device, an attacker may use various low-cost and open source tools to exploit the hard-coded encryption key, for example, by intercepting the encryption key moved from flash to memory during power up. Once the hard-coded encryption key is revealed to an attacker, security of the entire Zigbee network is potentially compromised. The attacker could intercept and alter data that is on this network.
- *Key attacks*: Key attacks are usually launched remotely. Zigbee applies over the air key delivery in some applications. In this case, an attacker may deploy a device that mimics a Zigbee node in the network and picks the transmissions exchanged among legitimate devices. Captured packets can be analyzed or decrypted off-line afterwards. Due to the nature of passive attacks, it is almost impossible to detect disclosure of packets. Even worse, attackers may equip with better devices that can extend the range of Zigbee coverage to stay further away.
- *Replay and injection attacks*: These attacks aim to dupe Zigbee devices to executing unauthorized actions, which may consume a lot of resources or even interrupt communications. Zigbee devices are particularly vulnerable to replay attacks since they are equipped with a lightweight design of the protocol with weak replay protection.

9.2 IEEE 802.15.4 Security Features

Zigbee and other IEEE 802.15.4 based WPANs are vulnerable to a number of security threats and attacks. The IEEE 802.15.4 specification provides security services as data confidentiality, message integrity, and protection to replay attacks. The IEEE 802.15.4 security layer is handled at the MAC layer, below application control. An application must explicitly enable security, otherwise security is not enabled by default. IEEE 802.15.4 defines four packet types for the MAC layer: beacon packets, data packets, acknowledgments packets, and control packets. The specification does not support security for acknowledgment packets. Other packet types can optionally support integrity protection and confidentiality protection for the data field of a packet [77].

9.2.1 Security Levels

The IEEE 802.15.4 specification introduces procedures and mechanisms for protecting MAC frames, through symmetric-key cryptography techniques based on the AES-CCM algorithm [78]. In particular, the IEEE 802.15.4 specification defines eight security levels to protect the frame generated at the MAC layer, as listed in Table 9.2.

The eight security levels can be broadly classified into *no security, data authenticity only, data confidentiality only, data confidentiality, and data authenticity*. Security level 0 provides no security, which is set by default. Security levels 1 to 3 are based on AES-CBC-MAC

Table 9.2 Security suites supported by IEEE 802.15.4.

Security level	Security attributes	Data confidentiality	Data authenticity	MIC length	Replay detection
0 (000)	Null	OFF	NO	0	YES
1 (001)	MIC-32	OFF	YES	4	YES
2 (010)	MIC-64	OFF	YES	8	YES
3 (011)	MIC-128	OFF	YES	16	YES
4 (100)	AES-CTR	ON	NO	0	YES
5 (101)	ENC-MIC-32	ON	YES	4	YES
6 (110)	ENC-MIC-64	ON	YES	8	YES
7 (111)	ENC-MIC-128	ON	YES	16	YES

that provides data authenticity only. Security level 4 is based on AES-CTR that provides data confidentiality only. Security levels 5 to 7 are based on AES-CCM that provides both data confidentiality and data authenticity. Each category that supports data authenticity comes in three variants with different message integrity code (MIC) sizes. The MIC can be either 4, 8, or 16 bytes long, depending on security levels. A longer MIC increases protection against authenticity attacks, with the trade-off of a larger packet size. For each security level that offers data confidentiality, the recipient can optionally enable replay protection. IEEE 802.15.4 specification only requires radio chips to support Null and the ENC-MIC-64 security levels.

9.2.2 IEEE 802.15.4 Frame Structure

IEEE 802.15.4 specification defines a MAC layer data packet as shown in Figure 9.3. Each data packet has variable length and is used by a node to send a message to a single node

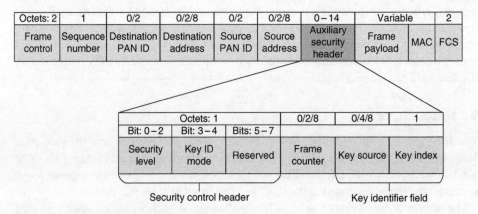

Figure 9.3 IEEE 802.15.4 MAC frame format.

or to broadcast a message to multiple nodes. A MAC frame is composed of a MAC header (7–23 bytes), a MAC payload (0–118 bytes), and a frame check sequence (FCS) footer (2 bytes). The FCS footer is a 16-bit cyclic redundancy check value. Each data packet has a flag field that indicates i) if security is enabled; ii) which addressing modes are in use; and iii) whether the sender requests an acknowledgment. If the security enabled flag of the frame control field is set to 0, then no security is provided. If the flag is set to 1, then one of the other seven security levels is applied. Security parameters are included within the auxiliary security header (ASH) field. The ASH field is a 5–14 bytes data structure composed of three fields as follows [73, 79]:

- The *Security control header* (1 byte): It specifies the *security level* (3 bits) and the *key identifier mode* (2 bits).
- The *frame counter* (4 bytes): It is to protect against replay attacks.
- The *key identifier field* (0–9 bytes): It contains *key source* (0/4/8 bytes) and *key index* (1 byte), which are needed to determine the key for the encryption. The key identifier field is optional.

The ASH is transmitted in clear text with authentication provided. The key identifier mode field is an unsigned integer that indicates whether the key can be derived implicitly or explicitly. Furthermore, it is used to indicate the particular representations of the key identifier field, if the key is derived explicitly. Values of the key identifier mode field (listed as mode) and key identifier field length field (listed as length) shall be set according to Table 9.3.

The formatting of the data field for the three main security categories is shown in Figure 9.4. When AES-CTR is applied, only the MAC payload is encrypted. When authentication is applied without confidentiality, the entire MAC header, ASH, and MAC payload are input to AES-CBC-MAC to generate a MIC. When both confidentiality and authentication are required, then the MIC is first generated with the inputs of MAC header, ASH, and MAC payload. Then the MAC payload and MIC are encrypted.

Table 9.3 Values of the key identifier mode field.

Mode	Description	Length
00	Key is determined implicitly from the originator and recipient(s) of the frame as indicated in the frame header	0
01	Key is determined from the Key Index field	1
10	Key is determined explicitly from the 4-octet key source field and the key index field	5
11	Key is determined explicitly from the 8-octet key source field and the key index field	9

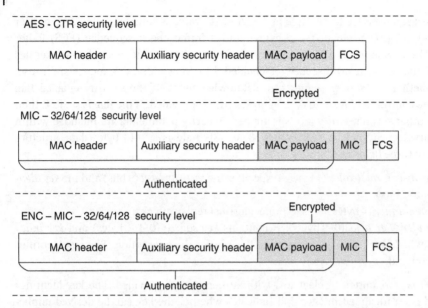

Figure 9.4 The formatting of the data field for the three main security categories.

9.3 Zigbee Upper Layer Security

Besides PHY and MAC layers, Zigbee specification implements two extra security layers at the *network layer* and the *application layer*. More specifically, it is the *application support sublayer*, which is a sub layer of the application layer, as shown in Figure 9.1. The Zigbee standard implemented complex security measures to ensure key establishment, secure networks, key transport, and frame security at both network and application support sublayers [80]. The Zigbee requires all protocol stack layers trust each other, thus each layer is responsible for the security of their respective frames.

9.3.1 Zigbee Security Models

Zigbee standard offers both *distributed* and *centralized* network architectures and corresponding security models to satisfy a wide range of applications. In order to ensure the optimal balance the security, ease of use, cost, and battery life with different network requirements, the two network architectures have different approaches in admitting new devices into the network and protecting messages on the network.

- *The Distributed Security Model*: This model is more appropriate for easier-to-configure Zigbee network systems. Such a distributed security model comprises of routers and end devices. If a Zigbee router detects an existing network when it is powered up, it may join the secure network. Otherwise, a Zigbee router can form a distributed secure network and start to issue network security keys. Other Zigbee routers and end devices will detect the secure network and join it by receiving the keys from the Zigbee network initiator. In a distributed Zigbee network, any router can issue network security keys.

- *The Centralized Security Model*: This model is more appropriate for Zigbee network systems that require higher security. In addition to routers and end devices, a centralized security model includes a *trust center*. A trust center is typically the network coordinator or a dedicated device that forms a centralized network. It is responsible for authentication and validation of each Zigbee device, which attempts to join the network. In a centralized Zigbee network, only the trust center can issue network security keys. The trust center also establishes a unique master key for each device on the network and link keys for each pair of devices as requested.

9.3.2   Security Keys in Zigbee

Since all Zigbee security policies rely on the same AES encryption algorithm, the 802.15.4 based hardware architecture deployed for the MAC layer is still valid. Including MAC layer, Zigbee standard applies three different kinds of keys to provide security in different layers. The keys used in Zigbee security are *master, link*, and *network* keys.

- **Master keys**: They are the keys pre-installed in each Zigbee device when manufacturing. A master key secures the link keys exchange between two nodes in the Key establishment procedure.
- **Link keys**: They are unique keys set for each pair of Zigbee nodes. Link keys are managed at the application support sublayer. They are used to encrypt data traffic between each pair of Zigbee nodes. Therefore, more memory resources are needed in each device if link keys are required.
- **Network key**: It is a unique 128-bit key shared among all the Zigbee devices in the network. The network key is generated by the *trust center* and regenerated at different intervals. Each Zigbee node must receive the network key in order to join the network. Once the trust center decides to update the network key, the new one is spread through the network using the old network key. Once this new key is updated in a device, its frame counter is initialized to zero.

Each pair of devices has both network key and link key, where only the link key is always used for better security. In a centralized security model with a trust center involved, there are two kinds of security policies, *residential mode* and *commercial mode* [81]. In the residential mode, a trust center issues the network key to the routers and end devices in the Zigbee network, as shown in Figure 9.5. This mode is normally chosen for wireless sensor networks, where the Zigbee nodes are embedded devices with limited resources.

In the commercial mode, a trust center issues all three types of keys for devices in the network, as shown in Figure 9.6. Besides the network key for all Zigbee devices, the trust

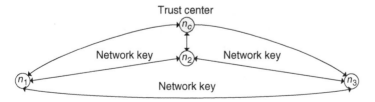

Figure 9.5   Trust center in Zigbee residential security mode.

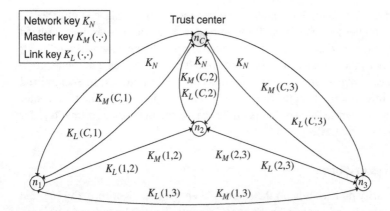

Figure 9.6 Trust center in Zigbee commercial security mode.

center also issues a master key and a link key for each pair of Zigbee devices (including the trust center itself) in the network. For example, between the trust center n_C and node 1 n_1 there exists a master key $K_M(C, 1)$, a link key $K_L(C, 1)$, and the network key K_N. Similarly, between nodes 1 and 2 there exists a master key $K_M(1, 2)$, a link key $K_L(1, 2)$, and the network key K_N.

9.3.3 Zigbee Network Layer Security

To protect a secure centralized network, Zigbee standard defines *install codes* at the network level to authenticate devices. An install code is a 128-bit unique code generated randomly for a new Zigbee device. To join a centralized secure network, the trust center will verify if the install code matches a code that is previously entered into the trust center without using a Zigbee message. For example, the install code may be printed in the packaging of the Zigbee device. The user can scan or manually enter the code into the trust center, or through a device (e.g. smart phone) that is connected to the trust center. Once the Zigbee device joins the secure network, the trust center derives a unique 128-bit link key from the install code using the Matyas–Meyer–Oseas (MMO) hash function [82].

Zigbee standard defines the same security environment at the network level for all Zigbee devices. Similar to the MAC layer security, Zigbee provides confidentiality and data authentication in the network layer by encrypting the transmitted frames with AES-CCM. AES-CCM is an authenticated encryption algorithm that combines both the counter mode of encryption and CBC-MAC mode of authentication. The key applied to the AES-CCM is the 128-bit network key, which is shared among all Zigbee devices in the network. In a centralized Zigbee network, the trust center usually generates and stores multiple network keys, where only one of the keys is the active network key. The initial network key is distributed to a Zigbee network via key-transport or pre-installation. An updated network key is usually encrypted with the link key and sent to each Zigbee device.

9.3.4 Zigbee Application Support Layer Security

Zigbee standard defines a secured link between each pair of Zigbee devices at the application support layer so that virtual private links can be supported between each pair of devices in the network. If security is provided for broadcast communications with the network key, the application support layer simply passed the encrypted frames to the network layer. If security is required for unicast communications, the frames are encrypted with link keys at the application support layer. A 128-bit link key is shared only between a pair of Zigbee devices in the network. The application support layer is also responsible for providing applications and the Zigbee device object with key establishment, key transport, and device management services. For example, in a smart home enabled with Zigbee network, all devices are connected in the same centralized network secured at the network level. Therefore communications in the network are protected from outsiders (e.g. neighbors). The application support layer security provides additional capability of protecting communications between two devices (e.g. a smart phone and a smart light) from others that are in the same network. The use of link key also limits the ability of an attacker that acquires the network key from intercepting or injecting messages that other devices would act upon.

9.3.5 Other Security Features in Zigbee

Besides the security definition in network layer and application support sublayer, Zigbee provides other security features. One of them is the usage of *application profiles*. Application profiles are agreements for messages, message formats, and processing actions that enable developers to create an inter-operable application with separate devices. Multiple application profiles can be created to allow devices of different vendors to properly and securely communicate with each other.

Another security feature is the *over-the-air (OTA) updates* defined for Zigbee applications. OTA updates allow a manufacturer to add new features, fix defects in its product, and apply security patches as new threats are identified. However, OTA updates also represent a potential security vulnerability if the protocol does not provide ample protections, or the device manufacturer does not use all available safeguards. As shown in Figure 9.7, Zigbee devices and associated silicon platforms provide multi-layered security to update devices in the field and assure that updated code images have not been modified maliciously. The OTA image may be encrypted during manufacturing so that only the end product contains the key to decrypt it. The OTA image may also be stored in on–chip memory that is configured with the debug read-back feature disabled preventing reverse engineering with standard debug tools, which is a common vulnerability of other solutions. Zigbee specification provides a method to encrypt all image transfers over the air with a unique key, e.g. k_1. The standard also provides a method to sign the OTA image with another unique key, e.g. k_2.

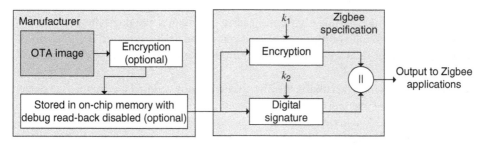

Figure 9.7 Security in Zigbee OTA updates.

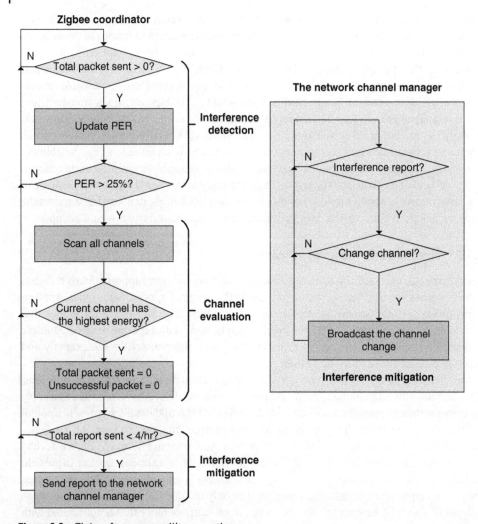

Figure 9.8 Zigbee frequency agility operation.

Zigbee standard also supports *frequency agility*, which enables a Zigbee network to switch operating channel. Frequency agility also improves the security of a Zigbee network against replay attacks. As shown in Figure 9.8 , frequency agility is conducted in three phases:

- *Interference detection*: In this phase, the Zigbee coordinator monitors the packet error rate (PER), i.e. the ratio of unsuccessful packets to the total number of packets sent. If the PER is greater than 25% with a minimum of 20 packets sent, then interference is confirmed.
- *Channel evaluation*: In this phase, the Zigbee coordinator scans all the channels to determine if the current channel has the highest energy.
- *Interference mitigation*: In this phase, the Zigbee coordinator reports to the network channel manager, which decides if a channel switch shall be conducted. If so, the decision is broadcast to all Zigbee nodes in the network.

9.4   Security-Related MAC PIB Attributes

The IEEE 802.15.4 standard defines the PAN information base (PIB) security-related attributes to serve multiple purposes. For example, some attributes determines the key for the encryption and decryption functionalities implemented within the outgoing frame security and the incoming frame security procedures. The configuration of the security-related attributes is a complex task but allows great flexibility in the use of the media access control security for various Zigbee application needs.

The security-related attributes defined in IEEE 802.15.4 specification are described in Table 9.4. In the *secKeyIdLookupList*, the *secKeyIdLookupDescriptors* contains:

- *secKeyIdMode*: the mode used for this descriptor.
- *secKeySource*: the originator of the key.
- *secKeyIndex*: information used to identify the key.
- *secKeyDeviceAddrMode*: the addressing mode for this descriptor.
- *secKeyDevicePanId*: the PAN ID for this descriptor.
- *secKeyDeviceAddress*: the address for this descriptor.
- *secKeyDescriptor*: an *secKeyDescriptor* associated with the parameters in this *secKeyId-LookupDescriptor*.

Table 9.4   The PIB security-related attributes.

PIB security-related attributes	Description
secKeyIdLookupList	A list of *secKeyIdLookupDescriptors* containing keys and security policy information that are useful for protecting an MAC frame
secDeviceList	A list of device information (i.e. *secDeviceDescriptors*) for each remote device with which this device securely communicates. Each of them contains the PAN ID, short MAC address, extended MAC addresses, as well as the counter of the latest packet received from the remote device and a boolean flag indicating if the considered node may override the minimum security level settings
secSecurityLevelList	A list of *secSecurityLevelDescriptors* that provides information about the minimal expected/required security level and the set of allowed security levels for each MAC frame type and subtype
secFrameCounter	The outgoing frame counter for this device to be used for keys which do not have *secFrameCounterPerKey* set to TRUE
secAutoRequestSecurityLevel	The security level used for automatic data requests
secAutoRequestKeyIdMode	The key identifier mode used for automatic data requests
secAutoRequestKeySource	The originator of the key used for automatic data requests
secAutoRequestKeyIndex	the index of the key used for automatic data requests

In the *secDeviceList*, the *secDeviceDescriptors* contains:

- *secPanId*: the PAN ID of the device in this DeviceDescriptor.
- *secShortAddress*: the short address of the device in this DeviceDescriptor.
- *secExtAddress*: the extended IEEE address of the device.
- *secExempt*: indication of whether the device may override the minimum security level settings.
- *secDeviceMinFrameCounter*: The smallest frame counter allowed to be sent by the other device for this key.

With the security-related attributes defined, a node that intends to secure a packet shall execute the outgoing frame security procedure in the following steps:

(1) identify the security level that has to be applied to the current MAC frame;
(2) check if the frame size is acceptable (i.e. 127 bytes maximum);
(3) identify the key to use during the encryption process;
(4) protect the MAC payload according to the selected security level, by using the corresponding algorithm;
(5) create the auxiliary security control field and include it within the protected frame;
(6) generate the FCS;
(7) reassemble the whole packet.

When a device receives an MAC frame, it should verify if it has been protected by the sender (i.e. if the security enabled flag is set to 1). In affirmative case, it will run the incoming frame security procedure with the following operations:

(1) verify the packet integrity through the check of the FCS;
(2) identify the key to exploit during decryption process;
(3) verify that the Security Level chosen by the sender is allowed for the message the packet contains;
(4) decrypt the payload;
(5) verify that all security constraints;
(6) deliver the message to the upper layer.

9.5 Mechanisms Used in Zigbee Security

Zigbee security applies AES-CTR for confidentiality, AES-CBC-MAC for data authentication, and AES-CCM for both confidentiality and data authentication at all the three layers, MAC layer, network layer, and application support sublayer. Zigbee also applies MMO hash function for link key generation.

9.5.1 AES-CTR

AES-CTR mode is a technique for encryption from a block cipher (i.e. AES in counter mode), as shown in Figure 9.9. To encrypt a plaintext, a nonce and a counter value CTR are first input to an AES encryption operation. The plaintext is then XORed to the output of the operation to generate the ciphertext. Note that the block operation is the same for decryption, i.e. only AES encryption is applied to the function.

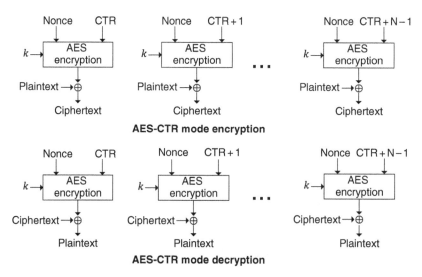

Figure 9.9 Overview of the AES-CTR mode.

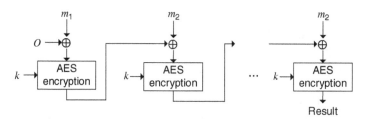

Figure 9.10 Overview of the AES-CBC-MAC mode.

9.5.2 AES-CBC-MAC

AES-CBC-MAC is a technique for constructing a message authentication code from a block cipher (i.e. AES in CBC mode), as shown in Figure 9.10. Note that the acronym *MAC* in AES-CBC-MAC stands for message authentication code. It should not be confused from the acronym *MAC*, which is defined as media access control in this chapter, while *MIC* is defined for the purpose of message integration (authentication) code. Zigbee security may choose 32, 64, or 128 bits from the 128-bit result.

9.5.3 Overview of the AES-CCM

AES-CCM mode is the mixture of AES-CTR mode and AES-CBC-MAC mode, as shown in Figure 9.11. A MIC is first generated using the AES-CBC-MAC mode. The plaintext and the MIC are encrypted using the AES-CTR mode after that.

9.5.4 Nonces Applied to the Security Mechanisms

In the security mechanisms described before, nonces are set differently for the Time-Slotted Channel Hopping (TSCH) mode and non-TSCH mode. For non-TSCH mode, the nonce is

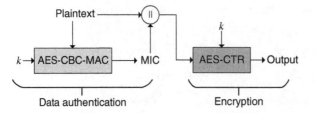

Figure 9.11 Overview of the AES-CCM mode.

Octets: 8	4	1
Source address	Frame counter	Nonce security level

Figure 9.12 Nonce for non-TSCH mode.

Octets: 8	5
Source address	ASN

Figure 9.13 Nonce for TSCH mode.

Octets: 3	1	2	2
IEEE 802.15.4 CID	0 x 00	PAN ID	Short address

Figure 9.14 Source address field for TSCH mode with short addressing.

formatted as shown in Figure 9.12. The source address field is set to the extended address of the device originating the frame. The frame counter field is set to the value of the respective field in the ASH. The nonce security level field is an unsigned integer that is set to the value of the security level of the security control field.

For TSCH mode, the nonce is formatted as shown in Figure 9.13. The source address field can be set to the extended address of the device originating the frame, or be formatted as shown in Figure 9.14. The IEEE 802.15.4 CID field contains the CID for IEEE 802.15.4. The PAN ID field contains the PAN ID. The short address field contains the short address of the device originating the frame. If short addresses are used in the nonce, then the coordinator must ensure the uniqueness of each short address.

The nonce for fragment frames is formatted as shown in Figure 9.15. Besides the source address security level fields, a nonce for fragment frames requires a PSDU counter and a fragment number to identify the fragment. The fragment indicator is set to one for this nonce.

9.5.5 Matyas–Meyer–Oseas Hash Function

MMO hash function is a single-block-length one-way compression function [82]. As shown in Figure 9.16, MMO hash function takes input the message m_i as the plaintext to be encrypted; and $g(H_{i-1})$ as the key for encryption. H_{i-1} is the hash value generated from the previous message and $g(\cdot)$ is a function that generates the key. The output ciphertext is then XORed with the same message block m_i to produce the next hash value H_i. In

Octets: 7	Bits: 0–25	26–31	32–35	36	37–39
Source address	PSDU counter	Fragment number	Reserved	Short address	Security level

Figure 9.15 Nonce for fragment frames.

Figure 9.16 General structure of MMO hash function.

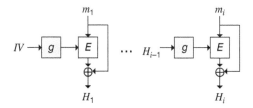

mathematical notation MMO can be described as follows:

$$H_i = \begin{cases} E_{g(IV)}(m_1) \oplus m_1, & i = 1, \\ E_{g(H_{i-1})}(m_i) \oplus m_i, & i \geq 2. \end{cases} \tag{9.1}$$

9.6 Summary

The Zigbee security is discussed in this chapter. The MAC layer security defined by the IEEE 802.15.4 standard is extended by Zigbee standard in network layer and application support sublayer. Due to the nature of symmetric-key cryptographic mechanisms, the level of security provided by the Zigbee security architecture depends on the safekeeping of the symmetric keys, on the protection mechanisms employed, and on the proper implementation of the cryptographic mechanisms and associated security policies involved. Therefore, the Zigbee security is based on the assumption that keys are securely stored and distributed. In practice, security provided by Zigbee standard is not enough. For example, if a Zigbee device joins a network, the key may be sent unprotected. Despite its short time frame, a potential key interception may happen. Another vulnerability to Zigbee is the low-cost nature of some types of Zigbee devices. An attacker may get physical access to such a device and extract the secret keying material as well as other privilege information. Zigbee has a great potential to be deployed in wireless sensor networks and Internet of things. Security must be carefully considered to provide those applications.

10

RFID Security

Traditional automatic identification and data capture (AIDC) technologies such as bar codes and quick response (QR) code use dedicated optical scanners and smart phone cameras to read labels. Radio-frequency identification (RFID) is a technological advancement in AIDC that support communications without optical line of sight and less human involvement in the identification process [83]. In this chapter, the security of RFID is introduced.

10.1 Introduction to RFID

10.1.1 Overview of RFID Subsystems

An RFID subsystem consists of two types of components, RFID tags and RFID readers [83], as shown in Figure 10.1. *RFID tags*, also known as transponders, are small electronic devices embedded in objects. Each RFID tag has a unique identifier (UID) and may also have other features such as memory to store additional data and security. The primary function of a tag is to provide an identifier to an RFID reader. *RFID readers* are devices that identify items associated with RFID tags. Both the RFID tag and the RFID reader are two-way radios that are capable of modulating and demodulating radio signals. Most RFID systems also contain an enterprise subsystem that makes the data acquired from RFID subsystem transactions useful to a supported business process. Each RFID tag has a UID. The Electronic Product Code (EPC) developed by the industry group EPCglobal is widely used across many industry sectors as the tag identifier. The EPC tag identifier format consists of four data fields, Header, EPC Manager ID, Object Class, and Serial Number. The *Header* specifies the EPC type. The *EPC Manager ID* identifies the organization that assigns the object class and serial number. The *Object Class* identifies a class of object. The *Serial Number* describes the specific instance of that class of objects.

10.1.2 Types of RFID Tags

There are different types of RFID tags in terms of the cost, size, performance, and security mechanisms. Based on the power source, RFID tags are categorized into passive, active, semi-active, and semi-passive.

Security in Wireless Communication Networks, First Edition. Yi Qian, Feng Ye, and Hsiao-Hwa Chen.
© 2022 John Wiley & Sons Ltd. Published 2022 by John Wiley & Sons Ltd.
Companion website: www.wiley.com/go/qian/sec51

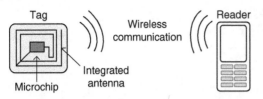

Figure 10.1 An example of a simple RF subsystem.

- *Passive tag*: A passive tag is powered by the electromagnetic energy from the transmission of a reader. Due to the limited power, a passive tag can only support lightweight data processing and has short operating range. Nonetheless, passive tags are usually cheaper and smaller than other types of tags.
- *Active tag*: An active tag is powered by an internal battery for communications and other functions. As a result, active tags can communicate over greater distance than other types of tags. However, active tags are usually more expensive and larger.
- *Semi-active tag*: A semi-active tag is also powered by an internal battery, similar to an active tag. In comparison, a semi-active tag only wakes up when it receives a signal from the reader. Otherwise, it remains dormant. As a result, semi-active tags generally have longer battery life.
- *Semi-passive tag*: A semi-passive tag has an internal battery. However, the battery only powers the on-board circuitry while not producing return signals. As a result, semi-passive tags are usually cheaper and smaller than active tags, whereas supporting more functionalities than passive tags.

10.1.3 RFID Transactions

RFID transactions can be initiated by either tags or readers, defined as *reader talks first (RTF)*, or *tag talks first (TTF)*. In an RTF transaction, the reader broadcasts a signal. The tags that receive the signal may then respond and continue transactions with the reader. In a TTF transaction, a passive tag initiates a transaction as soon as it gets power from the reader's signal. An active tag in a TTF transaction transmits periodically as long as its power supply lasts. In most RFID subsystems, readers and tags operate using only RTF or TTF transactions, not both types. Security wise, an active tag in a TTF transaction can be more vulnerable to eavesdropping as the attacker has no need to send signals.

10.1.4 RFID Frequency Bands

RFID systems may operate in different radio frequencies, ranging from low frequency (LF), high frequency (HF), ultra-high frequency (UHF), and microwave frequency [84]. The radio frequencies at which a tag transmits and receives signals have implications for tag performance characteristics, including operating range, and RFID data transfer rate [85]. An overview of the characteristics of each RFID frequency band is shown in Table 10.1.

Table 10.1 RFID frequency bands.

	LF	HF	UHF	Microwave
Frequency	30–300 KHz	3–30 MHz	300–1000 MHz	2–30 GHz
RFID RF	125–134 KHz	13.56 MHz	433 MHz (active) 865–956 MHz	2.45/5.8 GHz
Data rate	<1 Kbps	≈ 25 Kbps	≈ 30 Kbps	Up to 100 Kbps
Applications	Access control, animal tagging, inventory control, vehicle immobilizer	Smart card, contact-less access and security, item level tracking	Logistics case/pallet tracking, baggage handling	Railroad car monitoring, automated toll collection

10.2 Security Attacks, Risks, and Objectives of RFID Systems

10.2.1 Security Attacks to RFID Systems

RFID systems are vulnerable to several security attacks. The Smart Border Alliance assessed several security risks of RFID systems [85].

- *Counterfeit RFID tag attacks*: This type of attack seeks to duplicate legitimate tags through cloning or forgery. Since many low-cost RFID systems use unencrypted identifiers, they can be stolen by the counterfeit attacks.
- *Replay attacks*: An attacker can perform replay attacks especially when a counterfeit RFID tag is retrieved. Integrity of the legitimate RFID tag can be compromised by the replay attacks.
- *Eavesdropping attacks*: Since lightweight RFID tags are not required to communicate with RFID readers through encrypted channels, eavesdroppers can intercept the clear texts in the vicinity of the RFID systems.
- *Electronic collisions*: Electronic collisions may occur when multiple RFID tags and/or readers communicate to each other simultaneously. Collisions of RFID tags can result in failed transmissions, lost data, or faulty data integrity.

10.2.2 RFID Privacy Risks

RFID privacy risks reside in *data* and *location*. Data privacy involves user information contained on the RFID tag and the associated backend databases. Additional data privacy risks can arise if the RFID tags can be rewritten. Location privacy involves location information of a tag when its tag number is associated with a physical location and time. For example, the stored-value card for automated toll collection and public transportation may record the location information at the time the device is used. Disclosed location information may be further used in data mining or other value-added applications.

10.2.3 Security Objectives

The Smart Border Alliance listed the security objectives of RFID systems as *confidentiality*, *integrity*, *non-repudiation*, and *availability* [85].

- *Confidentiality*: Communication channels between RFID tags and legitimate readers should be protected from unauthorized access. All data stored on the RFID tags and the backend databases should be protected from unauthorized access. No unauthorized RFID readers should access data on the RFID tags.
- *Integrity*: User data stored on the RFID tags and the backend databases should be protected from unauthorized modification. Communication channels between RFID tags and legitimate readers should be protected from unauthorized modification and replay. The presence of multiple tags should not compromise system integrity. RFID tags should be prohibited from duplication.
- *Non-repudiation*: RFID tags and readers should achieve mutual authentication. Mutual authentication should also be achieved between RFID readers and any middleware in the system.
- *Availability*: All RFID system components should be available to users whenever needed. Neither the presence of multiple tags nor multiple readers should interrupt system operations. Data accessed from the backend systems should be available to multiple authorized entities at any time.

10.3 Mitigation Strategies and Countermeasures for RFID Security Risks

There are several mitigation strategies and countermeasure for RFID security risks, including cryptographic strategies, anti-collision algorithms, as well as other mitigation strategies. For clarification, Table 10.2 summarizes the RFID system risks and their impacts and countermeasures.

10.3.1 Cryptographic Strategies

10.3.1.1 Encryption
RFID signals can be encrypted so that rogue readers cannot intercept the messages. Therefore, encryption provides confidentiality to RFID systems. However, power consumption and cost limit the encryption capability of RFID tags.

10.3.1.2 One-Way Hash Locks
One-way hash locks provide access control based on a hash function [86]. The *lock state* is when an RFID tag stores the hash value of a unique key as its metaID. The key value and the metaID are sorted in a backend database. Per request from a reader, the RFID tag responds with its metaID. A legitimate reader would verify the metaID by consulting with the backend database. Once verified, the RFID tag enters the *unlocked state* and transmits its full functionality to the reader. However, one-way hash locks do not protect the systems from eavesdropping once the tag enters the unlocked state.

Table 10.2 RFID system risks and their impacts and countermeasures.

Risks	Security objectives	Countermeasures
Counterfeit attacks	Confidentiality Integrity	Encryption EPC tag PIN Physical shielding sleeve One-way hash locks Selective blocker tag
Replay attacks	Integrity Availability	EPC tag PIN Physical shielding sleeve One-way hash locks Selective blocker tag
Eavesdropping	Confidentiality	Encryption EPC tag PIN Physical shielding sleeve Selective blocker tag
Electronic collisions	Integrity Availability	Anti-collision algorithms Secure reader protocol 1.0

10.3.1.3 EPC Tag PINs

EPC Class 1 tags have personal identification number (PIN)-controlled access originally envisioned to allow readers to authenticate to RFID tags. The EPC tag PINs control several sensitive functions, including "write," "sleep," and "kill." This one-way authentication is extended to authenticate RFID tags to trusted readers [87]. Thus a mutual authentication of both RFID tags and readers can be achieved through a resultant challenge and response communication line.

10.3.2 Anti-Collision Algorithms

Anti-collisions algorithms provide *singulation* where a reader identifies a particular tag to avoid tag collisions. *Tree-walking* and *blocker tag* are common singulation methods in RFID systems.

10.3.2.1 Tree-Walking

RFID systems operating at 915 MHz generally implement the silent binary tree-walking algorithm as a singulation technique. As shown in Figure 10.2, the silent tree-walking protocol utilizes a binary algorithm that queries tags bit-by-bit. It resembles a depth-first search of a binary tree. RFID tags bear UIDs of a fixed bit-length, where each bit is represented as a "0" or a "1." An example is shown in Figure 10.2, the reader broadcasts current prefix. Each tag with *this* bit responds to the broadcast. If no collision, the reader adds one bit to current prefix. If responses collide, the reader tries both possibilities of the bit.

10.3.2.2 The Selective Blocker Tag

The selective blocker tag is an extension to the silent tree walking singulation protocol [88]. It is a form of jamming by broadcasting both "0" and "1" in response to any request from an

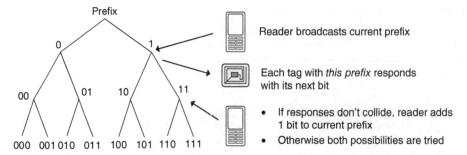

Figure 10.2 Illustration of tree walking.

RFID reader. It guarantees collision no matter what tags are present. To talk to an RFID tag, a reader must traverse every tree path, which is infeasible with a lengthy ID (e.g. 128 bits). To prevent illegitimate blocking, blocker tag shall be made selective for certain ID ranges.

10.3.3 Other Mitigation Strategies

10.3.3.1 Physical Shielding Sleeve (The Faraday Cage)
A Faraday cage is a physical shield made from metal mesh or foil that can protect RFID tags from eavesdropping. However, a physical shield must be removed to allow scans from legitimate readers. Without proper compliance, a physical shield may raise risks of availability and integrity. The integrity could be threatened since the RFID tag is not recorded by the legitimate reader if it is shielded in a Faraday cage.

10.3.3.2 Secure Reader Protocol 1.0
The EPCglobal developed Secure Reader Protocol 1.0 that defines communication between RFID middleware and RFID readers. As listed in Table 10.3, Secure Reader Protocol 1.0 is specified in three distinct layers. The *Transport Layer* provides operating system networking facilities. The *Messaging Layer* is allowed for multiple alternate implementations. One particular implementation is messaging/transport binding (MTB). Alternate implementations of MTB allow the RFID systems to support Ethernet, IEEE 802 local area network and metropolitan area network standards. The security of middleware and reader communication mode is indicated by the technology used to implement the MTB.

Table 10.3 Layers specified in Reader Protocol 1.0.

Layer	Definition
Reader layer	Message content/format
	Security services
Messaging layer	Message framing
	Connection establishment
Transport layer	Operating system networking facilities

10.4 RFID Security Mechanisms

The limited power supply and low processing power constraints dictate lightweight security mechanisms in RFID tag implementations. Several security protocols have been developed to protect RFID systems to a reasonable extent. The typical examples are hash locks [86] and the family of HB protocols [89–91].

10.4.1 Hash Locks

10.4.1.1 Default Hash Locking

Hash locking is an access control mechanism based on one-way hash functions [86]. Figure 10.3 illustrates an overview of the hash locking protocol. An RFID tag may enter the *locked state* and the *unlocked state* described as follows:

- *Lock a tag*: An RFID tag stores the hash of a random key, i.e. hash(key), as its metaID. In practice, a hardware-optimized cryptographic hash would suffice. The metaID generation may occur either over the RF communication channel or a physical contact channel for added security. The metaID also requires a portion of memory reserved on an RFID tag. The backend database stores both the key and metaID. The RFID tag enters the locked state. In this state, the RFID tag responds all queries with its metaID only and offers no other functionality.
- *Unlock a tag*: Once the metaID is queried from an RFID tag, a reader looks up the appropriate key in the backend database and transmits the key to the tag. The tag verifies the key by computing the hash value and comparing it to its stored metaID. If the values match, the RFID tag unlocks itself and enters the unlocked state, where all functionalities are open to nearby readers.

Analysis of Hash Locks

Due to the one-way hash function, hash locks prevent unauthorized readers from reading tag contents. However, metaID may be captured by an attacker who may spoof that tag and launch replay attacks to a legitimate reader. More importantly, the legitimate reader will retrieve and transmit the key to the attacker. To counter this matter, a legitimate reader may verify the ID and the associated metaID of a tag with the backend database. Hash locks may be extended to provide access control for other tag functionality and support multiple users.

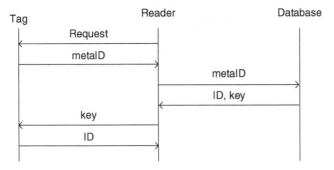

Figure 10.3 Hash locking.

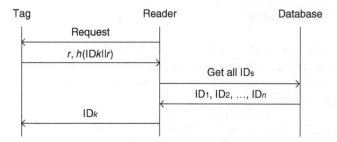

Tag   Reader   Database

Figure 10.4 Randomized hash locking.

Since the metaID acts as a UID of an RFID tag, location privacy risks may arise under hash locks even without knowing the real ID.

10.4.1.2 Randomized Hash Locking

Randomized hash locking provides a solution to tracking of individuals with a small number of RFID tags. The processes of tag locking and unlocking are illustrated in Figure 10.4.

- *Lock a tag*: Upon receiving queries from readers, an RFID tag picks a random value r and generates a hash value from its ID concatenated with r, i.e. hash$(ID\|r)$. Both r and hash$(ID\|r)$ are sent as responses to the queries. The RFID tag enters the lock state.
- *Unlock a tag*: A legitimate reader identifies an RFID tag by hashing each of its known IDs concatenated with r until it finds a match. Due to the complexity of brute-force search, the randomized hash locking is only feasible for a small number of RFID tags.

Analysis of Randomized Hash Locks The randomized hash locks may not be theoretically robust since a general one-way function does not guarantee secrecy. Therefore, it is possible that the bits of the input could be revealed. To counter this matter, a stronger primitive is needed to protect ID bits. In the protocol design, assuming that each RFID tag shares a unique secret key k with the reader and supports a pseudo-random function ensemble

$$F = \{f_n\}_{n\in\mathbf{N}}.$$

Upon receiving queries from readers, an RFID tag picks a random value r and computes $ID \oplus f_k(r)$. Both r and $ID \oplus f_k(r)$ are responses to the queries. The reader searches for a match through brute-force computation. To avoid storing tag IDs on the backend database, RFID tags may append their IDs to the computed hash values, i.e. $ID\|h(ID) \oplus f_k(r))$. The reader may identify tags by computing $f_k(r)$ for all their known keys and XORing it to the second part of the response. The value ending in the form $(x\|h(x))$ indicates a match.

10.4.2 HB Protocol and the Enhancement

10.4.2.1 HB Protocol

The HB protocol was proposed by Hopper and Blum [89]. An overview of a round of HB protocol is given in Figure 10.5. A secret x and a parameter η are known to both sides. The reader picks a random value a and queries the tag. The tag picks a random value v, computes, and replies $z = (a \cdot x) \oplus v$, where $a \cdot x$ and $a \oplus x$ represent scalar product and exclusive-or

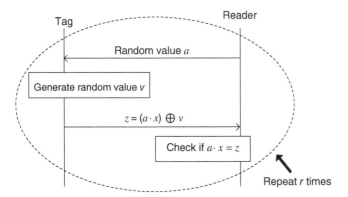

Figure 10.5 Illustration of HB protocol.

(XOR) of k-bit binary vectors a and x, respectively. The random value v is 1 with probability η, and 0 otherwise. The HB protocol repeats the round by r times and the tag is authenticated if check on the reader's side fails at most ηr times, or $a \cdot x \approx z$.

Analysis of HB Protocol

The HB protocol is based on the hardness of the learning parity with noise (LPN) problem, hence is deemed secure to the extent the related LPN problem is secure. The HB protocol uses an inner dot product that is bit-wise AND (&) of two strings and finds the parity of the result. Given several values of q and r, the value of k-bit x is determined using linear algebra as follows:

$$\begin{pmatrix} q_1 \\ q_2 \\ q_3 \\ \vdots \end{pmatrix} x = \begin{pmatrix} r_1 \\ r_2 \\ r_3 \\ \vdots \end{pmatrix}. \tag{10.1}$$

Note that the columns of the q matrix must be linearly independent. Adding *noise* by flipping the value of r randomly makes the LPN problem NP-hard [92]. Although this protocol works well under most circumstances against passive adversaries, it is vulnerable to active adversaries. As shown in Figure 10.6, an attacker (e.g. a rogue reader) can transmit a non-random value a to the tag. Each time, the tag replies a corresponding $z = (a \cdot x) \oplus v$. For example, if the queries are $a = 1000\ldots$, and the majority of the responses are 1, then the attacker could update the calculation of x as

$$\begin{pmatrix} 1000\ldots \\ q_2 \\ q_3 \\ \vdots \end{pmatrix} x = \begin{pmatrix} 1 \\ r_2 \\ r_3 \\ \vdots \end{pmatrix}. \tag{10.2}$$

After that, the attacker could initiate new queries $a = 0100\ldots$, and the majority of the responses are 0, then the attacker could update the calculation of x as

$$\begin{pmatrix} 1000\ldots \\ 0100\ldots \\ q_3 \\ \vdots \end{pmatrix} x = \begin{pmatrix} 1 \\ 0 \\ r_3 \\ \vdots \end{pmatrix}. \tag{10.3}$$

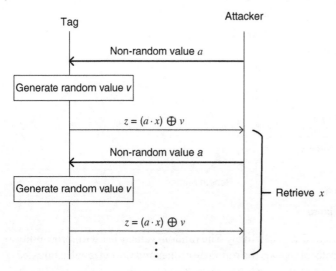

Figure 10.6 Active attacker to HB protocol.

The attacker could continue the process k times with corresponding a and finally retrieve the secret value x.

10.4.2.2 HB⁺ Protocol

The HB⁺ protocol modifies the HB protocol by adding a blinding value b so that active attacks that threat the HB protocol could be mitigated [90]. In HB⁺ protocol, two secrets (x, y) are pre-shared between the tag and the reader. As illustrated in Figure 10.7, in each round of the process, the tag sends a binding value b first to the reader. The reader replies with a random value a. Up on receiving a, the tag generates a random value v, where $v = 1$ with probability η, and $v = 0$ otherwise. Next computes

$$z' = (a' \cdot x) \oplus (b \cdot y) \oplus v.$$

The reader checks if $(a \cdot x) \oplus (b \cdot y) = z'$. The tag is authenticated if the checks fail at most ηr times.

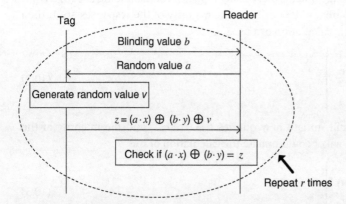

Figure 10.7 Illustration of HB⁺ protocol.

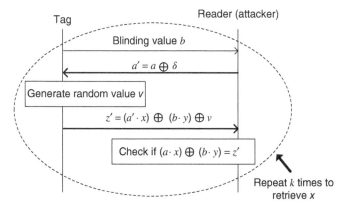

Figure 10.8 The man-in-the-middle attack on HB⁺ protocol to retrieve *x*.

Analysis of HB⁺

Although the HB⁺ protocol mitigates the active attacks that threat the original HB proto-col, it is vulnerable to man-in-the-middle attack. An attacker first manipulates challenges sent from a legitimate reader to retrieve the secret x from a legitimate tag. As illustrated in Figure 10.8. The attacker intercepts the binding value b from a legitimate tag. It then manipulate the challenge reply as

$$a' = a \oplus \delta,$$

where δ is a k-bit constant vector. The tag follows the protocol by generating the random value v and replies

$$z' = (a' \cdot x) \oplus (b \cdot y) \oplus v.$$

The attacker is assumed to have the capability to recognize if the authentication succeeds or not. In this case, a successful authentication indicates that $\delta \cdot x = 0$ with a high probability. A failed authentication indicates that $\delta \cdot x = 1$ with a high probability. The constant δ can be manipulated and run k times to reveal all bits of the secret value x.

Once x is identified, the attacker can impersonate the tag to retrieve the secret y from a legitimate reader. As shown in Figure 10.9, the attacker sends a given blinding value b to the reader. With the received challenge a, the tag replies $a \cdot x$ instead of computing a z value. If the authentication succeeds, the attacker knows that $b \cdot y = 0$ with high probability. If the authentication fails, $b \cdot y = 1$ with high probability. By repeating the process k times, the attacker could retrieve the secret value y. With both secrets (x, y) retrieved, the RFID system is under threat.

10.4.2.3 HB⁺⁺ Protocol

The HB⁺⁺ protocol further enhances the HB⁺ protocol to mitigate the threats from man-in-the-middle attacks [91]. As illustrated in Figure 10.10, the HB⁺⁺ protocol consists of a *preliminary stage* and a *round stage*. In this protocol, each tag has a unique secret Z. In the preliminary stage, the tag and the reader first exchange blinding values B and A. Based on the blinding values, both the tag and the reader derive secret values

$$(x, x', y, y') = h(Z, A, B),$$

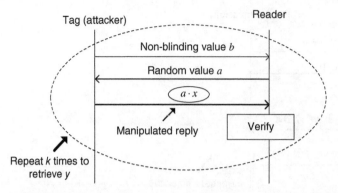

Figure 10.9 The man-in-the-middle attack on HB$^+$ protocol to retrieve y.

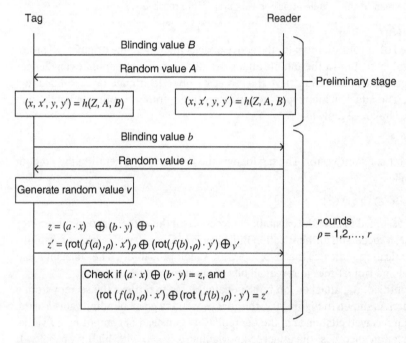

Figure 10.10 Illustration of HB^{++} protocol.

where $h(\cdot)$ can ba universal hash function. The secret keys are then used in the round stage for authentication. In each round, the tag and the reader first exchange blinding values b and a. The tag then generates a random value v and computes

$$z = (a \cdot x) \oplus (b \cdot y) \oplus v,$$
$$z' = (rot(f(a), \rho) \cdot x') \oplus (rot(f(b), \rho) \cdot y'),$$

where $rot(\cdot)$ is bit rotation, $\rho = 1, 2, \ldots, r$ is the round number, and $f(\cdot)$ is a lightweight function chosen to thwart the man-in-the-middle attack. Besides $f(\cdot)$, the hash function $h(\cdot)$ is also introduced for the same purpose. Please refer to the original paper for detailed discussion and example of the functions [91].

10.5 Summary

The RFID security is introduced in this chapter. With the pervasive application of RFID tags, it is imperative to provide security services such as confidentiality, integrity, privacy, and availability to RFID systems. However, due to the low cost and physical constraints of RFID tags, mitigation mechanisms to security risks are limited to lightweight cryptographic algorithms, anti-collision algorithms, and possible physical protection such as Faraday cage.

Part IV

Security for Wireless Wide Area Networks

11

GSM Security

Global System for Mobile (GSM) Communications is an international standard for one of the second generation (2G) digital cellular communication systems [93]. GSM is developed by the European Telecommunications Standards Institute (ETSI) [94]. Since early 1990s, GSM has been considered the most widely used cellular mobile phone system in the world. Services provided by GSM include voice communications, short messaging service, fax, data, etc. GSM provides security features for subscriber authentication, confidentiality on the radio links, and user anonymity. In this chapter, each of the security feature and selected security mechanisms will be demonstrated.

11.1 GSM System Architecture

A typical GSM system architecture is depicted in Figure 11.1. There are three parts in the GSM system architecture, *mobile station*, *base station subsystem*, and *network subsystem*.

11.1.1 Mobile Station

A Mobile Station (MS) is a subscriber's Mobile Equipment (ME, e.g. a cell phone) and a Subscriber Identity Module (SIM) card. Each mobile equipment has an International Mobile Equipment Identity (IMEI) as its unique global identity. For example, you may dial "*#06#" to check the IMEI of your cell phone. Most GSM equipment also has their IMEIs printed on the back. If a cell phone is lost, one can report the IMEI to the service provider for further actions (e.g. lock down the phone). A SIM card is a smart card that stores identities, keys, security algorithms, and other information (i.e. list of the last call attempts, list of preferred operators, and supplementary service data). The identities stored in a SIM card include an International Mobile Subscriber Identity (IMSI) and a Temporary Mobile Subscriber Identity (TMSI) if generated. Keys stored in a SIM card include a long term secret key and a session key. The long term secret key is used for the authentication process and the session key generation. The session key is used for data encryption over radio links. A SIM card is protected by a PIN code thus it is tamper-resistant.

Security in Wireless Communication Networks, First Edition. Yi Qian, Feng Ye, and Hsiao-Hwa Chen.
© 2022 John Wiley & Sons Ltd. Published 2022 by John Wiley & Sons Ltd.
Companion website: www.wiley.com/go/qian/sec51

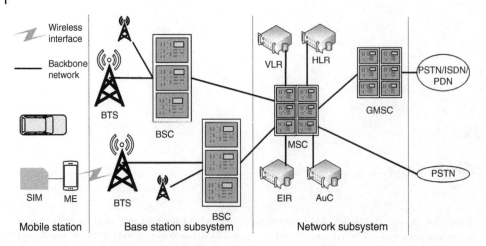

Figure 11.1 A typical GSM system architecture.

11.1.2 Base Station Subsystem

The region of a network operator is divided into *cells*. Each cell is served by one or more wireless transceivers for network services such as voice, text, and data. Multiple cells join together to cover a wide geographic area so that mobile users can roam freely with continuous mobile network services. The cells in the GSM system are supported by the base station subsystem. As shown in Figure 11.2, a base station subsystem comprises of several Base Transceiver Stations (BTSs) and Base Station Controllers (BSCs). A BSC is connected

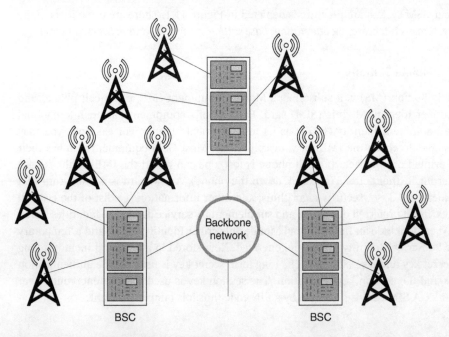

Figure 11.2 Connections of BTSs and BSCs.

to one or more BTSs through the backbone networks of the GSM operator. Each BTS forms a wireless cell that provides radio coverage for MSs. A BTS also handles the radio channels and the radio-link protocol in its cell. The MSs within the cell coverage connect the BTS through wireless links. A BSC manages the radio resources for one or more BTSs. A BSC also handles channel setup and handover of MSs. BSCs are connected to the Mobile Switching Center (MSC) through the backbone network.

11.1.3 Network Subsystem

Network subsystem provides the core functions of the GSM architecture, including telephone switching function, subscriber profiles function, and mobility management. There are several databases in a network subsystem for user and system management. They are MSC, Gateway MSC (GMSC), Home Location Register (HLR), Visitor Location Register (VLR), Equipment Identity Register (EIR), and Authentication Center (AuC).

The *MSC* is responsible for call signaling and processing, as well as coordinating the handover from one BTS to another as a MS roams in the network. The network traffic of the MSs in their respective cells is routed through the MSC. The MSC also manages the roles of inter-cellular transfer, mobile subscriber visitors, and interconnections with the Public Switched Telephone Network (PSTN). Each MSC is connected through a GMSC to the local PSTN or Integrated Service Digital network (ISDN) to provide the connectivity between the mobile and the fixed telephone users. An MSC may also connect to the Packet Data Networks (PDN) to provide mobiles with access to data services.

The *HLR* is a database in charge of the management of the operator's mobile subscribers. For a user that is registered with a network operator, the permanent data (i.e. the user's profile, subscriber's international identity number, and telephone number) and temporary data (i.e. the user's current location) are stored in the HLR. The main information stored in the HLR is utilized to route calls to the MSs that are managed by the HLR. When a call is placed to a user, the HLR is always queried first about the user's current location.

The *VLR* is responsible for a group of location areas, and stores the data of those users who are currently in its area of responsibility. This may include the user data that have been transmitted from the HLR to VLR for faster access. But the VLR may also assign and store local data such as a temporary identification. Concerning subscriber mobility, the VLR comes into play by verifying the characteristics of the subscriber and ensuring the transfer of location information. The VLR contains the current location of the MS and selected administrative information from the HLR. It is necessary for call control and provision of the services for each mobile currently located in the zones controlled by the VLR. A VLR is connected to one MSC and normally integrated into the MSC's hardware.

The *AuC* is responsible for the authentication process of a subscriber. It holds a copy of the 128-bit secret key and the IMSI that are stored in each subscriber's SIM card. The IMSI is retrieved to initiate an authentication process for a subscriber. The secret key is then applied to the rest of the authentication process.

The *EIR* registers equipment data in the GSM architecture. The GSM distinguishes explicitly between the user and the equipment, and deals with them separately. EIR is a database that contains a list of all valid MSs within the GSM network, where each MS is identified by its IMEI, which is a kind of serial number. Thus, the IMEI uniquely identifies an MS internationally. The IMEI is allocated by the equipment manufacturer and registered by the network operator who stores it in the EIR. The IMSI identifies uniquely each registered user and is stored in the SIM. An MS can only be operated if an SIM with a valid IMSI is inserted into equipment with a valid IMEI.

11.2 GSM Network Access Security Features

The GSM security features include *subscriber authentication*, *confidentiality* over radio links between MSs and BTSs, and *user anonymity*. Authentication and confidentiality are provided to ensure that only subscribers (paid users) can access the network and use the service. Cellular networks are provided by commercial service providers or network operators, thus no pay no access (except for emergency calls). Subscriber identity confidentiality is provided by GSM security schemes. Anonymity is also provided to protect users' identities by randomly choosing anonymous IDs for users and they always change. Moreover, GSM security provides key management and detection of compromised equipment. Key management in GSM security is independent of mobile equipment. It only involves the SIM card where the keys are stored. A SIM card is tamper-resistant and protected by a PIN code. As stated before, a SIM card is removable from the MS that contains all data specific to the end user which has to reside in the MS. SIM card can be used with different equipment. Data integrity is not one of the security goals. Because minor data loss or alteration would not affect a voice call in GSM. Readers should be aware that the GSM security specifications were designed by the GSM Consortium in secrecy. The details were only distributed on a need-to-know basis to hardware/software manufacturers and network operators. The enclosed authentication and enciphering algorithms were never exposed to the public. However, some of the algorithms have been revealed already.

11.2.1 GSM Entity Authentication

The entity authentication in GSM is to authenticate the subscriber and to defend against unauthorized user at the same time. The authentication involves an MS (specifically the SIM card that is attached to it), MSC/VLR, HLR, and AuC. In GSM, the MSC/VLR is responsible for call control and provides the services for each mobile located in the zones controlled by the MSC/VLR. The MSC/VLR tracks the user and allocates the calls so that the network knows where to route the call when a call is placed to a roaming user. When a call is placed from a mobile phone, the GSM network's VLR authenticates the individual subscriber's phone. The VLR immediately communicates with the original HLR of the visited MS, which in turn retrieves the subscriber's information from the AuC of the home network. The tasks of each component in the authentication process are listed in Table 11.1. The MSC/VLR has direct connection with the MS, the HLR in the home network of the MS is connected to the MSC/VLR over the network. The AuC performs the core functions to generate parameters for authentication.

The GSM authentication is a challenge–response scheme. There is a long term secret key (i.e. K_i) shared between the subscriber and the home network. In particular, K_i is a 128-bit key stored in both an SIM card and the AuC of the subscriber's home network. The key

Table 11.1 Components and their tasks in GSM authentication.

Component	Tasks
Home network AuC	Provides parameters for authentication and encryption functions of the original subscriber
Home network HLR	Provides MSC/VLR with authentication triplets (*RAND, SRES, K_c*); Handles MS location for the original subscriber
Visited network VLR	Stores generated triplets when a subscriber is not in the home network

K_i is not exposed to the visited network. When a subscriber roams to a visited network, the secret key should not be known by the service provider of the visited network. Therefore, in the authentication scheme, the subscriber identification IMSI is used for retrieving the subscriber's long term secret key (i.e. K_i) at the home network to verify the identity of the roaming MS. Because the authentication key K_i is the most essential component in the authentication process, it must not be transmitted over the air without protection to avoid possible interception. Therefore, K_i is placed only in the tamper-proved SIM card in the MS and in the AuC of the home MSC. As illustrated in Figure 11.3, the GSM authentication protocol follows the steps described as follows:

(1) The supplicant MS initiates the process by sending its IMSI to the visitor network.
(2) The VLR of the visited network forwards the IMSI to the home network of the supplicant according to the IMSI.
(3) In the home network, the HLR forwards the IMSI to the AuC.
(4) The AuC retrieves the corresponding K_i of the supplicant IMSI and generates authentication triplets (*RAND, SRES, K_c*).
(5) The home network forwards the triplets to the VLR of the current visited network of the MS.
(6) The VLR stores the *SRES* and the K_c and forwards the *RAND* to the MS.

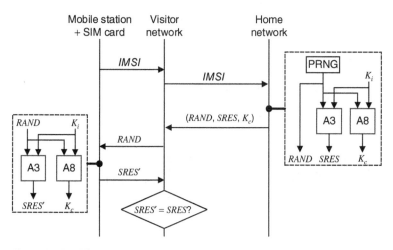

Figure 11.3 GSM authentication protocol.

(7) The MS uses the *RAND* to calculate the *SRES'* and the K_c.

(8) The MS sends the *SRES'* to the VLR of the visited network.

(9) The VLR compares the *SRES* generated by the AuC with the *SRES'* generated by the MS. If they are equal, the MS is authenticated, and admitted to the visited network.

The triplets $(RAND, SRES, K_c)$ are generated by three algorithms:

- *RAND* is generated by a pseudo-random number generator (PRNG);
- *SRES* is generated by algorithm $A3$;
- K_c is generated by algorithm $A8$.

When a MS has moved to a new MSC/VLR, the new MSC/VLR will normally establish the subscriber's identity by requesting the IMSI from the previous MSC/VLR. The previous VLR transfers, together with the IMSI, any unused triplets to the new VLR. This speeds up the authentication procedure, because the new VLR can only send a request for triplets to the subscriber's HLR/AuC after it has learned of the real identity of the subscriber, which is through this request to the previous VLR. It is worth noting that the keys (and triplets) are kept secret to one operator only. In other words, one operator does not have access to subscriber keys of another operator. An MS is not able to roam seamlessly between different operators.

11.2.2 GSM Confidentiality

Once an authentication process is completed, the session key K_c is shared between the MS and the visited network. The 64-bit K_c is used to encrypt the user data between the MS and the network over wireless links. As shown in Figure 11.4, the core function of GSM encryption is algorithm $A5$. $A5$ is a stream cipher that takes in two inputs: the session key K_c and the *FN*. *FN* is the 22-bit frame number. The output of $A5$ is a 114-bit keystream. The encryption is performed by XORing the user data (114-bit) and the keystream. At the receiver side, decryption is performed by generating the same keystream and XORing the ciphertext with it. The process is symmetric, thus either the MS or the BTS may initiate the transmission.

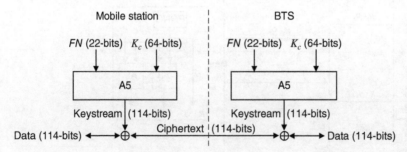

Figure 11.4 GSM encryption scheme.

11.2.3 GSM Anonymity

Anonymity is provided in GSM security by hiding the IMSI of each user. In GSM, each IMSI is associated with an user's identity. However, the IMSI should be used as seldom as possible to protect the user's identity from eavesdroppers. In GSM, anonymity is provided by assigning TMSI as a temporary identification number. The TMSI is used between an MS and the network. By using the TMSI, it prohibits tracing of the identity of a mobile subscriber by interception of the traffic on the radio link. TMSI is a 4-byte random number. The value cannot be all 1's because the SIM card uses this value to indicate that no valid TMSI is available. The network may assign a TMSI to a subscriber in several cases:

- *Case 1*: A TMSI is first assigned when a subscriber sends its IMSI to the AuC during authentication process. The TMSI is assigned for the duration that the subscriber is in the service area of the associated MSC. When a cell phone powers off, its TMSI is stored on SIM card to be reused.
- *Case 2*: The network operator can update the TMSI at any time during normal operation of a subscriber.
- *Case 3*: A new TMSI will be assigned every time a location is updated to a new MSC for the subscriber.
- *Case 4*: In a visited network, the VLR performs assignment, administration, and update of the TMSI.

11.2.4 Detection of Stolen/Compromised Equipment in GSM

When mobile equipment is lost or stolen, its IMEI can be used for detection of stolen or compromised equipment. The owner can report the loss of a cell phone to the network operator, who stores IMEI in the EIR. If required by law in the operator's jurisdiction, the operator can add the IMEI to the *blacklist* of stolen devices. The operator can optionally communicate the blacklist to a shared central EIR. Although a stolen device is still hard to trace, the network operators that share the central EIR can ban the stolen device from getting service. The IMEI number is not supposed to change easily. However, this is not the case. It is possible to clone an IMEI number to another device. Moreover, IMEI is an unauthenticated mobile identifier as opposed to IMSI. Therefore, a reported stolen IMEI may have been changed on the stolen device. As a result, it is unclear whether barring of IMEI has any positive effect for either the network operator or the subscriber.

11.3 GSM Security Algorithms

Algorithms $A3$, $A8$, and $A5$ are major algorithms applied in GSM security. $A3$ is the MS authentication algorithm. $A8$ is the session key (i.e. K_c) generation algorithm. And $A5$ is the algorithm for data encryption. Both $A3$ and $A8$ are implemented through function COMP128.

11.3.1 Algorithm *A*3

Algorithm *A*3 is applied to generate a signed response for MS authentication. As shown in Figure 11.5, *A*3 takes the inputs a 128-bit random number *RAND* and the 128-bit long term secret key K_i, then outputs a 32-bit response *SRES*.

11.3.2 Algorithm *A*8

Algorithm *A*8 is applied to generate the session key for each subscriber. As shown in Figure 11.6, *A*8 takes inputs a 128-bit random number *RAND* and the 128-bit long term secret key K_i, then outputs a 64-bit session key K_c. One session key is used until the MSC decides to authenticate the MS again, e.g. after a few days.

11.3.3 Algorithm COMP128

Both algorithms *A*3 and *A*8 are independent of hardware manufacturers and network operators. Instead, they are implemented on each SIM card and each AuC. In most GSM networks, *A*3 and *A*8 are derived from one algorithm COMP128 [95]. COMP128 is a keyed hash function, as shown in Figure 11.7. COMP128 takes inputs a 128-bit random number *RAND* and the 128-bit long term secret key K_i, then outputs an 128 bit string, which includes the 32-bit *SRES* and the 64-bit K_c.

COMP128 was a completely private set of algorithms originally. In 1997, a leaked document led to a successful reverse engineering in the algorithm. There are four versions of COMP128 algorithm as listed in Table 11.2. The full publication of COMP128-1 was

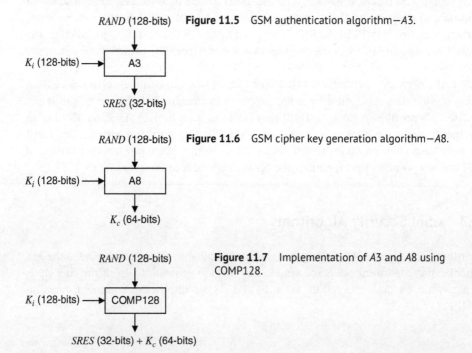

RAND (128-bits)   **Figure 11.5**   GSM authentication algorithm—*A*3.

K_i (128-bits) ⟶ A3

SRES (32-bits)

RAND (128-bits)   **Figure 11.6**   GSM cipher key generation algorithm—*A*8.

K_i (128-bits) ⟶ A8

K_c (64-bits)

RAND (128-bits)   **Figure 11.7**   Implementation of *A*3 and *A*8 using COMP128.

K_i (128-bits) ⟶ COMP128

SRES (32-bits) + K_c (64-bits)

Table 11.2 Four versions of COMP128 algorithm.

Version	Description
COMP128-1	The original algorithm with a 56-bit K_c generated
COMP128-2	A stronger algorithm, however, K_c is still 56 bits only
COMP128-3	The same algorithm as COMP128-2 with a 64-bit K_c generated
COMP128-4	Designed based on the 3GPP algorithm *Milenage* that uses AES [96]

Algorithm 11.1 COMP128-1

Input: *RAND* and K_i;
Output: *SREC* and K_c;
    Initialize $x[32]$; // 32 bytes
    $x[16 \text{ to } 31] \leftarrow RAND$;
    for $j := 0$ to 7 **do**
      $x[0 \text{ to } 15] \leftarrow K_i$;
      call Compression; // 5 rounds
      call FormBitsFromBytes
      if $j < 7$ **then**
        call Permute
      end if
    end for
    $SRES \leftarrow bit[0 \text{ to } 31]$;
    $K_c \leftarrow bit[74 \text{ to } 127] \| 0000000000_2$;
    return

---

published in 1998 [95]. The sketch of COMP128-1 algorithm is shown in Algorithm 11.1. COMP128-1 performs on a 32-byte array, i.e. $x[32]$. The initialization process sets $x[16 \text{ to } 31]$ to be *RAND*. The round function of COMP128-1 consists of three algorithms: *Compression*, *FormBitsFromBytes*, and *Permute*. The first 16 bytes of the array, i.e. $x[0 \text{ to } 15]$ is set to be the key K_i during each round function. COMP128-1 runs a total of eight times of its round function. The final output of COMP128-1 is a 128-bit string in which the first 32 bits for *SRES* and the last 54 bits with zero paddings to 64 bits for K_c. COMP128-1 has been discontinued in 2002 due to its weakness. The secret key of K_i can be recovered if an attacker has physical access to a SIM.

The core function of COMP128-1 is the *Compression* function (also known as the *Butterfly Structure*) [97]. The butterfly structure has five levels of compression, as shown in Figure 11.8. Details of the compression function are described in Algorithm 11.2. Two input bytes are used to determine the index of the look up table at each level. Two bytes of the array $x[32]$ are replaced by the two values from the corresponding look-up table in each compression. There are five look-up tables used for compression. The complete settings of the five tables are shown in Table 11.3. Each compression function applies the look-up

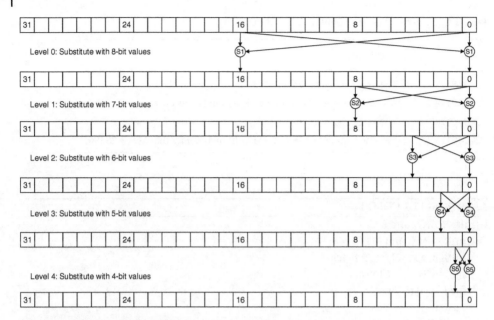

Figure 11.8 Illustration of COMP128-1 compression function.

Algorithm 11.2 COMP128-1 Compression

Input: $x[32]$;
Output: Compressed $x[32]$;

    for $j := 0$ to 4 **do**
      for $k := 0$ to $2^j - 1$ **do**
        for $l := 0$ to $2^{4-j} - 1$ **do**
          $m \leftarrow l + k \times 2^{5-j}$;
          $n \leftarrow m + 2^{4-j}$;
          $y \leftarrow (x[m] + 2 \times x[n]) \bmod 2^{9-j}$;
          $z \leftarrow (2 \times x[m] + x[n]) \bmod 2^{9-j}$;
          $x[m] \leftarrow T_j[y]$;
          $x[n] \leftarrow T_j[z]$;
        end for
      end for
    end for

---

table T_i at level i. Table T_i contains 2^{9-i} values, where each value is $(8 - i)$-bit long. For example, the look-up table at level 0 is T_0 which has 512 (i.e. 2^9) values, where each value is 8 bits. Similarly, the look-up table at level 4 is T_4, which has 32 values, where each value is 4 bits. All look-up tables are listed in Table. 11.4.

Function *FormBitsFromBytes* is to convert the result from bytes to bits, as detailed in Algorithm 11.3, where $\gg$ is the bitwise right shift, & is AND. Function *Permute* is to shuffle the array $x[32]$, as detailed in Algorithm 11.4, where $\ll$ is the bitwise left shift. The two

Table 11.3 Settings of look-up table T_j.

Level	Table name	Number of entries	Value
0	T_0	512	8-bit
1	T_1	256	7-bit
2	T_2	128	6-bit
3	T_3	64	5-bit
4	T_4	32	4-bit

Table 11.4 COMP128-1 look-up tables.

T_0:
```
66 B1 BA A2 02 9C 70 4B 37 19 08 0C FB C1 F6 BC
6D D5 97 35 2A 4F BF 73 E9 F2 A4 DF D1 94 6C A1
FC 25 F4 2F 40 D3 06 ED B9 A0 8B 71 4C 8A 3B 46
43 1A 0D 9D 3F B3 DD 1E D6 24 A6 45 98 7C CF 74
F7 C2 29 54 47 01 31 0E 5F 23 A9 15 60 4E D7 E1
B6 F3 1C 5C C9 76 04 4A F8 80 11 0B 92 84 F5 30
95 5A 78 27 57 E6 6A E8 AF 13 7E BE CA 8D 89 B0
FA 1B 65 28 DB E3 3A 14 33 B2 62 D8 8C 16 20 79
3D 67 CB 48 1D 6E 55 D4 B4 CC 96 B7 0F 42 AC C4
38 C5 9E 00 64 2D 99 07 90 DE A3 A7 3C 87 D2 E7
AE A5 26 F9 E0 22 DC E5 D9 D0 F1 44 CE BD 7D FF
EF 36 A8 59 7B 7A 49 91 75 EA 8F 63 81 C8 C0 52
68 AA 88 EB 5D 51 CD AD EC 5E 69 34 2E E4 C6 05
39 FE 61 9B 8E 85 C7 AB BB 32 41 B5 7F 6B 93 E2
B8 DA 83 21 4D 56 1F 2C 58 3E EE 12 18 2B 9A 17
50 9F 86 6F 09 72 03 5B 10 82 53 0A C3 F0 FD 77
B1 66 A2 BA 9C 02 4B 70 19 37 0C 08 C1 FB BC F6
D5 6D 35 97 4F 2A 73 BF F2 E9 DF A4 94 D1 A1 6C
25 FC 2F F4 D3 40 ED 06 A0 B9 71 8B 8A 4C 46 3B
1A 43 9D 0D B3 3F 1E DD 24 D6 45 A6 7C 98 74 CF
C2 F7 54 29 01 47 0E 31 23 5F 15 A9 4E 60 E1 D7
F3 B6 5C 1C 76 C9 4A 04 80 F8 0B 11 84 92 30 F5
5A 95 27 78 E6 57 E8 6A 13 AF BE 7E 8D CA B0 89
1B FA 28 65 E3 DB 14 3A B2 33 D8 62 16 8C 79 20
67 3D 48 CB 6E 1D D4 55 CC B4 B7 96 42 0F C4 AC
C5 38 00 9E 2D 64 07 99 DE 90 A7 A3 87 3C E7 D2
A5 AE F9 26 22 E0 E5 DC D0 D9 44 F1 BD CE FF 7D
36 EF 59 A8 7A 7B 91 49 EA 75 63 8F C8 81 52 C0
AA 68 EB 88 51 5D AD CD 5E EC 34 69 E4 2E 05 C6
FE 39 9B 61 85 8E AB C7 32 BB B5 41 6B 7F E2 93
DA B8 21 83 56 4D 2C 1F 3E 58 12 EE 2B 18 17 9A
9F 50 6F 86 72 09 5B 03 82 10 0A 53 F0 C3 77 FD
```

T_1:
```
13 0B 50 72 2B 01 45 5E 27 12 7F 75 61 03 55 2B
1B 7C 46 53 2F 47 3F 0A 2F 59 4F 04 0E 3B 0B 05
23 6B 67 44 15 56 24 5B 55 7E 20 32 6D 5E 78 06
35 4F 1C 2D 63 5F 29 22 58 44 5D 37 6E 7D 69 14
5A 50 4C 60 17 3C 59 40 79 38 0E 4A 65 08 13 4E
4C 42 68 2E 6F 32 20 03 27 00 3A 19 5C 16 12 33
39 41 77 74 16 6D 07 56 3B 5D 3E 6E 4E 63 4D 43
0C 71 57 62 66 05 58 21 26 38 17 08 4B 2D 0D 4B
5F 3F 1C 31 7B 78 14 70 2C 1E 0F 62 6A 02 67 1D
52 6B 2A 7C 18 1E 29 10 6C 64 75 28 49 28 07 72
52 73 24 70 0C 66 64 54 5C 30 48 61 09 36 37 4A
71 7B 11 1A 35 3A 04 09 45 7A 15 76 2A 3C 1B 49
76 7D 22 0F 41 73 54 40 3E 51 46 01 18 6F 79 53
68 51 31 7F 30 69 1F 0A 06 5B 57 25 10 36 74 7E
1F 26 0D 00 48 6A 4D 3D 1A 43 2E 1D 60 25 3D 34
65 11 2C 6C 47 34 42 39 21 33 19 5A 02 77 7A 23
```

T_2:
```
34 32 2C 06 15 31 29 3B 27 33 19 20 33 2F 34 2B
25 04 28 22 3D 0C 1C 04 3A 17 08 0F 0C 16 09 12
37 0A 21 23 32 01 2B 03 39 0D 3E 0E 07 2A 2C 3B
3E 39 1B 06 08 1F 1A 36 29 16 2D 14 27 03 10 38
30 02 15 1C 24 2A 3C 21 22 12 00 0B 18 0A 11 3D
1D 0E 2D 1A 37 2E 0B 11 36 2E 09 18 1E 3C 20 00
14 26 02 1E 3A 23 01 10 38 28 17 30 0D 13 13 1B
1F 35 2F 26 3F 0F 31 05 25 35 19 24 3F 1D 05 07
```

T_3:
```
01 05 1D 06 19 01 12 17 11 13 00 09 18 19 06 1F
1C 14 18 1E 04 1B 03 0D 0F 10 0E 12 04 03 08 09
14 00 0C 1A 15 08 1C 02 1D 02 0F 07 0B 16 0E 0A
11 15 0C 1E 1A 1B 10 1F 0B 07 0D 17 0A 05 16 13
```

T_4:
```
0F 0C 0A 04 01 0E 0B 07 05 00 0E 07 01 02 0D 08
0A 03 04 09 06 00 03 02 05 06 08 09 0B 0D 0F 0C
```

functions generate a new 128-bit array $bit[128]$ that is in the form of the final output. At last, COMP128-1 outputs:

$SRES$: $bit[0 \text{ to } 31]$;

K_c : $bit[74 \text{ to } 127]\|0000000000_2$.

Algorithm 11.3 COMP128-1 FormBitsFromBytes

Input: $x[32]$;

Output: $bit[128]$; // 128 bits

> **for** $j := 0$ to 31 **do**
>> **for** $k := 0$ to 3 **do**
>>> $bit[4j + k] \leftarrow [(x[j] \gg (3 - k)) \,\&\, 1]$;
>> **end for**
> **end for**

Algorithm 11.4 COMP128-1 Permute

Input: *bit*[128], *x*[32];
Output: Permuted *x*[32];
    for $j := 0$ to 15 **do**
      for $k := 0$ to 7 **do**
        $x[j + 16] \leftarrow [bit[(17(8j + k)) \bmod 128] \ll (7 - k)]$;
      end for
    end for

11.3.4 Algorithm A5

The wireless-link communication confidentiality is provided by algorithm A5 in the GSM cellular telephone standard. A5 has seven versions, all of which were initially kept secret. Because of leaks and reverse engineering, some of the A5 versions became public knowledge. Since then, a number of serious weaknesses of A5 have been identified. Versions 1 and 2 will be discussed as they are the most deployed A5 algorithms.

11.3.4.1 A5/1

The A5/1 algorithm (interchangeable with A5/1 hereafter for simplicity) was developed in 1987 to be used in Europe and the United States. The general design was leaked in 1994 and the algorithm was entirely reverse engineered by Marc Briceno in 1999. By 2011, confidentiality of 4 billion GSM customers was protected by A5/1. A5/1 is a stream cipher that takes in two inputs, frame number *FN* and session key K_c. The output of A5/1 is a 114-bit keystream. The encryption is by XORing the keystream with the message data.

The core of A5/1 consists of three Linear Feedback Shift Registers (LFSRs). The LFSRs are designed to produce pseudo-random bit sequence. Figure 11.9 illustrates a typical LFSR. As it shows, an LFSR consists of two parts:

- *Shift register*: a bit sequence feedback function;
- *Tap sequence*: bits that are input to the feedback function.

Shift register stores the current state of an LFSR. An *n*-bit LFSR has a total number of $2^n - 1$ internal states. The length of the output sequence before an LFSR repeating itself is defined as an LFSR *period*. The maximal period of an *n*-bit LFSR is also $2^n - 1$. The actual period is determined by the tap sequence. The polynomial formed by a tap sequence plus 1 must be a primitive polynomial (modulo 2).

The three LSFRs of A5/1 have different lengths. The combined length of the three LSFRs is 64 bits. As listed in Table 11.5, the three LSFRs are 19, 22, and 23 bits long with sparse feedback polynomials. An overview of the LSFR construction used in A5/1 is shown in Figure 11.10. All three registers are clocked, based on their middle bits (i.e. $R_1[8]$, $R_2[10]$, and $R_3[10]$). A register is clocked means that the tapped bits are XORed for a bit that

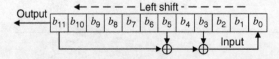

Figure 11.9 An example LSFR with feedback polynomial of $x^{11} + x^5 + x^3 + 1$.

Table 11.5 Specifications of the three LSFRs in A5/1.

LFSR	Length in bits	Characteristic polynomial	Clocking bit	Tapped bits
R_1	19	$x^{18} + x^{17} + x^{16} + x^{13} + 1$	8	13, 16, 17, 18
R_2	22	$x^{21} + x^{20} + 1$	10	20, 21
R_3	23	$x^{22} + x^{21} + x^{20} + x^7 + 1$	10	7, 20, 21, 22

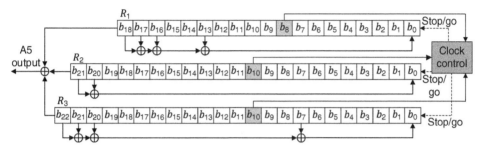

Figure 11.10 A5/1 LSFR construction.

is the input to the feedback. For example, if R_1 is clocked, then tapped bits are XORed as $R_1[18] \oplus R_1[17] \oplus R_1[16] \oplus R_1[13]$, the result value is input to the position of $R_1[0]$ and R_1 left shifts by 1 bit. A register is clocked if its middle bit agrees with the majority value of the three middle bits. The output of the three registers are XORed together, i.e. $R_1[18] \oplus R_2[21] \oplus R_3[22]$. The XORed result represents one bit for the keystream.

The clocking mechanism guarantees that at least two registers (i.e. the majority) are clocked after each round. To better understand the clocking mechanism in A5/1, three examples are shown in Figure 11.11. In the first example, the middle bits of the three registers are 1, 1, and 1, thus all three registers are clocked. In the second example, the middle bits are 0, 0, and 1, and then the first and the second registers are clocked. In the third example, the middle bits are 0, 0, and 0, then all three registers are clocked.

Example: producing three bits from the LSFRs. The current status of each LSFR and the detailed operating process are shown in Figure 11.12. The first bit output of the keystream is computed as:

$$R_1[18] \oplus R_2[21] \oplus R_3[22] = 1 \oplus 1 \oplus 1 = 1.$$

The middle bits are 1, 0, 1. By the majority rule, the first and the third registers are clocked. For the first register, tap sequence determines the bit that is input to the feedback

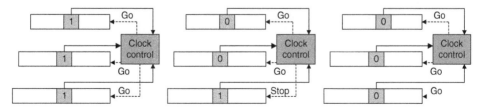

Figure 11.11 Examples of the clocking mechanism in A5/1.

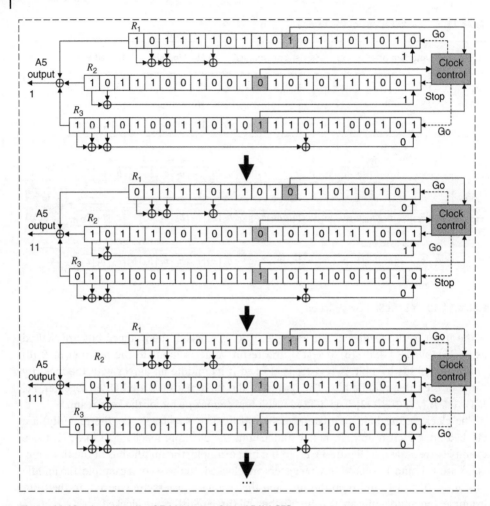

Figure 11.12 An example of 3-bit output from A5/1 LSFR.

as $R_1[18] \oplus R_1[17] \oplus R_1[16] \oplus R_1[13] = 1 \oplus 0 \oplus 1 \oplus 1 = 1$. The second register is not clocked. For the third register, the input to the feedback is $R_3[22] \oplus R_3[21] \oplus R_3[20] \oplus R_3[7] = 1 \oplus 0 \oplus 1 \oplus 0 = 0$. Thus, the second bit of the keystream is:

$$R_1[18] \oplus R_2[21] \oplus R_3[22] = 0 \oplus 1 \oplus 0 = 1.$$

The middle bits are 0, 0, 1, thus the first and the second registers are clocked. The inputs to the feedback are 0 and 1, respectively. The third bit of the keystream is:

$$R_1[18] \oplus R_2[21] \oplus R_3[22] = 1 \oplus 0 \oplus 0 = 1.$$

The three output bits are 111_2. This process continues to generate more outputs for keystreams.

Detailed A5/1 algorithm is shown in Algorithm 11.5. The three registers are initialized with the session key K_c and the frame number *FN*. The 64-bit K_c is first loaded into the register bit by bit. The Least Significant Bit (LSB) of the key is XORed into each of the registers. The registers are then all clocked. Note that the majority clocking mechanisms is disabled in this round. All 64 bits of the key are then loaded into the registers in the same way. The 22-bit frame number is also loaded into the register in the same way except that the majority clocking mechanisms applies from now on. After the registers have been initialized with the K_c and the current FN, they start the operation and clock one hundred times. The generated keystream bits are discarded. This is done in order to mix the frame number and keying material together. At this step, a total of 228 bits have been generated.

Algorithm 11.5 A5/1

Input: FN and K_c;
Output: Keystream;
    Set all LFSRs to 0 (i.e. $R_1 = R_2 = R_3 = 0$);
    for $i := 0$ to 63 **do**
        $R_1[0] \leftarrow R_1[0] \oplus K_c[i]$;
        $R_2[0] \leftarrow R_2[0] \oplus K_c[i]$;
        $R_3[0] \leftarrow R_3[0] \oplus K_c[i]$;
        Clock all three registers (i.e. for $j > 0$ $R_i[j] \leftarrow R_i[j-1]$, and $R_i[0]$ is set to the result of the primitive polynomial on the previous value of R_i);
    end for
    Set all LFSRs to 0 (i.e. $R_1 = R_2 = R_3 = 0$);
    for $i := 0$ to 21 **do**
        $R_1[0] \leftarrow R_1[0] \oplus FN[i]$;
        $R_2[0] \leftarrow R_2[0] \oplus FN[i]$;
        $R_3[0] \leftarrow R_3[0] \oplus FN[i]$;
        Clock all three registers
    end for
    for $i := 0$ to 99 **do**
        clock the cipher by its regular majority clocking mechanism, and discard the output;
    end for

The first 114 bits are used as the keystream to encrypt the message data from MS to BTS (i.e. uplink transmission), and the rest 114 bits are used as the keystream to encrypt the message data from BTS to MS (i.e. downlink transmission). For the next message data, the A5/1 algorithm will be initialized again with the same K_c and the next FN.

11.3.4.2 Algorithm A5/2

A5/2 algorithm was used for export instead of the relatively stronger (but still weak) A5/1. This algorithm is simpler than A5/1 and was developed by ETSI for use in Eastern European states that had restrictions to certain Western technologies. Similar to A5/1, A5/2 is also based on a combination of LFSRs. There are four LFSRs (i.e. R_1, R_2, R_3, and R_4) with irregular clocking and a non-linear combiner in A5/2. The length, characteristic polynomial, clocking bit, and tapped bits of each register are listed in Table 11.6.

The internal structure of A5/2 is shown in Figure 11.13. The clocking mechanism of A5/2 is different from A5/1. In A5/2, register R_4 controls the clocking of R_1, R_2, and R_3. The

Table 11.6 Specifications of the four LSFRs in A5/2.

LFSR	Length in bits	Characteristic polynomial	Clocking bit	Tapped bits
R_1	19	$x^{19} + x^5 + x^2 + x + 1$	8	1, 2, 5, 19
R_2	22	$x^{22} + x + 1$	10	1, 22
R_3	23	$x^{23} + x^{15} + x^2 + x + 1$	10	1, 2, 15, 23
R_4	17	$x^{17} + x^5 + 1$	10	5, 17

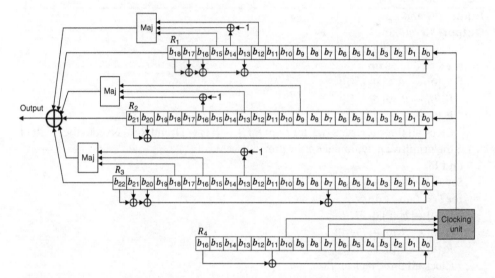

Figure 11.13 A5/2 LSFR construction.

clocking unit performs a majority function on the bits $R_4[3]$, $R_4[7]$, and $R_4[10]$. If $R_4[10]$ agrees with the majority, then R_1 is clocked. If $R_4[3]$ agrees with the majority, then R_2 is clocked. If $R_4[7]$ agrees with the majority, then R_3 is clocked. After R_1, R_2, and R_3 finish the clocking process, R_4 is clocked.

Once the clocking mechanism is performed, the output bit is computed as follows. In each register, the output is determined by the majority three inputs and the leftmost bit. For R_1, the majority function Maj($\cdot$) takes in $R_1[15]$, $R_1[14] \oplus 1$, and $R_1[12]$. The majority function is quadratic in its input, such that

$$\text{Maj}(a, b, c) = (a \cap b) \oplus (b \cap c) \oplus (c \cap a).$$

where $\cap$ is the AND operation. The output of the majority function is XORed with $R_1[18]$. For R_2, the majority function Maj ($\cdot$) takes in $R_2[16] \oplus 1$, $R_2[13]$, and $R_2[9]$. The output of the majority function is XORed with $R_2[21]$. For R_3, the majority function Maj($\cdot$) takes in $R_3[18]$, $R_3[16]$, and $R_3[13] \oplus 1$. The output of the majority function is XORed with $R_3[22]$. The final output for the keystream is the result of XORing all three outputs from R_1, R_2, and R_3.

The internal state initialization process of A5/2 is shown in Algorithm 11.6. The four registers are initialized with the K_c and FN. The LSB of K_c is XORed into each of the registers.

Algorithm 11.6 A5/2 internal state initialization

Input: *FN* and K_c;
Output: Keystream;

    Set all LFSRs to 0 (i.e. $R_1 = R_2 = R_3 = R_4 = 0$);

    for $i := 0$ to 63 **do**

      $R_1[0] \leftarrow R_1[0] \oplus K_c[i]$;
      $R_2[0] \leftarrow R_2[0] \oplus K_c[i]$;
      $R_3[0] \leftarrow R_3[0] \oplus K_c[i]$;
      $R_4[0] \leftarrow R_4[0] \oplus K_c[i]$;

    end for

    for $i := 0$ to 21 **do**

      $R_1[0] \leftarrow R_1[0] \oplus FN[i]$;
      $R_2[0] \leftarrow R_2[0] \oplus FN[i]$;
      $R_3[0] \leftarrow R_3[0] \oplus FN[i]$;
      $R_4[0] \leftarrow R_4[0] \oplus FN[i]$;

    end for

The *FN* is loaded into the registers in the same way. Once the initialization is finished, the keystream is generated as follows:

(1) Force the bits $R_1[15], R_2[16], R_3[18]$, and $R_4[10]$ to be 1;
(2) Run A5/2 for 99 clocks and ignore the output;
(3) Run A5/2 for 228 clocks and use the output as key-stream.

Note that A5/2 discards 99 bits of output while A5/1 discards 100 bits of output. The 228-bit output is separated in two halves. The first 114 bits are used as the keystream to encrypt the message data from the BTS to the MS (i.e. downlink transmission). The rest 114 bits are used as the keystream to encrypt the message data from the MS to the BTS (i.e. uplink transmission).

11.4 Attacks Against GSM Security

Despites its popularity, the GSM system has quite a few threats since the security algorithms are vulnerable to attacks.

11.4.1 Attacks Against GSM Authenticity

The authenticity of the GSM system is based on the secret key K_i. In fact, if K_i is compromised, the whole account of a user is compromised because both confidentiality and anonymity are provided based on K_i. If an attacker is able to retrieve the K_i, then the attacker can eavesdrop the subscriber's calls. The attacker may also place calls that are billed to the original subscriber's account. If the legitimate user is always registered to the network operator, then the attacker may not be able to steal the service because the GSM system has trip wires for two identical subscribers. The mechanism works as follows: if two MSs with the same ID (i.e. the same IMSI/TMSI) are powered on at the same time, the network operator notices this and makes a location query for the MSs. Once the operator confirms that the "same" MS is in two different locations at the same time, it will close the account to prevent the attacker and the legitimate subscriber from placing calls. However, this does not protect the subscriber from being eavesdropped by the attacker.

The secret key K_i is implemented in an SIM card. With physical access to an SIM card, K_i may be retrieved due to the flaw in COMP128-1 algorithm that was discovered by the Smartcard Developer Association and the ISAAC security research group [98]. The attack is based on a chosen-challenge attack. The COMP128-1 was broken in such a way that it reveals the information about the K_i when some *RANDs* are given to the algorithm $A8$. The attack was performed with physical access to the SIM card through a smartcard reader connected to a computer; however, it is applicable to launch the same attack over-the-air as well. Alternatively, it is also possible to retrieve the K_i from the AuC by lunching the same attack. Because the AuC answers each authentication request and return valid triplets. The attack procedure is similar to the procedure used in the MS to access the SIM card. Compared with the MS, the AuC computes much faster. The security of the AuC plays a critical role in whether this attack is possible or not.

11.4.1.1 Attacks Against GSM Confidentiality

The session key K_c and the encryption algorithm provide GSM confidentiality. If K_c is compromised, all traffic encrypted with this key is compromised to attackers. Fortunately, a real-time brute-force attack against the GSM security system is still infeasible regardless of the small key space. The time complexity of the attack is at least $O(2^{54})$, which makes it infeasible in eavesdropping on GSM calls in real time. However, it might be possible to have other ways of attacking the system by recording the frames between the MS and the BTS and launch the attack afterwards.

One possible attack is to retrieve K_c from its generating function. The COMP128-1 was broken by Marc Briceno et al. on 13 April 1998 [95]. The session key K_c was successfully retrieved because of the weakness in the compression function of COMP128-1. With the K_c, the attacker can clone an SIM card and impersonate the legitimate subscriber until the K_c expires. As a result, new versions (COMP128-2, COMP128-3, and COMP128-4) were developed. COMP128-1 was discontinued in 2002. Another possible attack to retrieve K_c is from algorithm A5. Despite of multiple versions, the original A5 (i.e. A5/1) is considerably stronger than most of the other A5 algorithms. Ironically, the other versions of A5 were developed because A5/1 was too strong to export GSM service to some countries. However, even A5/1 was not strong enough to protect confidentiality of a subscriber. Due to the small size of K_c, even brute-force attack only requires a time complexity of 2^{54} at most. A divide-and-conquer attack can further reduce the complexity from 2^{54} to 2^{45}.

The divide-and-conquer attack is to compute the third register by guessing the content of the first and the second registers. If the clocking of the first and the second registers is independent on the third register, then the time complexity would be $O(2^{40})$. However, due to the clocking bit of the third register, the attacker must guess half of the bits. This would increase the time complexity from $O(2^{40})$ to $O(2^{45})$. The attacker needs to determine the initial states of the registers from a known keystream sequence. To perform the attack, a total of 64 successive keystream bits needs to be retrieved if the attacker knows some ciphertext and the corresponding plaintext. This depends highly on the format of data frames. Each data frame contains a lot of constant information, e.g. frame headers. Although the 64 bits may not be known always, 32–48 bits are usually known.

A5/2 algorithm was extremely weak and could be broken by low end equipment in real time in the same month it was published in 1999 [99]. Since 1 July 2006, the GSM Association (GSMA) mandated that GSM mobile phones will not support the A5/2 algorithm. In July 2007, the 3GPP approved a change request to prohibit the implementation of A5/2 algorithm in any new cell phones. A5/1 is mandatory by the 3GPP association. If a GSM network does not support A5/1 or any other A5 algorithms implemented by the phone, an unencrypted connection can still be established.

11.4.2 Other Attacks against GSM Security

Besides attacks on the security algorithms, possible attacks may also be launched on signaling network, such as exploiting the unencrypted messages, false base station attacks, and denial of service attacks. As stated earlier, GSM security is only provided between the MS and the BTS. In the network subsystem, the data traffic is transmitted in plaintext without being encrypted. If an attacker gains direct access to the operator's signaling network, it would be able to eavesdrop everything that is being transmitted, including the triplets for authentication and the actual phone calls. It may not be easy to access the signaling network that is part of the wired backbone network. However, part of the signaling network may be connected to the BSC through a microwave or even a satellite link. The wireless links would be relatively easy to access with certain equipment. Another possible attack against GSM is *false base station attack*. This attack is possible in GSM system due to the one-way authentication. During the authentication process, only MS gets authenticated, leaving BTS a trusted party. IMSI catcher is a well-known device that exploits this security hole. In particular, the IMSI catcher launches man-in-the-middle attack: the attacker masquerades as a BTS to the MSs, but acts as an MS to the GSM network. The IMSI catcher thus can induce the MS to use no encryption by choosing A5/0 for no confidentiality, or to use A5/2 that can be easily broken. *Denial of Service (DoS) attack* is also possible to GSM network. Instead of getting "free" service or eavesdropping, an attacker that launches DoS attack simply wants to block the subscriber from getting network service. A typical DoS attack is by jamming the signal to prevent the MS from communications. For example, an attacker can launch a DoS attack by calling the subscriber again and again.

11.5 Possible GSM Security Improvements

Improvements can be made in several ways to enhance GSM security.

11.5.1 Improvement over Authenticity and Anonymity

GSM authenticity and anonymity depend on the secret key K_i. Existing GSM security is weak that a subscriber's K_i can be retrieved by attackers. Leakage of K_i causes the possibility of cloning an SIM card. To prevent an SIM card being cloned, another cryptographically secure algorithm is needed to replace A3 (or COMP128). It will require the network operators issuing new SIM card to all subscribers and update corresponding HLR software. This improvement is relatively easy for network operators since no third-party hardware or software manufacturers for the GSM Consortium will be involved in the process. However, the large population of GSM subscribers can cause difficulty in the upgrade.

11.5.2   Improvement over Confidentiality

While a real-time break of session key K_c is not possible, off-line approach is quick enough to jeopardize the subscribers' confidentiality. To enhance the confidentiality of GSM security, a larger key size and a better security algorithm (e.g. a new A5 implementation with stronger encryption) can be deployed so that a brute-force attack is infeasible in any case. It could be complicated to implement this improvement. First, the cooperation of the GSM Consortium will be needed. Second, third-party hardware and software manufacturers will be involved to upgrade their products that will comprise with the new encryption algorithm.

11.5.3   Improvement of the Signaling Network

Besides securing the wireless links between the MSs and the BTSs, it is critical to secure the signaling network in the network subsystem. To do so, it is necessary to encrypt the traffic on the backbone network. This would prevent the attacker from wiretapping on the signaling network. To implement the improvement, both hardware and software need upgrade from their manufacturers. However, the GSM Consortium may not be involved in the process.

11.6   Summary

Security of the GSM system is introduced in this chapter. GSM security provides authentication, confidentiality, and anonymity to protect the network operator and the subscribers. However, GSM security is vulnerable due to its short keys, weak algorithms, and unencrypted backbone network. Improvement should be made to enhance the GSM security. Unfortunately, very few improvements have been made even at the final stage of the GSM system before phasing out the GSM system recently. Security enhancement can be seen in the new generation cellular systems.

12

UMTS Security

In this chapter, security of the Universal Mobile Telecommunications System (UMTS) is introduced. UMTS is a third generation (3G) mobile cellular network system developed and maintained by the third Generation Partnership Project (3GPP). UMTS is a component of the International Telecommunications Union IMT-2000 standard set [100]. The architecture of UMTS is extended from the GSM standard. UMTS has made a great improvement to the GSM system in terms of security [101].

12.1 UMTS System Architecture

The UMTS system architecture is extended from the GSM system, as illustrated in Figure 12.1. It involves user equipment (UE), radio access network (RAN), and core network (CN). The CN of UMTS is extended from the GSM standard, which consists of visitor network and home network. The UE has a radio connection to the RAN. The RAN is connected to the CN through the operator's backbone network. According to the 3GPP Release 1999 [102], the components of the UMTS architecture have been named differently from that of the GSM system. Figure 12.2 depicts the most important elements in the 3GPP Release 1999. As it shows, the 3GPP UMTS architecture consists of three main parts: the UE, the UMTS Terrestrial Radio Access Network (UTRAN), and the CN.

12.1.1 User Equipment

A UE is a communication terminal that has a radio connection to the RAN. Each UE consists of two parts, the mobile equipment (ME) and the Universal Subscriber Identity Module (USIM). The ME is typically the cell phone that provides communication functionality with the network. The USIM is a smart card that is placed inside the ME. It contains all the operator-dependent data, including the identity (IMSI/TMSI), the long term shared key K_i, and other information of the subscriber. These parameters are stored in a tamper-resistant environment (i.e. in the Universal Integrated Circuit Card-UICC). The USIM card has the same appearance as the SIM card in the GSM system. However, the USIM card brings mutual authentication, which is not supported in the GSM system.

Security in Wireless Communication Networks, First Edition. Yi Qian, Feng Ye, and Hsiao-Hwa Chen.
© 2022 John Wiley & Sons Ltd. Published 2022 by John Wiley & Sons Ltd.
Companion website: www.wiley.com/go/qian/sec51

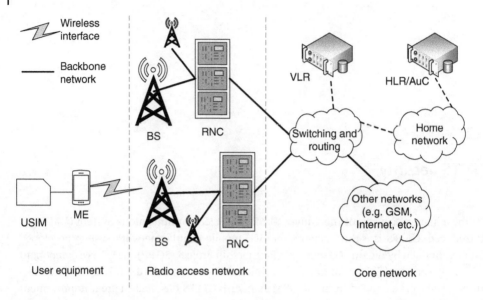

Figure 12.1 UMTS system architecture.

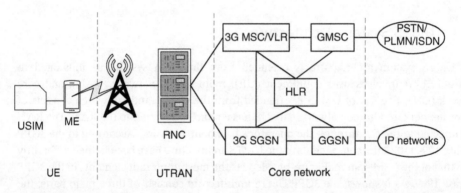

Figure 12.2 3GPP architecture—UMTS.

12.1.2 UTRAN

The RAN in the 3GPP system is the UTRAN, which is based on the wideband code division multiple access (W-CDMA) technology. UTRAN contains two types of elements: the base station and the radio network controller (RNC). The base station is the termination point of the radio interface on the network side. It is also named as NodeB in the 3GPP architecture. The base station is connected to the RNC. The RNC is the controlling unit of UTRAN. Different from the GSM system, a UE may be connected to the network via several RNCs simultaneously, thus better quality for the connection can be guaranteed. The RNC that maintains the connection to the CN for a UE is called the Serving RNC (SRNC), the other RNCs that connect to the UE are called the Drifting RNCs (DRNC).

12.1.3 Core Network

The CN consists of several elements, including the Mobile Switching Controller (MSC), the Visiting Location Register (VLR), The Gateway MSC (GMSC), the Home Location Register (HLR), the 3G Serving GPRS Support Node (SGSN), and the Gateway GPRS Support Node (GGSN). GPRS is General Packet Radio Service that provides packet oriented mobile data service in between GSM and UMTS systems. Two domains are served in the CN, including the *circuit switched* domain and the *packet switched* domain.

- In the circuit switched domain, the MSC and the VLR are part of the serving network. The two elements are typically integrated together. The VLR is a database that contains the users currently in the location area controlled by the MSC. The GMSC is responsible for connecting the mobile network to the Public Switched Telephone Network (PSTN).
- In the packet switched domain, the SGSN serves the function of MSC/VLR for the packet oriented mobile data service. The GGSN is responsible for connecting the mobile network to the IP network (e.g. the Internet).

The HLR and the Authentication Center (AuC) serve both domains in the CN, and they are part of the home network. The two elements are typically integrated together. The AuC has a database that holds the permanent security data related to subscribers (i.e. the identities and the secret keys) that can be used for security features in the serving network and in the access network.

12.2 UMTS Security Features

3GPP TS 33.102 is a 3GPP technical specification that defines the security architecture of the UMTS [103]. The security features defined in this technical specification are shown in Figure 12.3. According to the specification, there are five groups of security features in the UMTS, as stated in the following:

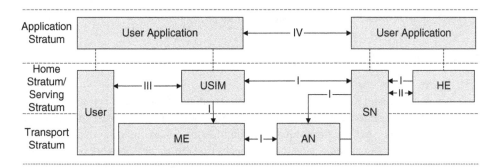

Figure 12.3 Security features defined by 3GPP TS 33.102.

- *Network access security (I)*: the set of security features that provide users with secure access to 3G services, in particular protect against attacks on the radio access link;
- *Network domain security (II)*: the set of security features that enable nodes in the provider domain to securely exchange signaling data, and protect against attacks on the backbone network;
- *User domain security (III)*: the set of security features that secure access to mobile stations;
- *Application domain security (IV)*: the set of security features that enable applications in the user and in the provider domain to securely exchange messages;
- *Visibility and reconfigurability of security (V)*: the set of features that enables the user to inform himself whether a security feature is in operation or not and whether the use and provision of services should depend on the security feature.

12.3 UMTS Network Access Security

The radio access technology of the UMTS is W-CDMA, which is changed from Time Division Multiple Access (TDMA). Nonetheless, requirements for access security do not change in the UMTS. Entity authentication is one of the security features that are provided in the UMTS. The identity of a subscriber must be authenticated before getting service. The authentication protects end users from fraudulent calls and packet data services that are made by others. The confidentiality is provided in the UMTS RAN to protect the transmission of user data. Confidentiality in the UMTS protects end users from data leakage to eavesdroppers. Moreover, confidentiality also protects other privacy of a user, for example, the location privacy. Although it may not be critical if a user is traceable, however, persistent tracking is certainly irritating thus should be avoided. The integrity of radio network signaling is also provided in the UMTS. Manipulation of all control messages can be avoided, thus enabling the network operator to control the network functions effectively. Besides, integrity also protects all data messages from being manipulated. Although some data loss does not affect a phone call too much, providing integrity certainly enhances the packet data service as well as the control efficiency from the network operator. In the rest of this section, the security features and ciphers in the first release of the 3GPP system specifications (i.e. Release 1999) are introduced. For the ease of illustration, the parameters used in the UMTS access control security are listed in Table 12.1.

12.3.1 Authentication and Key Agreement

12.3.1.1 The AKA Mechanism

The Authentication and Key Agreement (AKA) mechanism of the UMTS is extended from the GSM system. It is also a challenge-and-response authentication process. Different from the GSM system, the UMTS achieves mutual authentication, where the subscriber's identity is verified by the serving network, and the UE checks that the serving network is connected to the legitimate network. The mutual authentication protocol protects the UMTS from false base station attack.

The UTMS AKA mechanism is shown in Figure 12.4. The AKA mechanism has three major components: a USIM, the serving network, and the home network. A USIM is the

Table 12.1 Parameters used in UMTS access control security.

SQN	Sequence number
RAND	Random number
AMF	Authentication and key management field
K	Shared key
MAC	Message authentication code
XMAC	Expected MAC
CK	Cipher key
IK	Integrity key
AK	Anonymity key
XRES	Expected response
RES	Response
AUT	Authentication token
AV	Authentication vector

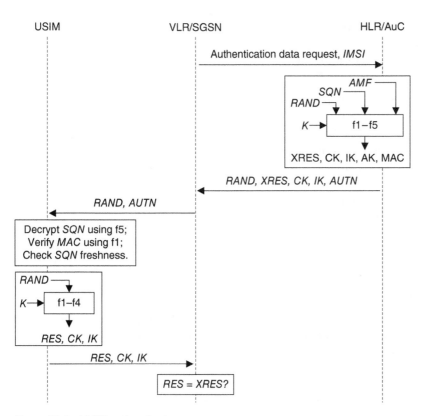

Figure 12.4 UMTS authentication.

subscriber to be connected to the serving network (i.e. the VLR/SGSN) that provides service. Since a USIM alone does not work properly without an ME (e.g. a cell phone), the UE is also referred to as a subscriber for better description. The home network (i.e. the HLR/AuC) is a part of the AKA mechanism. The VLR/SGSN has direct connection with the UE, the VLR/SGSN is contacting the HLR at the home network of the UE, and the AuC performs the core functions to generate parameters for authentication. The components and tasks are similar to the GSM system, with the addition of the SGSN that is responsible for the packet data service. The detailed steps of UMTS authentication protocol include:

(1) The UE initiates an authentication request by sending the IMSI to the serving network (i.e. the VLR/SGSN).
(2) The VLR/SGSN forwards the authentication request and the IMSI to the HLR of the home network of the UE.
(3) In the home network, the HLR forwards the IMSI to the AuC.
(4) The AuC retrieves the corresponding K of the supplicant and generates authentication vector ($RAND, XRES, CK, IK, AUTN$).
(5) The HLR forwards the authentication vector to the VLR/SGSN of the serving network.
(6) The VLR/SGSN stores the $XRES$, CK, and IK, and forwards the $RAND$ and $AUTN$ to the supplicant.
(7) The supplicant verifies the MAC and the SQN (part of the $AUTN$ that will be illustrated later); and generates RES, CK, and IK.
(8) The supplicant sends RES to the VLR/SGSN.
(9) The VLR/SGSN compares the $XRES$ generated by the AuC and the RES generated by the supplicant. If matches, then the authentication succeeds.

In the UMTS authentication protocol, the UE is authenticated by comparing the $XRES$ and the RES at the VLR/SGSN. The VLR/SGSN is also authenticated by verifying the $AUTN$ at the UE side. If the $AUTN$ is legitimate and fresh (i.e. was not sent beforehand to the USIM), then the UE can confirm that the serving network is connected to the legitimate network operator. The AKA mechanisms relies entirely on the long term secret key K, which is shared between the subscriber and the home network HLR/AuC. K is a 128-bit key stored in a USIM and the HLR/AuC. The secret key is never exposed to the serving network during the authentication process.

12.3.1.2 Authentication Vector Generation
The authentication vector (AV) consists of five fields, s.t.,

$$AV = RAND\|XRES\|CK\|IK\|AUTN,$$

where $RAND$ is a pseudo random number, $XRES$ is the expected response, CK is the ciphering key that will be used for encryption, and IK is the integrity key that will be used for identity protection. The $AUTN$ is the authentication token.

$$AUTN = (SQN \oplus AK)\|AMF\|MAC,$$

where SQN is a sequence number that is used to prevent replay attacks. AMF is the authentication and key management field that is used to direct a subscriber of the operator to ensure that a particular 3GPP authentication vector for that subscriber is used in the serving

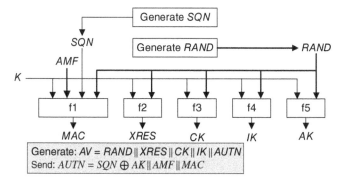

Figure 12.5 Generation of authentication vector.

network. *MAC* is the message authentication code. *AK*, *CK*, and *IK* are the three temporary keys derived during an authentication process. Technically, the keys are not part of the authentication, but are part of the Key Agreement mechanism. Therefore, the process is known as Authentication and Key Agreement (AKA) mechanism in the UMTS security. An illustration of the authentication vector generation is shown in Figure 12.5. The authentication vector is generated by five cryptographic functions:

- f1: the network authentication function;
- f2: the user authentication function;
- f3: the cipher key derivation function;
- f4: the integrity key derivation function;
- f5: the anonymity key derivation function.

All the five cryptographic functions are one-way functions. In other words, it is relatively easy to compute the output with the input, but practically impossible to invert the computation. Theoretically, nobody can derive the authentication vector without knowing the secret *K*.

The function f1 takes in four inputs, *K*, *AMF*, *SQN*, and *RAND*. The 64-bit *SQN* is generated to protect freshness of the AV. *RAND* is 128-bit long seed. A cryptographic pseudo-random number generator is assumed available to generate large amounts of unpredictable *RAND*. The output of f1 is the 64-bit *MAC*. The inputs for functions f2, f3, f4, and f5 are the same, i.e. *K* and *RAND*. The functions f2, f3, f4, and f5 are to generate the 32-bit *XRES*, 128-bit *CK*, 128-bit *IK*, and 64-bit *AK*, respectively. *SQN* is to protect the freshness of AV, thus transmitting it in clear text must be avoided. *AK* is generated to encrypt *SQN*.

There are two approaches to create *SQN*, one is to assign each user an individual *SQN*, the other is to generate *SQN* based on a global counter. One can also combine the two approaches together, where assigning the most significant part of the *SQN* as user-specific while using a global counter for the least significant part. 3GPP specification 33.102 describes three options for generating *SQN* [104]. The network operator is also free to choose other ways of generating *SQN*. One option that applies the combination of two parts define *SQN* as: $SQN = SEQ \| IND$. If AVs are sent in batches, the same *SEQ* is applied for all AVs in one batch to reduce the complexity. There are three ways to generate the value of *SEQ*.

- Generated based on an individual counter for each use.
- Generated based on a global counter and a deviation for each use. The deviation is ideally 0 for all users but could be different due to synchronization errors.
- Generated based on an individual counter (which generates 19-bit *SEQ1*) and a global counter (which generates 24-bit *SEQ2*). *SEQ1* is thus use-specific and stored in the database. *SEQ1* is kept constant until *SEQ2* wraps around.

As it shows, the first two ways are extreme cases of the third one. *IND* (5 bits) is based on a global counter. The value of *IND* is incremented by one for each new AV. When *IND* reaches the maximal value, it is cyclically counted to 0. Thus *SQN* is 48 bits in total.

12.3.1.3 AKA on the UE Side

The AKA on the UE side is illustrated in Figure 12.6. All the functions are performed on the USIM with parameters stored on the USIM as well. Therefore, the AKA on the UE side is actually handled by the USIM. However, the ME is a must to provide radio interface for data transmission. All five functions f1 to f5 are available at the UE side. As mentioned earlier, the authentication is completed by comparing *XRES* that is computed by the AuC and *RES* that is computed by the USIM. *RES* is the output of f2 with inputs *K* and *RAND*. Both *RAND* and *AUTN* are sent to the USIM. Therefore, *RES* can be easily computed by the subscriber. However, calculating *RES* alone only authenticates the subscriber to the network operator, which is the major flaw of the GSM AKA mechanism. In the UMTS AKA mechanism, the subscriber will first authenticate the accessing network. Based on the received *RAND* and *K*, the USIM computes *AK* through f5 first. Then, *SQN* is revealed to the USIM. Note that *SQN* is only known to the AuC and the USIM since *AK* cannot be generated without knowing *K*. The freshness of *SQN* is then checked by the USIM. If *SQN* is valid, then the USIM inputs *K*, *AMF*, *SQN*, and *RAND* to f1. The output is *XMAC*, which is compared with *MAC* received as part of *AUTN*. If *XMAC* matches *MAC*, then it implies that *RAND* and *AUTN* have been created by the AuC since it is the only entity that knows *K* besides the subscriber. In other words, the serving network is authenticated to the subscriber if *MAC = XMAC*. Once the

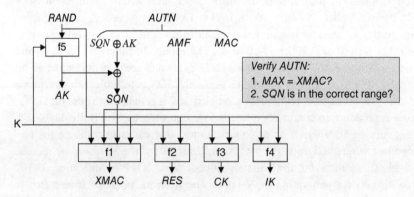

Figure 12.6 Verification on user equipment.

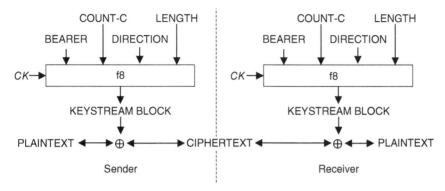

Figure 12.7 Stream cipher method in UMTS.

serving network is authenticated, the UE generates *RES* with f2 and replies it to the serving network for authentication. *CK* and *IK* are generated with f3 and f4, respectively. They are kept at the UE side for confidentiality and integrity of data transmission.

12.3.2 Confidentiality

Confidentiality is provided in the UMTS for data transmission over radio links between the UE and the RNC. As described earlier, once the user and the network have authenticated each other, a *CK* is shared between the serving network and the subscriber. 3GPP Release 1999 defined one encryption algorithm for secure communications over the radio link [102]. The UMTS encryption algorithm is based on a stream cipher f8. As illustrated in Figure 12.7, f8 takes in *CK* and a few other parameters to generate a keystream block that is to be XORed with the plaintext block. The decryption is the same by XORing the ciphertext with the keystream. The advantage of this encryption is that keystreams can be generated beforehand even without knowing the plaintext/ciphertext. Therefore the encryption/decryption operation can be done in real time.

The inputs of f8 include *CK*, BEARER, COUNT-C, DIRECTION, and LENGTH. *CK* is 128-bit long. For a stream cipher, a keystream should never be reused, thus, f8 takes in a few parameters besides *CK* so that the inputs change for each keystream output. BEARER is a 5-bit radio bearer identity that is to avoid generating the same keystream. Because different radio bearers associated with a single user may use the same *CK*.

COUNT-C is a 32-bit parameter that varies the input of f8. It is the combination of Connection Frame Number/Radio Link Control-Sequence Number (CFN/RCL-SN) and Hyper Frame Number (HFN). CFN is a counter in the medium access control layer, and RLC-SN is a counter in the radio link control layer. The two short counters change for every protocol data unit (PDU). HFN is a longer counter that increments whenever the short counter wraps around. If encryption occurs in the medium access control layer then CFN is applied. COUNT-C is computed as:

$$COUNT\text{-}C = HFN + CFN,$$

where HFN is 25 bits and CFN is 7 bits. If encryption occurs in the radio link control layer then RLC-SN is applied. COUNT-C is computed as:

COUNT-C = HFN + RLC-SN.

In this case, there are two modes: UM and HFN. In the UM mode, HFN is 25 bits and RLC-SN is 7 bits. In the AM mode, HFN is 20 bits and SN is 12 bits. The longer counter HFN is reset to zero whenever an AKA is processed. Thus, the reuse of COUNT-C with the same *CK* is almost impossible in practice.

DIRECTION is a 1-bit parameter that indicates whether the encryption is for uplink or downlink traffic. If LENGTH is 0, then encryption is for uplink (i.e. transmission from UE to RNC). If LENGTH is 1, then encryption is for downlink (i.e. transmission from RNC to UE). Applying direction indicator avoids using the same keystream to encrypt both uplink and downlink transmission, which is not achieved in the GSM system.

LENGTH is a 16-bit parameter that indicates the length of data to be encrypted. The limit of LENGTH is 20 000, or 20 000 bits as the maximal length of a message. Since the length of the plaintext block varies for a given BEARER and DIRECTION, LENGTH depends on the RLC PDU/MAC signaling data unit (SDU) size, number of RLC PDUs/MAC SDUs that may be sent in a single physical layer frame for a given BEARER and DIRECTION. The length of the output keystream is determined by LENGTH. The standard f8 algorithm is specified to support any lengths and length granularities.

KEYSTREAM is the output of f8. It has the length of the plaintext block, which consists of the payload of the particular RLC PDUs/MAC SDUs to be encrypted. The encryption is performed by XORing the PAINTEXT with the KEYSTREAM. The ciphertext has the length of the plaintext block (i.e. LENGTH). At the receiver side, the same *CK*, BEARER, COUNT-C, DIRECTION, and LENGTH are input to f8 to generate the same KEYSTREAM. Decryption is performed by XORing the CIPHERTEXT with the KEYSTREAM.

12.3.3 Data Integrity

The UMTS provides data integrity protection to authenticate individual control messages. Similar to confidentiality, integrity is also provided over the radio links between the UE and RNC. The integrity protection is based on a message authentication code. As mentioned in Chapter 6, a MAC takes in the message to be authenticated and a shared key. In the UMTS integrity mechanism, the key is IK that is generated during the AKA procedure. As shown in Figure 12.8, the core function of integrity protection is f9. The function takes in IK, COUNT-I, MESSAGE, DIRECTION and FRESH, and outputs 32-bit MAC-I. The MAC-I is appended to each message before sending to the receiver. Once the message (along with the MAC-I) is received, the receiver also computes an XMAC-I by running the same f9 function with the same inputs. If XMAC-I matches the received MAC-I, then the data integrity is verified. Any change in the input parameters will affect the output MAC in a random way.

IK is the 128-bit integrity key that is generated during the AKA procedure. IK is only known to the UE and the RNC. The AuC generates IK but it may not require it afterwards because the radio link transmissions are between the UE and the RNC only.

COUNT-I is a counter that is 32 bits. The counter changes for each message (incrementing by 1 for each message), thus it ensures a different input even for the same message. In other

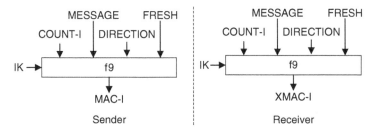

Figure 12.8 Message authentication in UMTS.

words, COUNT-I operates as a nonce that protects the system against replay attacks. The most significant part of COUNT-I is a 28-bit HFN. The least significant part of COUNT-I is a 4-bit RRC-SN. The COUNT-I is initialized using the same procedure as for COUNT-C. The Initial HFN is sent to the network by the user. The user stores the most used HFN from the previous connection and increments it for the new connection.

MESSAGE is the message to be transmitted. Unlike the 20 000-bit length limit set for the keystream of f8, there is no limit on the input message length of f9. Although, it was initially limited to 20 000 bits as well, however, it will most likely never be reached for f9 data input.

FRESH is a 32-bit parameter randomly chosen by the RNC and transmitted to the UE. The purpose of FRESH is to protect the system against a maliciously-chosen start value for COUNT-I, which may cause relay attack on an old MAC-I.

Radio bearer identity is not part of the inputs for f9 as it is for f8 (i.e. the encryption algorithm). However, radio bearer identity is always appended to the message. Thus, the output MAC-I is affected by the radio bearer identity, moreover, replay attacks based on recording different radio bearers are also unavailable.

12.3.4 User Identity Confidentiality

A UE needs to identify itself in many cases in the UMTS. For example, paging, location update, attach and detach all require an identity from the UE. The permanent identity of the user is IMSI, which is the same in GSM. In order to protect the user identity against passive eavesdroppers, the IMSI should not be transmitted very often. Therefore, the UMTS applies temporary identities for users in UTRAN. There are two types of temporary identities: *TMSI* in the circuit switched domain and *P-TMSI* in the packet switched domain.

The allocation of a temporary identity is shown in Figure 12.9. The SN first request the IMSI from the UE. The IMSI is then sent in clear text to the SN. Note that the permanent identity needs to be acquired during initial registration because a temporary identity is simply a random value to the SN at present. Once the SN acquires the IMSI, it stores the IMSI and allocates a unique temporary identity (a TMSI or a P-TMSI). The temporary identity is then sent back to the UE for further communications. An acknowledgment is sent by the UE to complete the allocation process.

If a temporary identity renew acknowledgment is not received by the SN, the SN keeps both the old and new temporary identities and accept either of them in uplink transmission (i.e. from UE to SN). However, in downlink transmission (i.e. from SN to UE), IMSI is used because the SN is not certain which temporary identity is stored in the UE. The UE will

UE SN **Figure 12.9** Temporary identity allocation process.

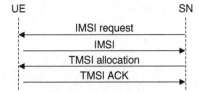

be informed to delete its stored temporary identity. A new temporary identity allocation follows after that.

Besides the temporary identity itself, a location identity is appended to it. For an IMSI, it is the location area identity (LAI) that stores the location information. For a P-TMSI, it is the routing area identity (RAI), respectively. When a UE is handed over to a new SN, the association between IMSI and temporary identity can be fetched from the previous SN without initiating another allocation. Moreover, unused AVs (for authentication) can also be transferred from the previous SN. If such location information is unknown, then the SN must initiate a new temporary identity allocation process as well as the AKA process.

Note that the user identity confidentiality mechanism does not offer a very good protection against active attackers. For example, an active attacker may pretend to be a SN and initiate a temporary identity allocation process to fetch the IMSI of a UE. Unfortunately, mutual authentication may not be applied here since the UE must be identified before the AKA process.

12.4 Algorithms in Access Security

As mentioned earlier in this chapter, f8 is the core function for the encryption mechanism, and f9 is the core function for the integrity mechanism. 3GPP TS 35.201 has the specifications publicly available for the functions [105]. Both f8 and f9 are based on a novel block cipher KASUMI, which is specified in 3GPP specification TS 35.202 [106]. In this section, the functions and algorithms of f8, f9, and KASUMI are illustrated.

12.4.1 Encryption Algorithm f8

Algorithm f8 is the core of the UMTS encryption mechanism to generate the keystreams. An overview of f8 is shown in Figure 12.10. Algorithm f8 makes use of the KASUMI block function, which operates on 64-bit input with a 128-bit key and 64-bit output block. Two 64-bit registers are used, one is a static register and the other is a counter. Inputs to the static register include a 64-bit initialization value IV, which concatenates 32-bit COUNT, 5-bit BEARER, 1-bit DIRECTION and a string of 26 zero bits. s.t.,

$$IV = COUNT\|BEARER\|DIRECTION\|0\ldots0.$$

Another input to the static register is $CK \oplus KM$, where KM is a 128-bit key modifier constant that is 01010101 repeated 16 times. For better illustration, let $KASUMI_{key}(input)$ be a KASUMI operation on a 64-bit $input$ with a 128-bit key. The output W of the static register is:

$$W = KASUMI_{CK \oplus KM}(IV).$$

The result W is stored in the static register to generate keystream bits.

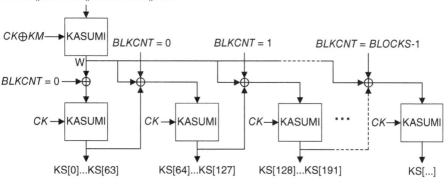

COUNT || BEARER || DIRECTION || 0...0

Figure 12.10 An overview of the f8 algorithm.

BLKCNT is the counter as the second register for f8. This register takes in W (the output of the static register), the counter BLKCNT, and the cipher key CK. The inputs change due to incrementation of BLKCNT after each operation, thus the register is not static. For each operation of the register, it generates a block of keystream bits, where each block has 64 bits. Some bits (between 0 and 63) of the least significant bits are discarded from the last block to match the total number of bits required by LENGTH (that is between 1 and 20 000 bits). Let BLOCKS denote the number of required blocks to generate the required keystream bits. The value of BLOCKS is determined by LENGTH as follows:

$$BLOCKS = \lceil LENGTH/64 \rceil.$$

The counter for the last block is $BLOCKS - 1$. Let KSB_i denote the output keystream block of the ith block. KSB_i has a total of 64 bits, e.g. $KSB_1 = KS[0] \ldots KS[63]$, then the current output keystream block is computed as follows:

$$KSB_n = KASUMI_{CK}(W \oplus (n - 1) \oplus KSB_{n-1}),$$

and the bit by bit keystream is:

$$KS[64(n - 1) + i] = KSB_n[i], \quad 0 \le i \le 63.$$

12.4.1.1 Integrity Algorithm f9
Algorithm f9 is the core function for 3GPP integrity mechanism. The specification of f9 is detailed in 3GPP TS 35.201 [105]. As shown in Figure 12.11, f9 is also based on the same block cipher KASUMI used by the confidentiality algorithm f8. The input of f9 is a padded string (PS), which is the concatenation of 32-bit COUNT, 32-bit FRESH, MESSAGE (with a length of LENGTH), 1-bit DIRECTION and a string in the form of 10 ... 0. The string has a single "1" bit followed by between 0 and 63 "0" bits so that the total length of the input is an integral multiple of 64 bits, i.e.

$$PS=COUNT||FRESH||MESSAGE||10 \ldots 0.$$

Unlike f8, there is no limitation on input length for f9. For any input to f9, it generates a 32-bit output MAC-I with the integrity key IK.

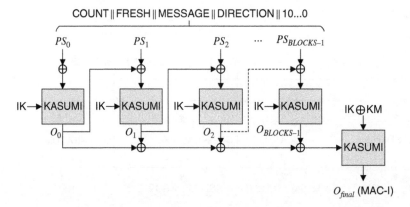

Figure 12.11 The f9 integrity mode.

Let *BLOCKS* be the number of blocks (64 bits each) of the initial input. Let PS_i be the input of the $(i + 1)$th block, the padded string is divided as follows:

$$PS = PS_0 \| PS_1 \| \dots PS_{BLOCKS-1}.$$

There are two parts in the f9 algorithm. The first part uses KASUMI in a form of CBC-MAC mode, where the output of each block operation is input to the next block operation. The operation of each block is:

$$O_i = \begin{cases} KASUMI_{IK}(PS_0), & i = 0, \\ KASUMI_{IK}(PS_i \oplus O_{i-1}), & 1 \leq i \leq BLOCKS - 1. \end{cases}$$

The outputs O_i for all block operations are XORed together and input to the second part of f9. The second part is a single block operation that generates the final MAC-I. In this block operation, the input key is $IK \oplus KM$, where KM is the 128-bit key modifier that is 10101010 (AA in HEX) repeating 16 times. Thus, the output of the final block operation is

$$O_{final} = KASUMI_{IK \oplus KM}(O_0 \oplus O_1 \oplus \dots \oplus O_{BLOCKS-1}).$$

The 32-bit MAC-I is the left-most 32 bits of O_{final}. The right-most 32 bits of O_{final} are discarded.

12.4.2 Description of KASUMI

Both confidentiality algorithm f8 and integrity algorithm f9 are based on the block cipher KASUMI. KASUMI was designed by the Security Algorithms Group of Experts (SAGE) for 3GPP to use in UMTS security system. The specification of KASUMI is detailed in 3GPP TS 35.202 [106]. An overview of the KASUMI algorithm is shown in Figure 12.12.

12.4.2.1 An Overview of KASUMI Algorithm

KASUMI is a block cipher that takes in 64-bit input and generates 64-bit output under the control of a 128-bit key. The core of KASUMI is an eight-round Feistel structure. In each round, the input is divided into two parts, where the 32 most significant bits are the input to the left hand side, and the 32 least significant bits are the input to the right hand side.

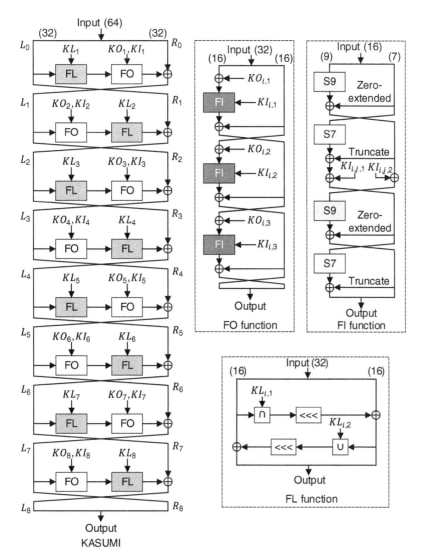

Figure 12.12 KASUMI block cipher.

A 16-bit round key is input to each round. The round keys are derived from the original 128-bit key using a fixed key schedule.

Let L_i and R_i be the two halves input to the ith round. The 64-bit input be I is divided into two 32-bit strings L_0 and R_0, s.t.,

$$I = R_0 \| L_0.$$

Let the round function be $F_i(\cdot)$ for the ith round ($1 \le i \le 8$), then the output of each round is computed as follows:

$$L_i = F_i(KL_i, KO_i, KI_i, L_{i-1}) \oplus R_{i-1},$$
$$R_i = L_{i-1},$$

where KL_i, KO_i, KI_i are round keys for the ith round. One can see that KASUMI follows Feistel structure: the right half of each round is XORed with the output of the round function $F_i(\cdot)$, and the halves are swapped for output. Note that the swapped outputs of the ith round are inputs to the $(i+1)$th round. The final output of a KASUMI function is the concatenation of the outputs of the last round as follows:

$$Output = L_8 \| R_8.$$

12.4.2.2 Round Function $F_i(\cdot)$

For the ith round, the round function $F_i(\cdot)$ takes in the 32-bit left input (i.e. L_{i-1}) under the control of a round key (K_i), and outputs a 32-bit output that is to be XORed to the right input R_{i-1}. The round key K_i is a composition of subkeys KL_i, KO_i, and KI_i. The round function comprises of two functions, i.e. $FL(\cdot)$ and $FO(\cdot)$. Depending on a specific round, the $F_i(\cdot)$ has two different forms.

- For rounds 1, 3, 5, and 7, the round function is defined as:

$$F_i(K_i, L_{i-1}) = FO(KO_i, KI_i, FL(KL_i, L_{i-1})).$$

- For rounds 2, 4, 6, and 8, the round function is defined as:

$$F_i(K_i, L_{i-1}) = FL(KL_i, FO(KO_i, KI_i, L_{i-1})).$$

12.4.2.3 Function FL

Function $FL(\cdot)$ takes in a 32-bit input I, which is split into two 16-bit halves $I = L \| R$. First, the left half of the input L is ANDed bitwise with subkey $KL_{i,1}$ and rotated left by one bit. Then, the results is XORed to the right half of the input R. The result is the right half of the output R'. Let $ROL(x, n)$ denote rotating left x of n bits, then R' is generated as follows:

$$R' = ROL(L \cap KL_{i,1}, 1) \oplus R.$$

To generate the left half of the output L', one need to bitwise OR the output R' with subkey $KL_{i,2}$ and rotate the result left by one bit. After that, the result is XORed to the left half of the input L to get the left half of the output L'. Mathematically, L' is generated as follows:

$$L' = ROL(R' \cup KL_{i,2}, 1) \oplus L.$$

The final output of function $FL(\cdot)$ is the concatenation of the left and right halves as $L' \| R'$.

12.4.2.4 Function FO

Function $FO(\cdot)$ also takes in a 32-bit input I, which is split into two 16-bit halves $I = L_0 \| R_0$. $FO(\cdot)$ has a Feistel-like structure with three rounds. In each round, the input of the left half goes through a function to generate the right half of the output. The input of the right half is directly applied as the left half of the next round. Let R_j and L_j $(1 \leq j \leq 3)$ denote the right and left halves of the output of the jth round, respectively. They are generated as

$$R_j = FI(KI_{i,j}, L_{j-1} \oplus KO_{i,j}) \oplus R_{j-1},$$
$$L_j = R_{j-1}.$$

where $KI_{i,j}$ and $KO_{i,j}$ are subkeys for the i-round of the KASUMI algorithm, and the jth round of $FO(\cdot)$. $FI(\cdot)$ is the round function applied in each round of $FO(\cdot)$. The final output of the function is 32-bit value $L_3 \| R_3$.

12.4.2.5 Function FI

The function $FI(\cdot)$ takes in a 16-bit input I and subkey $KI_{i,j}$. The input I is split to two unequal halves $I = L_0 \| R_0$. The left half L_0 is 9 bits and the right half R_0 is 7 bits. Subkey $KI_{i,j}$ is also split into 7-bit $KI_{i,j,1}$ and 9-bit $KI_{i,j,2}$. $FI(\cdot)$ also has a Feistel-like structure with four rounds. The round function in $FI(\cdot)$ is performed with two substitution box (S-box) 9-bit S9 and 7-bit S7. In rounds 1 and 3, S9 is applied; in rounds 2 and 4, S7 is applied. The four rounds in $FI(\cdot)$ are different from each other.

(1) In the first round, the inputs are L_0 and R_0. The output of the left half $L_1 = R_0$. To generate the output of the right half R_1, bits of L_0 are first shuffled by S9. The result is then XORed with the zero-extended R_0. Mathematically, R_1 is computed as follows:

$$R_1 = S9(L_0) \oplus (00 \| R_0),$$

where $S9(x)$ is shuffling x by S9.

(2) In the second round, R_1 is XORed with subkey $KI_{i,j,2}$ to generate L_2, s.t.,

$$L_2 = R_1 \oplus KI_{i,j,2}.$$

Then, bits of R_0 are shuffled by S7. The result is then XORed with the seven least significant bits (LS7) of R_1 and subkey $KI_{i,j,2}$ to get the output R_2. Mathematically, R_2 is computed as follows:

$$R_2 = S7(L_1) \oplus LS7(R_1) \oplus KI_{i,j,1}.$$

(3) In the third round, output L_3 is R_2. To generate the output of the right half R_3, bits of L_2 are first shuffled by S9, the result is then XORed with zero-extended R_2. Mathematically, R_3 is computed as follows:

$$R_3 = S9(L_2) \oplus (00 \| R_2).$$

(4) In the last round, bits of L_3 are shuffled by S7, the result is XORed with R_3 to generate L_4, s.t.,

$$L_4 = S7(L_3) \oplus LS7(R_3).$$

The output of the right half $R_4 = R_3$.

Finally, the output of $FI(I, KI_{i,j}) = L_4 \| R_4$.

12.4.2.6 S-boxes S7 and S9

S-boxes S7 and S9 are the heart of function FI. They are designed to shuffle 7 and 9-bit inputs, respectively. Let two 6-bit digits $[x_6, x_5, \ldots, x_0]$ and $[y_6, y_5, \ldots, y_0]$ be the input and output of S7. Let $x_i x_j$ indicate AND operation such that

$$x_i x_j = x_i \cap x_j.$$

Table 12.2 Decimal look-up table of S-box *S7*.

54,	50,	62,	56,	22,	34,	94,	96,	38,	6,	63,	93,	2,	18,	123,	33,
55,	113,	39,	114,	21,	67,	65,	12,	47,	73,	46,	27,	25,	111,	124,	81,
53,	9,	121,	79,	52,	60,	58,	48,	101,	127,	40,	120,	104,	70,	71,	43,
20,	122,	72,	61,	23,	109,	13,	100,	77,	1,	16,	7,	82,	10,	105,	98,
117,	116,	76,	11,	89,	106,	0,	125,	118,	99,	86,	69,	30,	57,	126,	87,
112,	51,	17,	5,	95,	14,	90,	84,	91,	8,	35,	103,	32,	97,	28,	66,
102,	31,	26,	45,	75,	4,	85,	92,	37,	74,	80,	49,	68,	29,	115,	44,
64,	107,	108,	24,	110,	83,	36,	78,	42,	19,	15,	41,	88,	119,	59,	3

Then the gate logic of S7 is as follows:

$$y_0 = x_1x_3 \oplus x_4 \oplus x_0x_1x_4 \oplus x_5 \oplus x_2x_5 \oplus x_3x_4x_5 \oplus x_6 \oplus x_0x_6 \oplus x_1x_6$$
$$\oplus x_3x_6 \oplus x_2x_4x_6 \oplus x_1x_5x_6 \oplus x_4x_5x_6,$$

$$y_1 = x_0x_1 \oplus x_0x_4 \oplus x_2x_4 \oplus x_5 \oplus x_1x_2x_5 \oplus x_0x_3x_5 \oplus x_6 \oplus x_0x_2x_6 \oplus x_3x_6$$
$$\oplus x_4x_5x_6 \oplus 1,$$

$$y_2 = x_0 \oplus x_0x_3 \oplus x_2x_3 \oplus x_1x_2x_4 \oplus x_0x_3x_4 \oplus x_1x_5 \oplus x_0x_2x_5 \oplus x_0x_6$$
$$\oplus x_0x_1x_6 \oplus x_2x_6 \oplus x_4x_6 \oplus 1,$$

$$y_3 = x_1 \oplus x_0x_1x_2 \oplus x_1x_4 \oplus x_3x_4x_0x_5 \oplus x_0x_1x_5 \oplus x_2x_3x_5 \oplus x_1x_4x_5$$
$$\oplus x_2x_6 \oplus x_1x_3x_6,$$

$$y_4 = x_0x_2 \oplus x_3 \oplus x_1x_3 \oplus x_1x_4 \oplus x_0x_1x_4 \oplus x_2x_3x_4 \oplus x_0x_5 \oplus x_1x_3x_5$$
$$\oplus x_0x_4x_5 \oplus x_1x_6 \oplus x_3x_6 \oplus x_0x_3x_6 \oplus x_5x_6 \oplus 1,$$

$$y_5 = x_2 \oplus x_0x_2 \oplus x_0x_3 \oplus x_1x_2x_3 \oplus x_0x_2x_4 \oplus x_0x_5 \oplus x_2x_5 \oplus x_4x_5 \oplus x_1x_6$$
$$\oplus x_1x_2x_6 \oplus x_0x_3x_6 \oplus x_3x_4x_6 \oplus x_2x_5x_6 \oplus 1,$$

$$y_6 = x_1x_2 \oplus x_0x_1x_3 \oplus x_0x_4 \oplus x_1x_5 \oplus x_3x_5 \oplus x_6 \oplus x_0x_1x_6 \oplus x_2x_3x_6$$
$$\oplus x_1x_4x_6 \oplus x_0x_5x_6.$$

In practice, one may apply the decimal look-up table of S7, as detailed in Table 12.2. Similarly, let $[x_8, x_7, \ldots, x_0]$ be the input and $[y_8, y_7, \ldots, y_0]$ be the output of S9, the gate logic of S9 is as follows:

$$y_0 = x_0x_2 \oplus x_3 \oplus x_2x_5 \oplus x_5x_6 \oplus x_0x_7 \oplus x_1x_7 \oplus x_2x_7 \oplus x_4x_8 \oplus x_5x_8$$
$$\oplus x_7x_8 \oplus 1,$$

$$y_1 = x_1 \oplus x_0x_1 \oplus x_2x_3 \oplus x_0x_4 \oplus x_1x_4 \oplus x_0x_5 \oplus x_3x_5 \oplus x_6 \oplus x_1x_7 \oplus x_2x_7$$
$$\oplus x_5x_8 \oplus 1,$$

$$y_2 = x_1 \oplus x_0x_3 \oplus x_3x_4 \oplus x_0x_5 \oplus x_2x_6 \oplus x_3x_6 \oplus x_5x_6 \oplus x_4x_7 \oplus x_5x_7$$
$$\oplus x_6x_7 \oplus x_8 \oplus x_0x_8 \oplus 1,$$

$$y_3 = x_0 \oplus x_1x_2 \oplus x_0x_3 \oplus x_2x_4 \oplus x_5 \oplus x_0x_6 \oplus x_1x_6 \oplus x_4x_7 \oplus x_0x_8 \oplus x_1x_8$$
$$\oplus x_7x_8,$$

$$y_4 = x_0x_1 \oplus x_1x_3 \oplus x_4 \oplus x_0x_5 \oplus x_3x_6 \oplus x_0x_7 \oplus x_6x_7 \oplus x_1x_8 \oplus x_2x_8 \oplus x_3x_8,$$

$$y_5 = x_2 \oplus x_1x_4 \oplus x_4x_5 \oplus x_0x_6 \oplus x_1x_6 \oplus x_3x_7 \oplus x_4x_7 \oplus x_6x_7 \oplus x_5x_8 \oplus x_6x_8,$$
$$\oplus x_7x_8 \oplus 1$$

Table 12.3 Decimal look-up table of S-box 9.

```
167,239,161,379,391,334,   9,338,  38,226,  48,358,452,385,  90,397,
183,253,147,331,415,340,  51,362,306,500,262,  82,216,159,356,177,
175,241,489, 37,206, 17,   0,333,  44,254,378,  58,143,220,  81,400,
 95,  3,315,245,  54,235,218,405,472,264,172,494,371,290,399,  76,
165,197,395,121,257,480,423,212,240,  28,462,176,406,507,288,223,
501,407,249,265,  89,186,221,428,164,  74,440,196,458,421,350,163,
232,158,134,354,  13,250,491,142,191,  69,193,425,152,227,366,135,
344,300,276,242,437,320,113,278,  11,243,  87,317,  36,  93,496,  27,

487,446,482,  41,  68,156,457,131,326,403,339,  20,  39,115,442,124,
475,384,508,  53,112,170,479,151,126,169,  73,268,279,321,168,364,
363,292,  46,499,393,327,324,  24,456,267,157,460,488,426,309,229,
439,506,208,271,349,401,434,236,  16,209,359,  52,  56,120,199,277,
465,416,252,287,246,   6,  83,305,420,345,153,502,  65,  61,244,282,
173,222,418,  67,386,368,261,101,476,291,195,430,  49,  79,166,330,
280,383,373,128,382,408,155,495,367,388,274,107,459,417,  62,454,
132,225,203,316,234,  14,301,  91,503,286,424,211,347,307,140,374,

 35,103,125,427,  19,214,453,146,498,314,444,230,256,329,198,285,
 50,116,  78,410,  10,205,510,171,231,  45,139,467,  29,  86,505,  32,
 72,  26,342,150,313,490,431,238,411,325,149,473,  40,119,174,355,
185,233,389,  71,448,273,372,  55,110,178,322,  12,469,392,369,190,
  1,109,375,137,181,  88,  75,308,260,484,  98,272,370,275,412,111,
336,318,   4,504,492,259,304,  77,337,435,  21,357,303,332,483,  18,
 47,  85,  25,497,474,289,100,269,296,478,270,106,  31,104,433,  84,
414,486,394,  96,  99,154,511,148,413,361,409,255,162,215,302,201,

266,351,343,144,441,365,108,298,251,  34,182,509,138,210,335,133,
311,352,328,141,396,346,123,319,450,281,429,228,443,481,  92,404,
485,422,248,297,  23,213,130,466,  22,217,283,  70,294,360,419,127,
312,377,   7,468,194,   2,117,295,463,258,224,447,247,187,  80,398,
284,353,105,390,299,471,470,184,  57,200,348,  63,204,188,  33,451,
 97,  30,310,219,  94,160,129,493,  64,179,263,102,189,207,114,402,
438,477,387,122,192,  42,381,   5,145,118,180,449,293,323,136,380,
 43,  66,  60,455,341,445,202,432,   8,237,  15,376,436,464,  59,461
```

$$y_6 = x_0 \oplus x_2 x_3 \oplus x_1 x_5 \oplus x_2 x_5 \oplus x_4 x_5 \oplus x_3 x_6 \oplus x_4 x_6 \oplus x_5 x_6 \oplus x_7 \oplus x_1 x_8$$
$$\oplus x_3 x_8 \oplus x_5 x_8 \oplus x_7 x_8,$$

$$y_7 = x_0 x_1 \oplus x_0 x_2 \oplus x_1 x_2 \oplus x_3 \oplus x_0 x_3 \oplus x_2 x_3 \oplus x_4 x_5 \oplus x_2 x_6 \oplus x_3 x_6 \oplus x_2 x_7$$
$$\oplus x_5 x_7 \oplus x_8 \oplus 1,$$

$$y_8 = x_0 x_1 \oplus x_1 x_2 \oplus x_3 x_4 \oplus x_1 x_5 \oplus x_2 x_5 \oplus x_1 x_6 \oplus x_7 \oplus x_2 x_8 \oplus x_3 x_8.$$

The decimal look-up table of S9 is detailed in Table 12.3.

12.4.2.7 Key Schedule

The input key K to the KASUMI algorithm is 128 bits. All the round keys and sub keys are derived from K. First, the 128-bit key K is divided into eight 16-bit sub keys K_i, $1 \leq i \leq 8$ as follows:

$$K = K_1 \| K_2 \| \dots \| K_8.$$

In addition to K, a modified key K' is also used. K' is derived from the original key by XORing with 128-bit HEX value:

$$C = 123456789ABCDEFFEDCBA9876543210.$$

Thus the modified key is:

$$K' = K \oplus C$$

The modified key is also divided into eight 16-bit parts as follows:

$$K' = K'_1 \| K'_2 \| \dots \| K'_8.$$

All subkeys of the eight round in KASUMI algorithm are derived from the K_i and K'_i. For the ith round, subkeys KL are derived as follows:

$$KL_{i,1} = ROL(K_i, 1),$$
$$KL_{i,2} = K'_{i+2}.$$

Subkeys KO are derived as follows:

$$KO_{i,1} = ROL(K_{i+1}, 5),$$
$$KO_{i,2} = ROL(K_{i+5}, 8),$$
$$KO_{i,3} = ROL(K_{i+6}, 13).$$

Subkeys KI are derived as follows:

$$KI_{i,1} = K'_{i+4},$$
$$KI_{i,2} = K'_{i+3},$$
$$KI_{i,3} = K'_{i+7}.$$

Note that the sub key index additions are cyclic, i.e. $i + j \leftarrow (i + j) \mod 8$.

12.4.3 Implementation and Operational Considerations

UMTS security requires the algorithms to accommodate a range of implementation options, including both hardware and software implementations. It is required to implement one instance of the algorithm using at most 10 000 gates for hardware implementation. It is also required to implement the algorithms to achieve an encryption speed that satisfies 2 Mbps transmission rate on both the downlink and the uplink. In addition, the encryption throughput requirements are based on a clock speed of 20 MHz. In practice, a typical clock speed is much greater than 20 MHz thus the algorithm should support much higher throughput requirements. Depending on different RLC modes, the implementation and operational requirements of new keystream block generation per frame, maximum/minimum number of bits per frame, and granularity are listed in Table 12.4.

Table 12.4 Implementation and operational requirements.

	RLC - transparent	RLC - UM	RLC - AM
New keystream block per frame	10 ms	$\geq 156\ \mu s$	$\geq 156\ \mu s$
Maximum number of bits per frame	5114 bits	1016 bits	1024 bits
Minimum number of bits per frame	1 bit	16 bits	24 bits
Granularity	1 bit	8 bits	8 bits

12.5 Other UMTS Security Features

12.5.1 Mobile Equipment Identification

The same ME identification method is carried over from the GSM system to the UMTS system. An international mobile equipment identity (IMEI) is the unique identity of an ME. IMEI may be helpful for a user by tracking the location of the phone. Or, the IMEI of a stolen ME may be reported to the network operator, who will deny network access of the stolen device. There are no particular protection methods for IMEI from the network operator. Only the terminal side can provide protection to its IMEI, for example, making it difficult to modify the IMEI when requested by the network. IMEI does not participate in the UMTS access security except for some important features in the network that can only be based on the value of IMEI, e.g., making emergency calls with a cell phone that is without a USIM.

12.5.2 Location Services

UMTS tracks user location to apply network service with mobility. Authentication parameters and temporary identities can be quickly allocated if location of a user is tracked and stored. In addition, many services can benefit from location service. For example, automatically update weather information for user's current location. However, location information is private to the user. Security mechanisms are needed to protect against leakage of location information to unauthorized parties. In practice, users are in charge of who know about their whereabouts. Users can control applications on an ME whether they can track and distribute the location information.

12.5.3 User-to-USIM Authentication

USIM is a critical part of the UMTS security. Therefore, user-to-USIM authentication is a must before a user gets access to the network. The UMTS carries over the feature from the GSM system by using a Personal Identification Number (PIN) that is known only to the user and the USIM. The PIN is 4–8 digits long, given to the USIM before network access is granted. However, once the authentication is done for a USIM (with an ME), user-to-USIM authentication does not provide further protection against theft of the ME.

12.6 Summary

In this chapter, the UMTS security is introduced. Since the UMTS is developed based on the GSM system, several security mechanisms are reused with modifications. The authentication and key agreement of the UMTS is a challenge-response scheme that achieves mutual authentication. Both cipher key and integrity key are generated during the AKA process. Data confidentiality and integrity are both provided in the UMTS security. Compared with the GSM system, integrity protection is added in the UMTS. The confidentiality and integrity are provided by algorithms f8 and f9 respectively. Both algorithms are based on another block cipher - KASUMI. Besides network access security, other security features such as ME identification, location service, and user-to-USIM authentication are also provided in the UMTS security. While GSM security has flaws, UMTS is considered much better improved in security.

13

LTE Security

The International Telecommunication Union Radio Communication Sector (ITU-R) established standards for 4G connectivity in March 2008. 4G requires a connection speed for mobile/stationary user to have a peak data rate of at least 100 Mbps/1 Gbps, respectively. Long-term evolution, commonly referred to 4G LTE, is a standard for wireless communications of high-speed data for mobile devices developed by the Third Generation Partnership Project (3GPP). LTE is based on the global system for mobile (GSM) and Universal Mobile Telecommunications System (UMTS) network technologies. The increase of capacity and speed is because of improvements on different radio interface and core network. LTE are the most popular upgrade path for carriers with both GSM/UMTS and CDMA2000 networks. LTE is also known as evolved packet system (EPS) and system architecture evolution (SAE). The three terms are interchangeable in this chapter. LTE security is based on the evolution of UMTS security, but has in fact improved much further. In this chapter, LTE security as well as some other related 4G security issues are discussed.

13.1 LTE System Architecture

An overview of the LTE/EPS basic architecture is given in Figure 13.1. It consists of user equipment (UE), evolved UMTS Terrestrial Radio Access Network (E-UTRAN), and evolved packet core (EPC). UE is connected to the EPC over E-UTRAN, i.e. the LTE access network. The EPC is connected to Internet Protocol (IP) network or the Internet [107]. UE is a cellular device that contains a mobile equipment (ME), i.e. a cell phone and a Universal Integrated Circuit Card (UICC). A UICC card serves the same role as a subscriber identity module (SIM) card in GSM or a universal subscriber identity module (USIM) card in UMTS. Despite their technical difference inside the cards, the same outer appearance of the three modules makes them widely recognized as SIM cards. Personal information such as contacts and text messages may be stored on a UICC card. It stores more importantly several security related parameters, such as international mobile subscriber identity (IMSI)

Security in Wireless Communication Networks, First Edition. Yi Qian, Feng Ye, and Hsiao-Hwa Chen.
© 2022 John Wiley & Sons Ltd. Published 2022 by John Wiley & Sons Ltd.
Companion website: www.wiley.com/go/qian/sec51

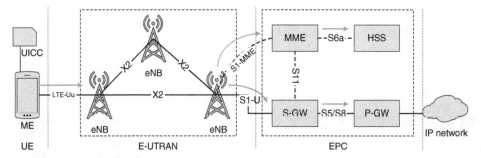

Figure 13.1 LTE architecture.

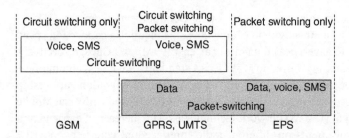

Figure 13.2 Evolution of services provided by GSM, UMTS, and EPS.

(i.e. the subscriber identifier), keys, and authentication algorithms. Because of backward compatibility, a UICC is also responsible for running the SIM and USIM applications.

An E-UTRAN comprises of multiple evolved NodeBs (eNBs) that are the base stations for LTE radio network. An eNB is a radio component of LTE network that connected to the core network that communicates directly through wireless links with UEs, like a base transceiver station in GSM networks. An eNB demodulates radio frequency (RF) signals and transmits IP packets to the core network. It also modulates IP packets and transmits RF signals to UE. The interface that connects eNBs is X2. Note that there is no separate controller element for eNBs. The simplification allows lower response time in LTE. The EPC is the latest evolution of the 3GPP core network architecture. The GSM architecture relies on circuit-switching to provide services. Though GPRS added packet-switching to transport packets, circuits still transport voice and short messages. The UMTS system architecture introduces a dual-domain concept to the core network side. LTE/EPS uses IP as the key protocol to transport all services. Therefore, the EPC did not have a circuit-switched domain anymore as an evolution of the packet-switched architecture used in GPRS/UMTS. Figure 13.2 illustrates such evolution from GSM to EPS. The EPC comprises of four major components, mobility management entity (MME), home subscriber server (HSS), serving gateway (S-GW), and packet data network gateway (P-GW). An access security management entity (ASME) is also a component of the EPC. However, it is assumed by the MME for E-UTRAN access networks.

- The **MME** deals with the control plane (signaling data). It handles signaling related to mobility and security for E-UTRAN access. The MME is also responsible for tracking and

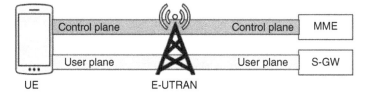

Figure 13.3 LTE communications planes.

paging a UE that is in idle-mode. The MME also plays an important role in LTE mobility, both in handover between eNBs and handover between LTE and other 3GPP accesses. An MME is a termination point of the non-access stratum (NAS).

- The **ASME** is an entity that receives the top-level keys in an access network from the HSS. For EPS, the role of the ASME is assumed by the MME.
- The **HSS** is based on the home location register (HLR) and authentication center (AuC) that are in previous generations of cellular networks. Basically, the HSS is a database that contains user-related and subscriber-related information, e.g. IMSI and keys. It provides support functions in user authentication and access authorization, mobility management, and call/session setup.
- The **S-GW** deals with the user plane (user data). It interconnects the radio access network and the EPC. Incoming and outgoing IP packets to a UE are routed by the S-GW. A S-GW is logically connected to P-GW.
- The **P-GW** deals with the user plane. It interconnects the EPC and the external IP networks. The networks are also known as packet data network (PDN). The P-GW routes packets to and from the PDNs. The P-GW also performs various functions such as IP address/IP prefix allocation, policy control, and charging. While S-GW and P-GW are specified independently, they may be combined together as they are logically connected by network vendors.

LTE has a flatter architecture compared with UMTS. LTE separates control plane (i.e. for signaling data) and user plane (i.e. for user data) as shown in Figure 13.3. The two logical planes are multiplexed into the same RF signal that is transmitted between a UE and the E-UTRAN. Data from different planes is routed to different end points. Signaling data is routed to the MME, while user data is routed to S-GW.

13.2 LTE Security Architecture

LTE security reuses UMTS authentication and key agreement (AKA) mechanisms. Therefore the use of USIM is required (however, GSM SIM is excluded). Compared with USIM, LTE further extends the key hierarchy (details will be given in Section 13.3). LTE provides greater protection for backhaul networks to ensure the security of EPC. LTE also integrates inter-networking security for legacy and non-3GPP networks [108]. LTE has a trust model that is similar to that of UMTS. The EPC is the secure core network and the E-UTRAN is a radio access network that is vulnerable to attack. Since a radio network controller does not exist in LTE, the UE user plane security is terminated in the eNB. There are five set of security features defined by 3GPP for LTE [109], as shown in Figure 13.4.

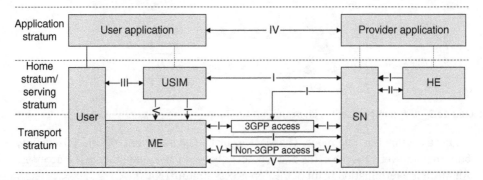

Figure 13.4 Overview of LTE security architecture.

- *Network access security (I)*: The set of security features that provides the UEs with secure access to the EPC and protect against various attacks on the radio link.
- *Network domain security (II)*: The set of security features that protects against attacks on the wire line network and enable nodes to exchange signaling data and user data in a secure manner.
- *User domain security (III)*: The set of security features that provides a mutual authentication between the USIM and the ME before the USIM access to the ME.
- *Application domain security (IV)*: The set of security features that enables applications in the UE and in the provider domain to securely exchange messages.
- *Non-3GPP domain security (V)*: The set of features that enables the UEs to securely access to the EPC via non-3GPP access networks and provides security protection on the radio access link.

Security over the LTE air interface is provided through strong cryptographic techniques. The backhaul link from the eNB to the core network makes use of Internet key exchange (IKE) and the IP security (IPsec) protocol when cryptographic protection is needed. Strong cryptographic techniques provide end-to-end protection for signaling between the core network and UE. Therefore, the main location where user traffic is threatened by exposure is in the eNB.

Moreover, to minimize susceptibility to attacks, the eNB needs to provide a secure environment that supports the execution of sensitive operations, such as encryption or decryption of user data, and the storage of sensitive data like keys for securing UE communications, long-term cryptographic secrets, and vital configuration data. Likewise, the use of sensitive data must be confined to this secure environment.

Even with the aforementioned security measures in place, one must consider attacks on an eNB, because, if successful, they could give attackers full control of the eNB and its signaling to UEs and other nodes. To limit the effect of a successful attack on one eNB, attackers must not be able to intercept or manipulate user and signaling plane traffic that traverses another eNB – for example, after handover.

13.3 LTE Security

13.3.1 LTE Key Hierarchy

As mentioned earlier, LTE has extended key hierarchy of UMTS. More keys are applied in LTE security. In this section, an overview of LTE keys is given for better understanding in further security mechanisms. 3GPP TS 33.401 specification [110] defines key distribution and key derivation scheme for EPS. The key hierarchy is shown in Figure 13.5. K is the permanent key shared between the UICC (the USIM on the UICC) and the AuC (in the HSS). CK and IK are the pair of keys derived in the AuC and on the USIM during an AKA process. Since EPS AKA is based on UMTS AKA, those three keys are shared between UE and HSS in EPS. Although CK and IK are derived in EPS security, they are handled differently. The two keys are used to generate an intermediate key K_{ASME} between UE and ASME for further EPS AKA. Besides K_{ASME}, EPS security also includes the following keys: K_{eNB}, $K_{NAS_{int}}$, $K_{NAS_{enc}}$, $K_{UP_{enc}}$, $K_{RRC_{int}}$, and $K_{RRC_{enc}}$.

- K_{eNB} is a 256-bit key derived by UE and MME from K_{ASME} when the UE goes to ECM-CONNECTED state or by UE and target eNB during eNB handover.
- $K_{NAS_{enc}}$ 256-bit is an encryption key that is derived by UE and MME from K_{ASME} using the key derivation function (KDF). It is used to provide confidentiality for NAS traffic with a particular encryption algorithm.
- $K_{NAS_{int}}$ is a 256/128-bit integrity key that is also derived by UE and MME from K_{ASME} using the KDF. It is used to provide integrity for NAS traffic with a particular integrity algorithm.
- $K_{UP_{enc}}$ is a 256/128-bit encryption key that is derived by UE and eNB from K_{eNB}, as well as an identifier for the encryption algorithm (i.e. NAS-enc-alg and Alg-ID) using the KDF.

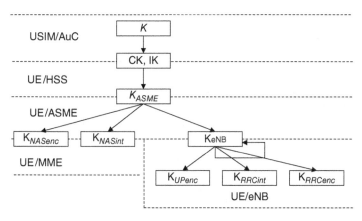

Figure 13.5 Key hierarchy in LTE.

This key is used to provide confidentiality for user plane traffic with a particular encryption algorithm.

- $K_{RRC_{enc}}$ is a 256/128-bit integrity key that is derived by UE and eNB from K_{eNB}, as well as an identifier for the encryption algorithm (i.e. UP-enc-alg and Alg-ID) using the KDF. This key is used to provide confidentiality for radio resource control (RRC) traffic with a particular encryption algorithm.
- $K_{RRC_{int}}$ is a 256/128-bit integrity key that is derived by UE and eNB from K_{eNB}, as well as an identifier for the integrity algorithm (i.e. UP-int-alg and Alg-ID) using the KDF. This key is used to provide integrity for RRC traffic with a particular integrity algorithm.

Figure 13.6 shows the dependencies among the different keys, and how they are derived from the network nodes point of view. *K*, *CK*, and *IK* are the three keys that exist in UMTS. Despite those three keys, all the others keys in EPS are derived from the core function KDF, which is defined in the 3GPP TS 33.220 specification. The KDF is an HMAC function that is based on SHA-256. Therefore the output of the KDF is 256 bits, which is the length of the EPS keys. Nonetheless, in case the encryption or integrity algorithm used to protect NAS, UP, or RRC requires a 128-bit key as input, the key is truncated by function Trunc to 128 bits. The function Trunc takes as input a 256-bit string and returns the 128 least significant bits of that string as output. The keys are also derived on the user side. Instead of deriving them with different components, the ME simply derives all the keys with the same process performed on the network side. EPS key hierarchy ensures that keys used for one access network cannot be used in another access network. It also ensures that the same key is not used for multiple purposes or with different algorithms. Moreover, the key hierarchy makes it possible to frequently change the keys used between a UE and eNBs without changing the master secret key *K*.

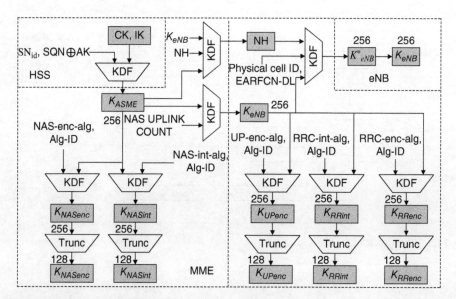

Figure 13.6 Key derivation in LTE.

13.3.2 LTE Authentication and Key Agreement

Mutual authentication between the UE and the EPC is an important security feature in the LTE security framework. The UMTS AKA protocol has already provided mutual authentication, as a good foundation for LTE AKA protocol. LTE AKA provides mutual authentication between the UE and the EPC, ensuring that no fraudulent entities can pose as a valid network node. Besides entity authentication, LTE AKA also produces keying material forming a basis for UP RRC and NAS ciphering keys, as well as RRC and NAS integrity keys. One must keep in mind that although LTE is backward compatible with GSM service, however, SIM cards are not allowed in LTE for AKA because they do not provide adequate security due to one-way authentication and weak keys in GSM system. The LTE AKA process is sketched in Figure 13.7. The 11 steps can be seen as three parts: request, challenge, and response.

Request:

(1) The MME (serving network) initiates by requesting the user identity, which is the same as UMTS.

(2) A UE replies with an identity response that is further relayed to the HSS (home network). There are two possibilities for a UE identity response. First, if a UE has authenticated itself to the MME (serving network) before, then a temporary identity globally unique temporary identifier (GUTI) exists. GUTI will be sent back as a response to the request. On the other hand, if a GUTI is not available, a UE sends its permanent identity IMSI as a response.

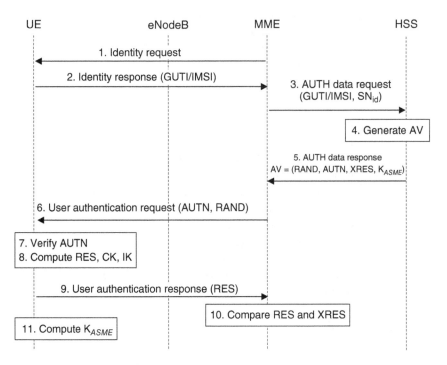

Figure 13.7 LTE authentication and key agreement.

Challenge:

(3) The MME forwards the GUTI/IMSI to the HSS. If GUTI is applied, the network retrieves user context (i.e. IMSI and K) based on GUTI in HSS [111]. In both cases, it is the IMSI of a UE that confirms the identity. A serving network's identity (SN_{id}) is also sent to the HSS to authenticate the serving network.

(4) Once the HSS retrieves the IMSI of the supplicant (i.e. a UE) and the corresponding long term secret key K, it generates an authentication vector AV as the challenge part of LTE AKA. The AV generation is almost the same as UMTS AKA process, i.e.

$$AV = (RAND, AUTN, XRES, K_{ASME}),$$

where $RAND$ is a random number, $AUTN$, and $XRES$ are the same parameters applied in the UMTS AKA. Please refer to the previous chapter for more details. The difference between LTE and UMTS AV is the key. Keys CK and IK are generated in both LTE and UMTS AKA processes, however, LTE AKA does not include them in the AV. Instead, the two keys are binded together through a function and create K_{ASME}, s.t.,

$$K_{ASME} = KDF(SN_{id}, CK\|IK).$$

(5) HSS sends AV to the MME.

(6) The MME sends to the UE the authentication token $AUTN$ and the random challenge $RAND$. Note that K_{ASME} is not sent to the UE. It will be generated at the UE side in the response part.

Response:

(7) USIM verifies whether $AUTN$ can be accepted.

(8) Then computes CK, IK, and RES. Please refer to the previous chapter for details.

(9) The RES is sent to the MME for comparison with the $XRES$.

(10) The authentication process succeeds if $RES = XRES$.

(11) After a success response, the UE computes K_{ASME} from CK, IK, and SN_{id} using the KDF function.

Note that the response part from the UE side follows the same process as UMTS AKA. It is mentioned in step 3 that SN_{id} is applied for the authentication of the SN's identity. It is because the generation of K_{ASME} takes input SN_{id} at both UE and HSS, thus if the derived keys from K_{ASME} are successfully used, the SN's identity is implicitly authenticated. To expedite future AKA, a UE may store CK and IK from a run of EPS AKA. The UE shall not store the two keys on the USIM, instead, the two keys shall be stored in a non-volatile part of the ME memory. It is because that the USIM does not have some of the required fields (e.g. NAS COUNT values) to store EPS security context. Moreover, storing CK and IK for EPS AKA on a USIM may cause overwriting keys resulting from an earlier run of UMTS AKA. Therefore, when a UICC is inserted into another ME, an EPS AKA authentication is required to be initiated. This is in contrast to UMTS where the USIM may store the security context to avoid another UMTS AKA run.

13.3.3 Signaling Protection

LTE separates control plane and user plane, thus, security protection provided for radio-specific signaling and user-plane traffic is also separated.

13.3.3.1 Protection of Radio-Specific Signaling

LTE provides *confidentiality* and *integrity* between the UE and eNB. The backhaul signaling between the eNB and MME is protected by IKE/IPsec (which is beyond the scope of this book). Specific protocols are also available in LTE to provide end-to-end protection of signaling between the MME and UE. In particular, *Confidentiality* may be provided to both RRC signaling and NAS signaling. It is applied at Packet Data Convergence Protocol (PDCP) layer for the protection of RRC signaling, which can prevent UE tracking based on certain radio signaling information, such as cell level measurement reports, handover message mapping, etc. It is an operator's option whether confidentiality needs to be provided for RRC signaling. Confidentiality is usually provided for the NAS signaling by the NAS protocol. In fact, if the NAS security has not been activated, the UE would not send international mobile equipment identity (IMEI) to the network. Both RRC and NAS signaling confidentiality protection is recommended to be used. *Integrity* shall be provided to both RRC and NAS signaling. For the RRC signaling integrity, it is applied at the PDCP layer only. There is no layer below PDCP being integrity protected. For the NAS signaling integrity, it is provided by the NAS protocol.

13.3.3.2 Protection of User-Plane Traffic

LTE provides *confidentiality* between the UE and eNB, while integrity is not provided. Nonetheless, support for both confidentiality and integrity is mandatory in the eNB. The backhaul traffic between the eNB and the S-GW is protected by IKE/IPsec. *Confidentiality* is applied at PDCP layer and protects the user-plane traffic between the UE and eNB. Confidentiality is an operator's option but is recommended to be used. *Integrity* is NOT provided for the user-plane traffic between the UE and eNB. On one hand, integrity leads to expensive bandwidth overhead. On the other hand, it is almost impossible for an attacker to intelligently manipulate user plane packets.

13.3.4 Overview of Confidentiality and Integrity Algorithms

Both confidentiality and integrity algorithms in LTE security are depending on a 128-bit input key. 256-bit input key is another option for LTE confidentiality and integrity algorithms.

13.3.4.1 Confidentiality Mechanism

The confidentiality of both radio signaling and user-plane traffic is provided by the same EEA, as illustrated in Figure 13.8. There are five input parameters to the encryption algorithm: KEY, COUNT-C, BEARER, DIRECTION, and LENGTH.

- KEY is the 128-bit ciphering key (i.e. $K_{RRC_{enc}}$, $K_{NAS_{enc}}$, and $K_{UP_{enc}}$).
- COUNT-C is a 32-bit bearer specific, time and direction dependent value that prevents replay attacks.
- BEARER is an 8-bit radio bearer identity.
- DIRECTION is a 1-bit value that indicates the transmission direction, i.e. 1 for uplink and 0 for downlink.
- LENGTH is the length of the keystream required. Readers shall note that it is the length of keystream block, but not the actual bits in it.

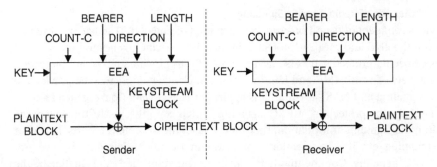

Figure 13.8 LTE ciphering mechanism.

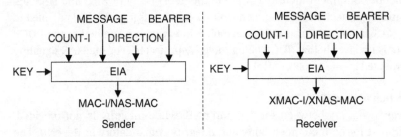

Figure 13.9 LTE integrity mechanism.

The EEA generates the output keystream block KEYSTREAM based on the five input parameters. The encryption is performed by XORing the plaintext block and the keystream block. At the receiver side, the plaintext is recovered by generating the same KEYSTREAM using the same input parameters, and XORing KEYSTREAM with the ciphertext. Although the same algorithm is applied to both signaling traffic and user-plane data, no collision will occur between input parameters. Because the BEARER identities are different between signaling radio bears used for RRC messages and the radio bearers for user-plane data.

13.3.4.2 Integrity Mechanism
Both RRC and NAS signaling integrity are protected by the same evolved packet system integrity algorithm (EIA), as illustrated in Figure 13.9. There are five input parameters to the integrity algorithm: KEY, COUNT-I, MESSAGE, DIRECTION, and BEARER.

- KEY is the 128-bit integrity key (i.e. $K_{RRC_{int}}$ and $K_{NAS_{int}}$).
- COUNT-I is a 32-bit value that prevents replay attacks.
- MESSAGE is the message itself.
- DIRECTION is a 1-bit value that indicates the transmission direction, i.e. 1 for uplink and 0 for downlink.
- BEARER is an 8-bit radio bearer identity.

The output of EIA is the message authentication code or MAC for integrity protection. The MAC is then appended to the message when sent. Based on different applications, the outputs have different names. That is to say, MAC-I/NAS-MAC for radio/NAS signaling integrity protection respectively. The XMAC-I/XNAS-MAC are the corresponding message

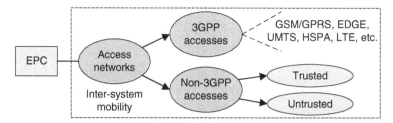

Figure 13.10 3GPP and non-3GPP access networks.

authentication code generated at the receivers' sides, respectively. A receiver verifies the data integrity of the message by comparing the XMAC to the received MAC.

13.3.5 Non-3GPP Access

LTE supports non-3GPP access, thus E-UTRAN is not the only access technology for a UE to reach the EPC. For example, some service operators have already support voice over Wi-Fi in addition to their LTE access. In fact, since code division multiple access (CDMA) or CDMA2000 is non-3GPP technologies, service providers that are based on those technologies shall promote non-3GPP access to LTE. The idea is to use a unique core network that provides various IP-based services over multiple access technologies.

An overview of non-3GPP access to EPC is illustrated in Figure 13.10. Non-3GPP accesses are generally categorized into *trusted* and *untrusted* groups. Any *trusted* non-3GPP accesses direct interact with the EPC, while any *untrusted* non-3GPP accesses interact with the EPC via the evolved packet data gateway (ePDG). Security tunneling is provided by the ePDG to connect UE to the EPC over untrusted non-3GPP access. It is the operator's decision of which non-3GPP technologies being trusted or untrusted. For example, a CDMA based operator may trust CDMA access while a WiMAX based operator may trust WiMAX access. A trusted access shall have EPS security contexts already to process AKA with the EPC. Nonetheless, the access-level authentication is optional for untrusted access. Authentication of an untrusted access is provided by IKE/IPsec security association between the ePDG and UE using EAP AKA. When UE's mobility is based on the dual-stack mobile IPv6 protocol (DSMIPv6), EAP AKA is optional. Because EAP AKA is already applied between UE and PDN-GW by default in DSMIPv6.

13.4   Handover Between eNBs

13.4.1   Overview

Handover from an eNB to another eNB includes two cases. One is handover over X2 interface, the other is handover over S1 interface. X2-handover occurs between two eNBs directly. As shown in Figure 13.11, the source eNB (i.e. eNB_1) initiates handover by sending a request to the target eNB (i.e. eNB_2). The UE security capabilities are included in the request message. The target eNB selects both ciphering and integrity algorithms with highest priority according to the locally configured list of algorithms. The target eNB

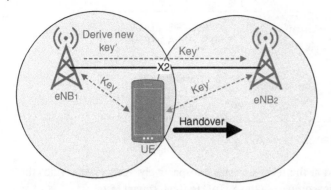

Figure 13.11 Handover from a source eNB over X2 to a target eNB.

then acknowledges handover request with the source eNB and establishes connection with UE. A path switch request is sent from the target eNB to the MME after the handover process to verify UE's security capabilities received from the source eNB. If a handover is legit, then the MME shall confirm. Otherwise, the MME may log the event and take additional measures.

S1-handover involves MME to the process. In the case of MME handover, S1-handover is definitely the choice for process. In S1-handover, a transparent container is created by the MME that contains the UE security capabilities. The container is sent to the target eNB in the handover request. The target eNB selects both ciphering and integrity algorithms with highest priority according to the locally configured list of algorithms. Acknowledgment and command messages are exchanged in the same way. Different from X2-handover, a *handover notify message* is sent from the target eNB to the MME to verify the UE's security capabilities. If a handover is legit, then the MME shall confirm. Otherwise, the MME may log the event and take additional measures.

Note that there might be a need for the target eNB to get a new key that is unknown to the source eNB for security after handover. Therefore, even if an attacker may compromise one of the eNBs and obtain its key, the other eNB can still have its keys secure and continue providing service to UE. This process is illustrated in Figure 13.12. Details of the key handling process is introduced later in this section.

13.4.2 Key Handling in Handover

13.4.2.1 Initialization
Key handling in LTE handover starts with the initialization of K_{eNB} and Next Hop (NH) parameter. The initialization process has three steps, as depicted in Figure 13.13.

(1) K_{eNB} is calculated from the K_{ASME} and the NAS uplink COUNT. Whenever a fresh K_{eNB} is calculated, both UE and MME calculate the NH parameter according to process **Proc-NH0**:

Proc-NH0:

$$NCC = 0; \quad NH = KDF(K_{ASME}, K_{eNB}); \quad K_{eNB}^* = K_{eNB}. \tag{13.1}$$

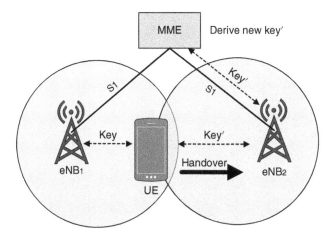

Figure 13.12 Handover from a source eNB over S1 to a target eNB with new key derived by the MME.

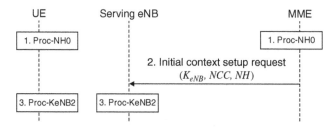

Figure 13.13 Initialization of NH key derivation parameter and K_{eNB}.

(2) MME sends initial context setup request to serving eNB providing K_{eNB}, NCC, and NH in the AS security context. Serving eNB stores NCC and NH as NH^*, and K_{eNB} as K^*_{eNB} for further calculation.

(3) Both UE and serving eNB run process **Proc-KeNB2** that creates new K_{eNB} and RRC/UP keys.

Proc-KeNB2:

If index increases from previous handover

$$K_{eNB} = NH^*.$$

If index does not increase from previous handover:

$$K_{eNB} = KDF(K^*_{eNB}, \text{target C-RNTI});$$

Derive RRC and UP keys from K_{eNB}; Delete K_{eNB^*}.

C-RNTI is the cell radio network temporary identifier that is used for identifying RRC connection and scheduling which is dedicated to a particular UE. C-RNTI is assigned by the serving eNB to UE.

After initialization of K_{eNB} and NH, key handling can be performed in handover.

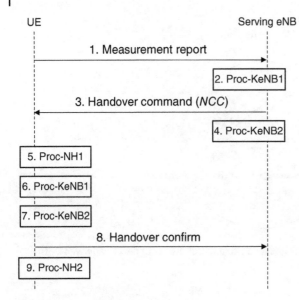

Figure 13.14 Key refresh with intra-eNB handover.

13.4.2.2 Intra-eNB Key Handling

The most intuitive handover is intra-eNB handover, where the NCC and NH parameters are the same as in previous handover. The 9-step key derivation is described for handover procedure in Figure 13.14.

(1) UE sends measurement report to the serving eNB.
(2) The serving eNB runs process **KeNB1** to create target eNB key (K_{eNB}^*).
    Proc-KeNB1: create target eNB key (K_{eNB}^*) – run in UE, eNB, and MME as follows,

---

Proc-KeNB1:
If index increases from previous HO: $K_{eNB} = NH^*$,

$$K_{eNB}^* = KDF(K_{eNB}, PCI). \tag{13.2}$$

---

(3) The serving eNB transfers handover (NCC included) command to UE.
(4) The serving eNB runs process **Proc-KeNB2** to create new K_{eNB}.
(5) UE runs process **Proc-NH1** that updates the NH depending on UE's current NCC value and the NCC value included in the handover command.

Proc-NH1:

Temp-NCC = NCC; $NH^* = NH$.

In UE repeatedly update NH if needed (e.g. $Temp = NCC < Received - NCC$). UE shall try at least MIN-NH-UPDATE (at least 04 in HEX) times before giving up with an error:

$$NH^* = KDF(K_{ASME}, NH^*), + + \text{Temp-NCC}.$$

---

(6) UE runs process **Proc-KeNB1**.

(7) UE runs process **Proc-KeNB2**.

(8) UE sends handover confirmation to the serving eNB.

(9) UE runs process **Proc-NH2** to update current NH.

Proc-NH2:

$$NH^* = NH; \quad NCC = \text{Temp-NCC};$$

$$\text{Delete } NH^*; \quad \text{Delete Temp-NCC.}$$

13.4.2.3 Intra-MME Key Handling

Hanover between eNBs may occur without changing the MME, or it is called intra-MME. X2-handover only applies to intra-MME handover. Figure 13.15 illustrates intra-MME key handling process.

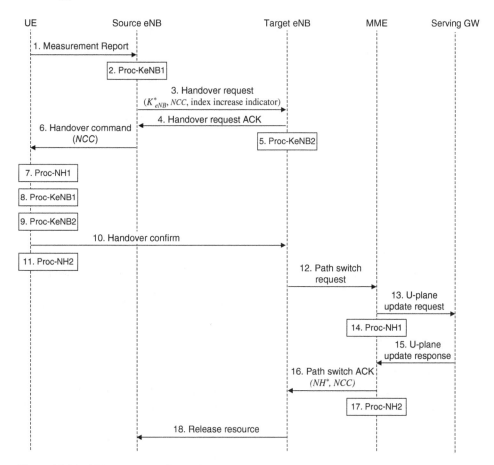

Figure 13.15 NH based key refresh with intra-MME handover.

(1) UE sends measurement report to the source eNB.

(2) The source eNB runs process **Proc-KeNB1**.

(3) The source eNB sends handover request to the target eNB. The request also includes security context NCC, index increase indicator, and K^*_{eNB}.

(4) The target eNB sends handover request acknowledgment to the source eNB.

(5) Target eNB runs process **Proc-KeNB2**.

(6) The source eNB sends handover command (NCC included) to UE.

(7) UE runs process **Proc-NH1** that updates the NH depending on UE's current NCC.

(8) UE runs process **Proc-KeNB1**.

(9) UE runs process **Proc-KeNB2**.

(10) UE sends handover confirmation to the target eNB.

(11) UE runs process **Proc-NH2**.

(12) The target eNB sends path switch request to MME. Note that the path switch message is transmitted after the handover process, it is only used to provide keying material for the next handover procedure and the target eNB. Since the source eNB has the target eNB keys, key separation happens only after two hops.

(13) MME sends user-plane update request to S-GW.

(14) MME runs process **Proc-NH1** and may update NCC and NH^*.

(15) S-GW sends user-plane update response to MME.

(16) MME sends path switch acknowledgment together with NCC and NH^*. Target eNB updates NCC and NH^* with the received NCC and NH^* respectively.

(17) MME runs process **Proc-NH2**.

(18) Target eNB sends release resource message to source eNB and completes the entire handover process.

13.4.2.4 Inter-MME Key Handling

If handover is required between different MMEs, it is called inter-MME handover. Figure 13.16 describes key derivation steps for inter-MME handover procedure.

(1) UE sends measurement report to the source eNB.

(2) The source eNB runs process **Proc-KeNB1**.

(3) The source eNB sends handover required message to the source MME. Included in the request are AS level security context NCC, K^*_{eNB}, and index increase indicator.

(4) The source MME sends forward relocation request to the target MME. Included in the request are AS level security context from the source eNB, plus NH and NCC.

(5) The target MME updates NH by running process **Proc-NH1**.

(6) The target MME runs process **Proc-NH1** again.

(7) The target MME sends handover request to the target eNB. Included in the request are the AS level security context from the source eNB, plus fresh NH^* and NCC. The target eNB updates NCC and NH with fresh NH^* and NCC in the handover request if MME included them in addition to the security context from the source eNB.

(8) The target eNB sends handover request acknowledgment to the target MME.

(9) The target eNB runs process **Proc-KeNB2** based on the context received from the source eNB.

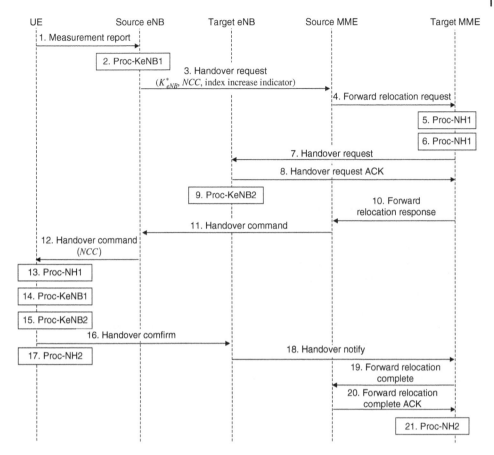

Figure 13.16 Key refresh with inter-MME handover.

(10) The target MME sends forward relocation response to the source MME.

(11) The source MME sends handover command to the source eNB. Included in the message are NCC and *NH** if updated.

(12) The source eNB sends handover command to UE. Included in the message is NCC from the source MME if present, otherwise using the NCC that is in its memory.

(13) UE runs process **Proc-NH1** that updates the NH depending on UE's current NCC and the received NCC in the handover command.

(14) UE runs process **Proc-KeNB1**.

(15) UE runs process **Proc-KeNB2**.

(16) UE sends handover confirmation to the target eNB.

(17) UE runs process **Proc-NH2**.

(18) The target eNB sends handover notification to the target MME.

(19) The target MME sends forward relocation completion notice to the source MME.

(20) The source MME sends forward relocation completion acknowledgment to the target MME.

(21) The source MME runs process **Proc-NH2**.

Table 13.1 Security algorithms in LTE.

Identifier	Code name	Core algorithm
0000	128-EEA0	Null ciphering algorithm
0001	128-EEA1	SNOW 3G
0010	128-EEA2	AES
0001	128-EIA1	SNOW 3G
0010	128-EIA2	AES

13.5 Security Algorithms

Confidentiality of LTE signaling traffic and user plane traffic is protected by EEA. Integrity of LTE signaling traffic is protected by EIA. As mentioned earlier in this chapter, algorithm identifiers are involved in LTE AKA as well as LTE handover processes. Therefore, there are options when implementing EEA and EIA besides different key size. Algorithm identifier is a 4-bit value. The security algorithms in LTE are listed in Table 13.1. Algorithms EEA1 and EIA1 are identical to the UMTS algorithms with a mapping of LTE parameters to UMTS parameters. Both EEA2 and EIA2 will be illustrated in this section with 128-bit key input [112]. Readers may have noted that the algorithm identifier has reserved a few spaces for future options. For example, EEA3/EIA3 based on ZUC may be the next security algorithm choice for LTE. EEA3/EIA3 will also be illustrated in this section with 128-bit key input [113].

13.5.1 128-EEA2

As shown in Figure 13.17, EEA2 uses block cipher AES in CTR mode to generate key stream. Encryption/decryption are performed by XORing the plaintext/ciphertext with a key stream. AES is chosen as the kernel for the following reasons. First eNB needs to support network domain security (NDS)/IP which uses AES. Moreover, the KASUMI based 3GPP security algorithm has a licensing fee if not used for 3GPP access protection. The AES-CTR uses 128-bit counters (T_i). Counters are initialized for AES block operation to generate key stream. Figure 13.18 illustrates the structure of the first counter block T_1. Subsequent counter blocks are obtained by applying the standard integer incrementing function [114].

> **Standard Integer Incrementing Function**: let n be the number of blocks, b be the block size, and m be the number of bits to be incremented. The standard incrementing function takes $[x]_m$ and returns $[x + 1 \bmod 2^m]_m$.

In this EEA2 implementation, $b = 128$ and $m = 64$ for standard integer incrementing. A key stream block is generated by encrypting a counter using AES, s.t., $KS(i) = AES(T_i)$. A plaintext is partitioned into 128-bit blocks for encryption and decryption. The operation is simply XORing a key stream with a plaintext block.

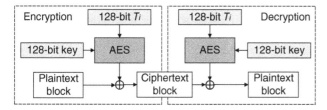

Figure 13.17   Overview of EEA2.

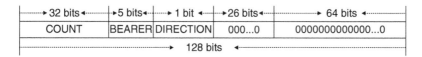

Figure 13.18   First counter block.

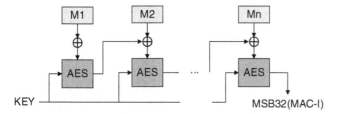

Figure 13.19   Overview of EIA2.

32 bits	5 bits	1 bit	26 bits	64 bits
COUNT	BEARER	DIRECTION	000...0	First 64 bits of MESSAGE

128 bits

Figure 13.20   First input block (M_1) for 128-EIA2.

13.5.2 128-EIA2

EIA2 is based on AES-CMAC mode to generate the message authentication code. An overview of EIA2 is given in Figure 13.19. The input message and other parameters are partitioned into 128-bit blocks for processing. The 32 most significant bits of the final output is the message authentication code MAC-I.

The first input block includes COUNT, BEARER, DIRECTION, zero paddings, and the first 64 bits of MESSAGE, as shown in Figure 13.20. The following blocks are partitioned MESSAGE except for the last message block M_n. Two subkeys K_1 and K_2 are derived from the input key for the last message block. If the last block from the original message (e.g. M_n^*) contains 128 bits (a complete block), then $M_n = K_1 \oplus M_n^*$. If M_n^* needs padding to be complete, then $M_n = K_2 \oplus (M_n^* \| 10 \ldots 0)$. Subkeys K_1 and K_2 are derived using Algorithm 13.1, where function $\lll$ is left rotation by bits, i.e. $(L \lll n)$ indicates n-bit left rotation of L.

Algorithm 13.1 EIA2 subkey derivation

Input: K;

Output: K_1, K_2;

    $R \leftarrow 0 \dots 010000111$;

    $L \leftarrow \text{AES}(K, 0 \dots 0)$;

    if $MSB_1(l) == 0$ **then**

        $K_1 \leftarrow L \lll 1; K_2 \leftarrow K_1 \lll 1$;

    else

        $K_1 \leftarrow (L \lll 1) \oplus R; K_2 \leftarrow (K_1 \lll 1) \oplus R$;

    end if

    Return K_1, K_2;

13.5.3 EEA3

3GPP TS 35.221 [115] and 3GPP TS 35.222 [116] specify another set of algorithms, EEA3 and EIA3 to be applied for LTE security. The two algorithms are based on ZUC, a stream cipher algorithm that is named after ZU, Chongzhi, a famous ancient Chinese mathematician [113]. The overview of 128-EEA3 is shown in Figure 13.21. The inputs are cipher key *CK* and a 128-bit IV value. The ciphering key is used in bytes as follows:

$$CK = CK[0]\|CK[1]\|CK[2]\|, \dots, \|CK[15].$$

Each block operation generates a 128-bit key stream to encrypt/decrypt a data block. EEA3 is able to encrypt/decrypt a data block between 1 and 20 000 bits long. The 128-bit IV is initialized based on COUNT, BEARER, and DIRECTION. The initialization process is illustrated in Algorithm 13.2.

Algorithm 13.2 EEA3 initialization

Input: CK;

Output: *Key*, *IV*;

    for $i := 0$ to 15 **do**

        $Key[i] \leftarrow CK[i]$;

    end for

    for $i := 0$ to 3 **do**

        $IV[i] \leftarrow COUNT[i]$;

    end for

    $IV[4] \leftarrow BEARER\|DIRECTION\|00$;

    for $i := 5$ to 7 **do**

        $IV[i] \leftarrow 0$;

    end for

    for $i := 8$ to 15 **do**

        $IV[i] \leftarrow IV[i - 8]$;

    end for

Figure 13.21 Overview of 128-EEA3.

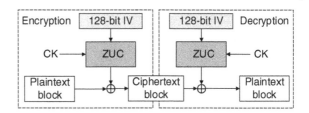

Figure 13.22 Overview of 128-EEA3.

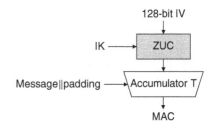

Once IV is generated, it can be applied to ZUC to generate key words $z[i]$, where $i = 1, 2, \ldots, [LENGTH/32]$, according to Algorithm 13.3. The encryption/decryption of EEA3 is the same as other stream ciphers by XORing a keystream with a data block.

Algorithm 13.3 EEA3 keystream generation

Input: *LENGTH, Key, IV*;
Output: *KEYSTREAM*;
   $L \leftarrow [LENGTH/32]$;
   Allocate memory space for the *KEYSTREAM*;
   Generate *KEYSTREAM*;

13.5.4 EIA3

EIA3 is a universal hash function using ZUC as its core, as shown in Figure 13.22. The final output is a 32-bit MAC value for integrity protection. An initialization vector IV is also generated for EIA3, however, it is different from the IV of EEA3. The initialization process of 128-EIA3 is illustrated in Algorithm 13.4.

ZUC is applied to generate L key words $KS[1], KS[2], \ldots, KS[L]$, where $L = [N/32]$ and $N = LENGTH + 64$. The process of 128-EIA3 keystream generation is illustrated in Algorithm 13.5. Finally, the accumulator T generates the 32-bit MAC based on the message and the keystream based on Algorithm 13.6. Note that $k[0], k[1], \ldots, k[31], k[32], \ldots, k[N-1]$ are the key bit stream corresponding to the previous key words *KS*.

Algorithm 13.4 EIA3 initialization

Input: *CK*;
Output: *Key, IV*;
    for $i := 0$ to 15 **do**
      $Key[i] \leftarrow CK[i]$;
    end for
    for $i := 0$ to 3 **do**
      $IV[i] \leftarrow COUNT[i]$;
    end for
    $IV[4] \leftarrow BEARER\|000$;
    for $i := 5$ to 7 **do**
      $IV[i] \leftarrow 00000000$;
    end for
    $IV[8] = IV[0] \oplus (DIRECTION << 7)$
    for $i := 9$ to 13 **do**
      $IV[i] \leftarrow IV[i - 8]$;
    end for
    $IV[14] = IV[6] \oplus (DIRECTION << 7)$;
    $IV[15] = IVP[7]$;

Algorithm 13.5 EIA3 keystream generation

Input: *LENGTH, Key, IV*;
Output: *KEYSTREAM*;
    $N = LENGTH + 64$;
    $L \leftarrow \lceil LENGTH/32 \rceil$;
    Allocate memory space for the *KEYSTREAM*;
    Generate *KEYSTREAM*;

Algorithm 13.6 EIA3 MAC Calculation

Input: *LENGTH, plaintext, KEYSTREAM*;
Output: *MAC*;
    for $i := 0$ to $LENGTH - 1$ **do**
      if $M[I] == 1$ **then**
        $T = T \oplus k_i$;
      end if
      $ciphertext[i] \leftarrow plaintext[i] \oplus KEYSTREAM[i]$;
    end for
    $T \leftarrow T \oplus k_{LENGTH}$;
    $MAC \leftarrow T \oplus k_{N-32}$;

13.6 Security for Interworking Between LTE and Legacy Systems

LTE has been deployed by many operators in the world, however, it is not the only wireless service provided for most of the operators. For example, some operators still provide GSM, UMTS, and LTE to their subscribers simultaneously. Seamless transition between LTE and legacy systems must be achieved not only for services but also for security. This section illustrates the security for interworking between LTE and UMTS systems, and between LTE and GSM systems for both idle mode and handover mode.

13.6.1 Between LTE and UMTS

Security for interworking between LTE and UMTS takes place in their serving networks, i.e. E-UTRAN and UTRAN correspondingly. As illustrated earlier in this chapter and the previous chapter, LTE NAS attaches request messages during AKA process include the UTRAN security capabilities of UE. All parameters needed for UMTS security are sent back to the UE in the corresponding integrity protected response messages. However, since the two systems are not exactly the same, security interworking is thus not as intuitive.

13.6.1.1 Idle Mode Mobility from E-UTRAN to UTRAN

If UE has been previously authenticated with the serving GPRS support node (SGSN) that is part of an UTRAN, then UE has valid SGSN security context already. As shown in Figure 13.23, a routing area update (RAU) request shall be sent to SGSN together with P-TMSI and a key set indicator (KSI). Note that LTE only supports packet service, thus P-TMSI is applied as user identity. KSI is to indicate cached keys in the SGSN. If the network does not have valid cached security context, then the UE must run UMTS AKA with the serving network.

If UE has not been previously authenticated with the SGSN, thus it does not have valid cached SGSN security context. As shown in Figure 13.24, UE sends a RAU request, its P-TMSI, and a KSI to the target SGSN. In this case, the P-TMSI is mapped from its GUTI. A NAS-token is included in the P-TMSI signature. The KSI only indicates KSI_{ASME}, which is the key indicator of K_{ASME}. Both the UE and the target SGSN will assign the value of KSI_{ASME} to KSI, i.e. $KSI = KSI_{ASME}$, thus to locate K_{ASME}. After that, the target SGSN forwards the P-TMSI signature and NAS-token to the MME. The MME verifies the signature by comparing it with a NAS-token. Sometimes the NAS-token needs to be truncated some of the most significant bits for comparison. Once the verification succeeds, the security context request message is authenticated and authorized. Both the UE and the MME will derive

Figure 13.23 Interworking from E-UTRAN to UTRAN with cached context.

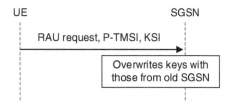

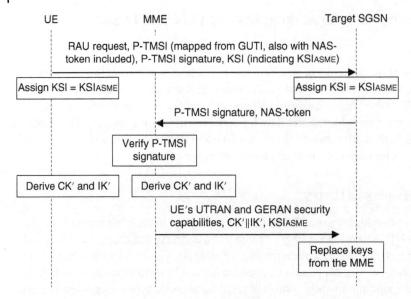

Figure 13.24 Interworking from E-UTRAN to UTRAN without cached context.

CK' and IK' based on K_{ASME} and the current NAS downlink COUNT value indicated by the NAS-token, s.t.,

$$CK'\|IK' = KDF(K_{ASME}, COUNT).$$

Note that the UE may not derive the keys simultaneously with the MME in practice. It is logical for the UE to perform the operation at this step. After that, the MME sends the security context, including UE's UTRAN and GERAN security capabilities, $CK'\|IK'$, and KSI_{ASME} to the target SGSN. Finally, the target SGSN replaces the keys with the ones from the MME, K_{ASME} is located with KSI_{ASME}.

13.6.1.2 Idle Mode Mobility from UTRAN to E-UTRAN

Interworking from UTRAN to E-UTRAN is to transfer $CK\|IK$ from UE to MME so that both MME and UE can derive K'_{ASME} for further LTE service. *Mapped context* and *cached context* are two specific cases.

- *Mapped context*: In this case, UE does not have cached context available. As shown in Figure 13.25, UE sends TAU request, together with source system temporary identity, KSI_{SGSN}, and $NONCE_{UE}$ to MME. Since no cached security context is available, the TAU request is sent unprotected. MME then generates a $NONCE_{MME}$ and fetches $CK\|IK$ from SGSN based on KSI_{SGSN} and the temporary identity. After that, MME generates a fresh mapped K'_{ASME} with $NONCE_{UE}$, $NONCE_{MME}$, and $CK\|IK$. The final TAU accept message is protected using the NAS keys based on K'_{ASME}
- *Cached context*: In this case, UE has cached E-UTRAN security context. As shown in Figure 13.26, UE sends the tracking area update (TAU) request, together with the temporary identity, KSI_{SGSN}, KSI_{ASME}, and $NONCE_{UE}$ to MME. The cached security context

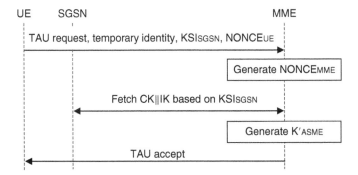

Figure 13.25 Interworking from UTRAN to E-UTRAN with mapped context.

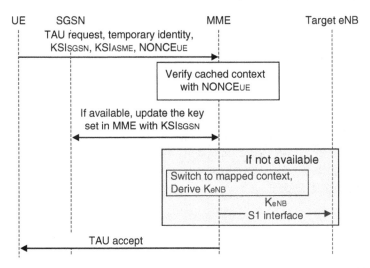

Figure 13.26 Interworking from UTRAN to E-UTRAN with cached context.

algorithms are applied to protect the TAU request message. $NONCE_{UE}$ is for MME to verify if cached context is available. If available, MME is pointed to the source SGSN and updates the key set based on KSI_{SGSN}. If cached security context is not available in the network, the network will switch to mapped security context and derive K_{eNB}. K_{eNB} is delivered to the target eNB on the S1 interface.

13.6.1.3 Handover Mode from E-UTRAN to UTRAN

In handover mode, interworking from E-UTRAN to UTRAN always uses mapped context. As shown in Figure 13.27, once MME receives RAU request, it shall derive CK' and IK' from K_{ASME} and the current NAS downlink COUNT value using KDF function. Then, MME transfers $CK'\|IK'$ and KSI_{ASME} to SGSN. KSI_{ASME} is assigned to KSI since UTRAN does not support K_{ASME}. SGSN transfers RAU acknowledgment and the algorithms to RNC. The selected algorithms are indicated to UE in the handover command message.

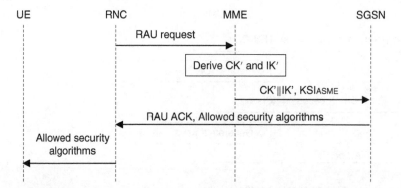

Figure 13.27 Interworking from E-UTRAN to UTRAN in handover mode.

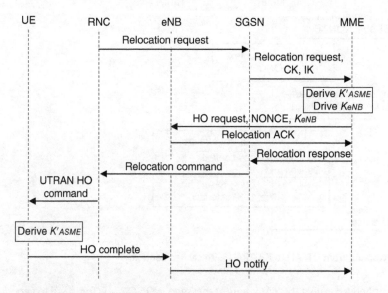

Figure 13.28 Interworking from UTRAN to E-UTRAN in handover mode.

13.6.1.4 Handover Mode from UTRAN to E-UTRAN

The interworking from E-UTRAN to UTRAN in handover case is illustrated in Figure 13.28. RNC sends a relocation request to SGSN. SGSN forwards the relocation request, together with UE's E-UTRAN security capabilities (i.e. *CK* and *IK*) to MME. MME derives K'_{ASME} based on *CK*, *IK*, and $NONCE_{MME}$. K_{eNB} is derived at MME based on K'_{ASME}. Then, MME sends handover request, including $NONCE_{MME}$ and K_{eNB} to eNB. KSI_{ASME} and KSI_{SGSN} are also sent to eNB for the purpose to support both mapped and cached context based K_{eNB}. eNB is responsible of selecting NAS, UP, and RRC algorithms. They are sent to MME in the relocation acknowledgment. MME shall include the information in a relocation response and send it to SGSN. It is then forwarded to RNC in a relocation command. RNC shall forward it to UE in UTRAN HO command. UE then derives K'_{ASME} and notifies eNB of handover completion. Finally, eNB notifies MME of the completion.

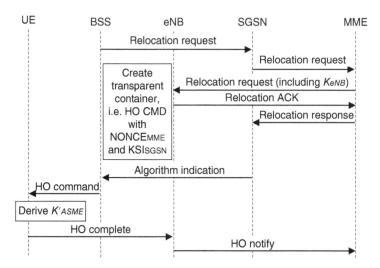

Figure 13.29 Interworking from GERAN to E-UTRAN in handover mode.

13.6.2 Between E-UTRAN and GERAN

Interworking between E-UTRAN and GERAN is to ensure backward capability from LTE to GSM (GPRS).

13.6.2.1 Idle Mode
Similar to interworking between E-UTRAN and UTRAN, interworking from E-UTRAN to GERAN also supports *cached context* and *mapped context*.

- *Cached context*: If UE has cached valid SGSN security context, it sends LAU request together with P-TMSI and *KSI* to SGSN. Keys indicated by *KSI* shall overwrite the ones in the target SGSN. If no valid cached security context is available in the network, then an AKA shall be initiated.
- *Mapped context*: If UE does not have cached SGSN security context, it also sends LAU request together with P-TMSI and *KSI* to SGSN. However, this P-TMSI is mapped from GUTI, and *KSI* indicates KSI_{ASME}. MME shall transfer UE's UTRAN and GERAN scrutiny capabilities as well as CK' and IK' with KSI_{ASME} to SGSN. Keys from MME shall replace the ones in the target SGSN.

Interworking process from GERAN to E-UTRAN is the same as the process from GERAN to E-UTRAN. Please refer to Section 13.6.1 for more information.

13.6.2.2 Handover Mode
The interworking process in handover mode from E-UTRAN to GERAN is similar to the process from E-UTRAN to UTRAN. MME derives CK' and IK' from K_{ASME} and transfer them to the SGSN. SGSN derives K_c (ciphering key in GERAN) from CK' and IK'. In the meantime, MME shall also transfer UE's UTRAN and GERAN security capabilities to SGSN. It is SGSN's responsibility to select the encryption algorithm to use in GERAN after handover. Note that GERAN does not support integrity. Figure 13.29 illustrates

interworking process from GERAN to E-UTRAN. BSS sends relocation request to SGSN. SGSN forwards the request to MME. Then, MME selects the NAS security algorithms and the K_{eNB}. The information is included in a relocation request sent from MME to the target eNB. eNB shall select the RRC and UP algorithms and reply to MME a relocation acknowledgment with the algorithm indication. Once the acknowledgment is received by MME, it sends a relocation response to SGSN with the selected NAS, UP, and RRC algorithms. SGSN shall transfer those algorithms to BSS through a specific message. A transparent container that includes handover command, $NONCE_{MME}$ and KSI_{SGSN} is created by eNB and transferred to the target BSS. UE is informed with the algorithms and other information through the handover command. The rest of the process is the same as that from UTRAN to E-UTRAN.

13.7 Summary

In this chapter, LTE security is illustrated. LTE architecture is developed from UMTS system, thus several security mechanisms are carried over with modifications, especially the AKA protocol. The major difference between LTE security and the previous generation is that LTE separates control plane and user plane. More keys are involved in LTE security to serve the more complicated system. Several security algorithms have been proposed and implemented to protect LTE security. Algorithms such as AES and ZUC are applied as core to LTE security algorithms EEA and EIA. Moreover, LTE also provides security for interworking with legacy systems as well as non-3GPP access. LTE has a strong security implemented comparing with the previous generation system. LTE will continue to serve as an important part with the next generation wireless system. Even although LTE has been deployed for a few years already, it is still necessary to continue enhancing the security.

Part V

Security for Next Generation Wireless Networks

14

Security in 5G Wireless Networks

The fifth generation (5G) wireless network systems are the next generation mobile wireless telecommunications beyond the current fourth generation (4G)/International Mobile Telecommunications (IMT)-Advanced Systems [117]. The advanced features of the 5G wireless network systems bring new security requirements and challenges. Since 5G wireless network systems are still under development, this chapter introduces the current solutions and research results on security of 5G wireless network systems.

14.1 Introduction to 5G Wireless Network Systems

14.1.1 The Advancement of 5G

The 5G wireless network systems (interchangeable with 5G systems for the rest of this chapter) are more than a direct improvement over the legacy 4G cellular networks. The new systems will include several advanced features to support demanding services [118]. As listed in Table 14.1, the 5G wireless network systems will have huge improvement in communication bandwidth, transmission latency, coverage, and energy efficiency (EE). The advanced features are introduced to better support the implementation of high-density mobile broadband networks, device-to-device (D2D) communications, massive machine-type communications, low-latency and low-energy Internet of things (IoT), vehicular communication networks, and many other types of applications [119–121]. Various new technologies have been developing to support all the advanced features of 5G systems [122]. The most significant ones include heterogeneous networks (HetNets), massive multiple-input multiple-output (MIMO), millimeter wave (mmWave) [123], D2D communications [124], software defined network (SDN) [125], network functions visualization (NFV) [126], and networking slicing [127]. A generic architecture of 5G systems is shown in Figure 14.1. As it shows, 5G systems can provide not only traditional voice and data communications, but also many new use cases, such as vehicle-to-vehicle and vehicle-to-infrastructure communications, industrial automation, health services, smart cities, smart homes, critical services, and many other new industry applications, as well as a multitude of devices and applications to connect society at large [128–130].

Security in Wireless Communication Networks, First Edition. Yi Qian, Feng Ye, and Hsiao-Hwa Chen.
© 2022 John Wiley & Sons Ltd. Published 2022 by John Wiley & Sons Ltd.
Companion website: www.wiley.com/go/qian/sec51

Table 14.1 Advanced features of 5G wireless systems.

High data rate	1–10 Gbps connections to end points in the field
Low latency	1 ms latency
Large connectivity	10–100 times the number of connected devices
High availability	99.999% availability
High energy efficiency	90% reduction of network energy usage

Source: Based on GSMA Intelligence [119].

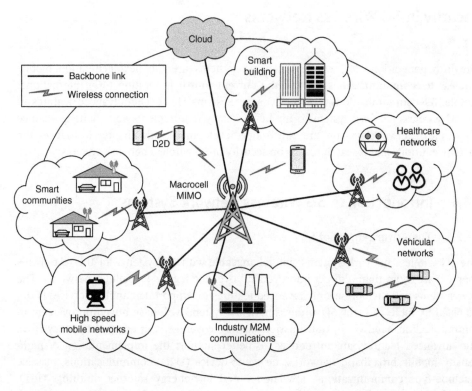

Figure 14.1 A generic 5G wireless system.

14.1.2 5G Wireless Network Systems

The standardization process for 5G wireless systems is in progress. 5G systems certainly introduce new perspectives and changes to legacy systems. In a 5G wireless network, user equipment (UE) and services are not limited to traditional mobile phones, voice and data services. More than 70 different use cases are included in user interface. The use cases are classified into four different groups by 3GPP. The four groups are: massive IoT, critical communications, network operation, and enhanced mobile broadband [131].

Figure 14.2 shows a general 5G wireless network system [132]. The radio access network in a 5G wireless network combines virtualization, centralization, and coordination techniques for efficient and flexible resource allocation. In addition to 3GPP access (e.g. LTE)

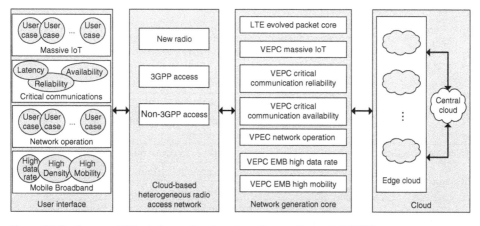

Figure 14.2 A general 5G wireless network system. Source: Fang et al. [132].

and non-3GPP access (e.g. CDMA and Wi-Fi), new radio access from certain user cases will also be supported for efficient spectrum utilization. Moreover, there are several new technologies, such as massive MIMO, HetNet, and D2D communications, to be deployed to the access network to improve the network performance. The core of 5G systems will inherit the LTE and evolved packet core (EPC) in the first stage of implementation. In later stages, eEPC will be virtualized with network slicing, SDN, and NFV implementation to be flexible and service-oriented for different use cases. The virtualized evolved packet core (VEPC) is expected to be access-independent by fully separating control plane and user plane for flexibility and scalability. Different types of cloud computing will be supporting the VEPC. Edge cloud is distributed to improve the service quality. Central cloud can implement global data share and centralized control.

14.2 5G Security Requirements and Major Drives

The new architecture, new technologies, and new use cases in 5G systems, nonetheless, will bring new challenges to security and privacy protection [133].

14.2.1 Security Requirements for 5G Wireless Networks

The security requirements of 5G wireless networks are widely discussed among the communities [120, 127, 134]. 5G wireless networks need to be backward compatible to legacy LTE security. For example, mutual authentication and user traffic encryption are provided between UE and base station; key hierarchy and handover key management are provided to secure access and mobility management in legacy LTE.

In addition to those have been applied to legacy LTE systems, new security mechanisms are needed to comply with the overall 5G advanced features such as low latency and high energy efficiency [135]. As shown in Table 14.2, 5G wireless networks will be service-oriented that emphasize on security and privacy requirements regarding to corresponding services [133]. Due to the rapid increase of network capacity and number of user devices,

Table 14.2 Security requirements for 5G wireless networks.

4G compliance	Improve resilience and availability of the network against signaling based threats including overload caused maliciously or unexpectedly
	Specific security design for extremely low latency
	Comply with security requirements defined in 3GPP 4G standards
	Provide Public Safety and Mission Critical Communications (resilience and high availability)
5G network traffic	Be resilient to significant changes in network traffic patterns
5G radio access	Improve system robustness against smart jamming attacks
	Improve security for 5G small cell nodes
5G signaling plane	Improve overload controls to mitigate possible attacks from huge number of compromised M2M/IoT devices
5G user plane	Possible integrity protection to user plane data in special cases
5G security measures	Possible mandates on security measures in 5G wireless networks

5G network traffic can suffer from sudden and significant changes in network traffic pattern. It is required for 5G security to be resilient to those significant changes. Security of the 5G radio access requires to consider small cell nodes in the design. Also, 5G radio access needs to be robust against smart jamming attacks. Moreover, huge number of compromised machine-to-machine (M2M) and IoT devices may bring down the 5G signaling plane. Therefore, overload controls must be improved to mitigate the possible effects. Some special cases in 5G wireless networks such as battery power IoT devices or low latency 5G devices may not afford application level integrity protection. Protection can be added by providing integrity to the 5G user plane. However, most user cases should rely on upper layers to provide data integrity. Moreover, it may be helpful to mandate the use of security measures in 5G wireless networks to mitigate hazards due to the optional implementation in the legacy LTE systems.

14.2.2 Major Drives for 5G Wireless Security

There are three major drives for 5G wireless security: *supreme built-in-security*, *flexible security mechanisms*, and *automation*, as illustrated in Figure 14.3.

14.2.2.1 Supreme Built-in-Security

Security must be considered as an integral part of the overall architecture and from the very beginning of 5G wireless network systems design. On one hand, security requirements introduced from new use cases and new technologies need to be addressed in 5G security design [136]. For example, HetNet in 5G systems may need delay sensitive and frequent authentications due to complex network structures [137]. SDN and NFV in 5G systems will support new service delivery models and thus require new security aspects [138, 139]. On the other hand, new technologies also introduce new security approaches in 5G system design. For example, massive MIMO in 5G systems is considered effective against

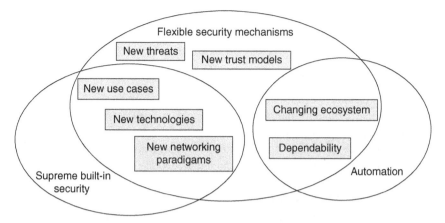

Figure 14.3 Major drives for 5G wireless security.

eavesdropping with proper implementation [140]. SDN and NFV certainly introduce more flexible and power software implementation of security mechanism in 5G systems.

14.2.2.2 Flexible Security Mechanisms
Security of more than 70 use cases must be supported in 5G wireless networks. Thus trust models in 5G systems needs to be refreshed from the one used in legacy LTE systems. As shown in Figure 14.4, the trust models are enabled between users/services, and users/networks, while trust models are extended to services/networks in 5G systems [133]. Security mechanisms design must be flexible for vertical industries use cases [141]. For instance, mobile devices require relatively lightweight security mechanisms due to power constraint. High-speed services also require efficient security services due to latency requirements. Moreover, the authentication management in 5G systems is more complex due to various types and massive number of connected devices.

14.2.2.3 Automation
Automated and intelligent security controls need to be combined with automated holistic security management in 5G wireless security [135]. Automation in 5G systems may introduce more challenges since services closely interact to each other. For example, the fixed telephone line, Internet access, and TV service could be interrupted simultaneously given an outage of a major network [133]. Moreover, privacy needs extra attention in 5G wireless

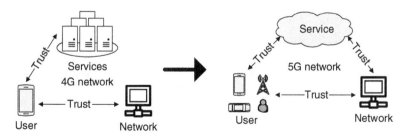

Figure 14.4 Trust models of 4G and 5G wireless networks.

security design since more personal information would be appeared to network and service management.

14.2.3 Attacks in 5G Wireless Networks

Security attacks in wireless networks can be classified into passive and active ones. Passive attacks tend to make use of the information from legitimate users, while not altering active communications. Active attacks can involve modification of the data, and/or interruption of legitimate communications. Figure 14.5 illustrates four typical attacks at physical (PHY) and medium access control (MAC) layers in 5G wireless networks.

14.2.3.1 Eavesdropping and Traffic Analysis
Eavesdropping is a passive attack that uses an unintended receiver to intercept messages by tapping into communication channels, as shown in Figure 14.5a. Due to the nature of wireless communications, eavesdropping can be easily deployed, whereas detection of such attack is rather difficult. Traffic analysis is another type of passive attack that intends to intercept information such as location and identity of the communication parties by analyzing the traffic of the received signal without understanding the contents carried by the signal.

14.2.3.2 Jamming
Jamming is an active attack that tends to disrupt on-going communications between legitimate users, as shown in Figure 14.5b. Such attack is conducted by generating high interference in the communication channel thus normal transmission cannot be identified at the receiver side. In some MAC layer protocols, e.g. carrier sense multiple access-collision

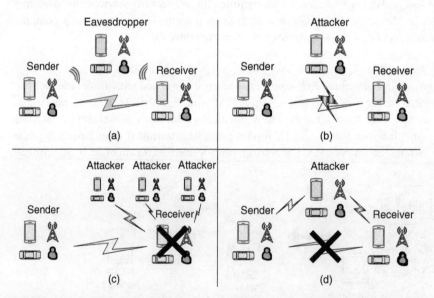

Figure 14.5 PHY/MAC layer attacks in 5G wireless networks. (a) Eavesdropping and traffic analysis. (b) Jamming. (c) DoS and DDoS. (d) Man-In-The-Middle.

avoidance (CSMA/CA), jamming attack may block authorized users from accessing radio resources.

14.2.3.3 DoS and DDoS

DoS and DDoS are active attacks that tend to exhaust resources from legitimate users and/or network systems, as shown in Figure 14.5c. Success of a DoS/DDoS attack can compromise the availability of 5G wireless networks, posing threats to both operators and users [134]. A DoS/DDoS attack against the network infrastructure can strike the signaling plane, user plane, management plane, support systems, radio resources, logical, and PHY resources. A DoS attack against device/user can target on battery, memory, disk, CPU, radio, actuator, and sensors. Currently, detection is mostly used to recognize DoS and DDoS attacks.

14.2.3.4 Man-In-The-Middle (MITM)

MITM is an active attack that tend to intercept, modify, and replace communication messages between the two legitimate users, as shown in Figure 14.5d. Success of an MITM attack may compromise data confidentiality, integrity, and availability of the 5G wireless networks. In a legacy cellular network, false base station based MITM forces a legitimate user to create a connection with a rogue base transceiver station [142]. Mutual authentication between the mobile device and the base station is normally used to prevent the rogue base station based MITM.

14.3 A 5G Wireless Security Architecture

A new security architecture is needed in 5G networking paradigms [143]. New elements will be added to achieve all those required security services.

14.3.1 New Elements in 5G Wireless Security Architecture

Typical new elements at the control plane of 5G wireless security architecture are cloud based computing systems. As shown in Figure 14.6, the newly added computing systems are edge clouds and a central cloud. Edge clouds are applied to improve the network performance by reducing the communication delay. Central cloud connects the edge clouds for data sharing and centralized control.

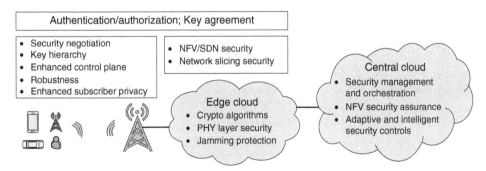

Figure 14.6 Elements in a 5G security architecture.

The 3GPP Technical Report (TR) 23.799 defines the major network functions at the control plane of the VEPC of 5G systems [144]:

- *Access and mobility management function (AMF):* This function is applied to manage access control and mobility, similar to mobility management entity (MME) for the 4G network. AMF can vary with different use cases in 5G systems. Note that mobility management function is not necessary for fixed access applications.
- *Session management function (SMF):* This function is set up based on network policy to manage sessions. Multiple SMFs can be assigned to a single AMF for each user.
- *Unified data management (UDM):* This function manages subscriber data and profiles, such as authentication data of users, for both fixed and mobile accesses in VEPC.
- *Policy control function (PCF):* This function provides roaming and mobility management, quality of service, and network slicing. AMF and SMF are controlled by PCF. Differentiated security can be provided with PCF.

14.3.2 A 5G Wireless Security Architecture

A 5G wireless security architecture is shown in Figure 14.7 [132]. With the new characteristics of VEPC, a separation of data plane and control plane of VEPC is proposed, where the data plane can be programmable for flexibility. AMF and SMF are integrated in the 4G network as MME. The separation of AMF and SMF in 5G can support a more flexible and scalable architecture. In the control plane, network functions can be applied based on different use cases. Similar to the 4G network, four security domains are defined in the 5G wireless security architecture.

14.3.2.1 Network Access Security (I)

The set of security features that provide the user interface to access VEPC securely and protect against various attacks on radio access links. New PHY layer communication technologies such as massive MIMO, HetNet, D2D communications, and mmWave are applied to the radio access network with new challenges and opportunities in network access security. Current research on network access security focuses on providing user identity and location confidentiality, user data and signaling data confidentiality, and entity authentication.

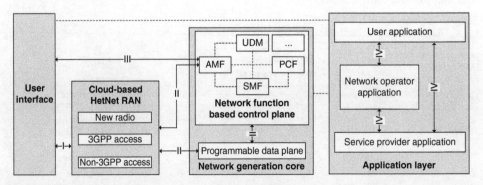

Figure 14.7 A 5G wireless network security architecture. Source: Fang et al. [132].

14.3.2.2   Network Domain Security (II)

The set of security features that protect against attacks between access network and VEPC for secure signaling data and user data exchange. New security vulnerabilities introduced by technologies such as cloud computing, network slicing and NFV need to be addressed in this domain. Nonetheless, the separation of control plane and user plane can greatly reduce the overhead from signaling data synchronization. Entity authentication, data confidentiality, and data integrity are the main security services in this level. With the independent characteristics of access technologies of AMF, the network domain security performance can be simplified and improved.

14.3.2.3   User Domain Security (III)

The set of security features that provide mutual authentication between the user interface and VEPC before the control plane access to the user interface. Based on each use case, the authentication may be needed for more than two parties. For example, the authentication can be required between a user and network operator as well as between a user and application service provider. Moreover, different service providers may need to authenticate each other to share the same user identity management. Compared with the device-based identity management in legacy cellular networks, new identity management methods are needed to improve the security performance.

14.3.2.4   Application Domain Security (IV)

The set of security features that ensure the security message exchange between applications on the interfaces, between user interface and service provider, as well as between a user and the network operator.

14.4   5G Wireless Security Services

14.4.1   Cryptography in 5G

Cryptography will be used to 5G security services including authentication, availability, confidentiality, and key management. Recent development and solutions can be mainly divided into two categories: *cryptographic approaches* with new networking protocols and PHY *layer security approaches*. The cryptographic techniques, both symmetric-key and public-key based, are commonly deployed at the upper layers of the 5G wireless networks for new networking protocols. The performance of a cryptography-based security service depends on the key length and computational complexity of the corresponding algorithms. Key management of symmetric key algorithms is well designed in the traditional cellular networks. Due to heterogeneous architecture and various access networks, 5G systems require a re-visit to key management designs in the new protocols [132]. In comparison, PHY layer security has relatively low computational complexity and high scalability [145]. Different from cryptographic algorithms, security performance on PHY layer security is evaluated based on *secrecy capacity* and *secrecy outage probability*.

As shown in Figure 14.8, the secrecy capacity C_s is defined as:

$$C_s = C_m - C_e, \tag{14.1}$$

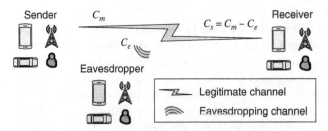

Figure 14.8 Channel capacities in PHY layer security.

where C_m is the main channel capacity of the legitimate user, and C_e is the channel capacity of the eavesdropper. The secrecy outage probability is an extension of the secrecy capacity by computing the probability of an instantaneous C_s being less than a target secrecy rate R_s, i.e.

$$P_{out} = P(C_s < R_s). \tag{14.2}$$

Besides these two metrics, transmission power can also be involved. For example, secrecy energy efficiency can be defined as the ratio of C_s to the corresponding power consumption [146].

14.4.2 Identity Management

In the 4G cellular networks, the identity management is device-based, through the universal subscriber identity module (USIM) cards. However, USIM cards may not be available to devices such as IoT sensors and smart home controllers in 5G systems. Therefore, the UDM in 5G devices will handle the identity management based on cloud. Moreover, anonymity service is required in many use cases in 5G wireless networks. Therefore, the identity management will be different in 5G wireless networks compared with that in the 4G networks. New identity management is required. Based on the characteristics of the use cases, different identity managements can be applied as shown in Table 14.3.

The 5G system user-based identity management will be more efficient to let the user determine what devices are allowed to access the network and services. One user may have multiple device identities. Besides only considering the device identity, service identity can be added with device identity as device and service identity management. The device

Table 14.3 Identity management in 5G wireless networks – from USIM to UDM.

USIM	Device-based identity management
	⇓
	Device-based identity management
	User-based identity management
UDM	Device and service identity management
	User-based and service identity management
	Federated identity management

identity is unique and service identity can be assigned by a service provider in certain session. With service identity, revocation process will be simplified. Moreover, for the trusted service providers, federated identity management can be applied to simplify the identity management and also improve the user experience. The identity management in 5G wireless networks is not unified for all use cases.

14.4.3 Authentication in 5G

Authentications in the legacy cellular networks are normally symmetric-key based. For example, the UMTS and LTE systems provide mutual authentication between a mobile station and the network. The authentication also includes a key generation and distribution process. Besides supporting the legacy systems, 5G authentication must also be flexible with new technologies to support many use cases.

14.4.3.1 Flexible Authentication

In the legacy cellular networks, mutual authentication is applied between a user and the network. However, the authentication between a user and the services provider is not implemented by the network. In 5G wireless network systems, some use cases may require both the service provider and network provider to carry out authentication with the users. Therefore, 5G wireless networks require flexible authentication to meet security requirements while satisfying the quality of services requirements. As shown in Figure 14.9, the authentication mechanism in 5G systems is selected based on UE, access technique, service requirement, and security requirement.

The chosen mechanism leads to the choices of trust model and cryptographic function. The input information can be included in the PCF, which then controls the AMF in the VEPC to perform the authentication procedure. Although non-3GPP access is supported in legacy LTE systems, a full authentication and key agreement (AKA) is required once a user changes the access technology. In the 5G security architecture, a full AKA can be skipped during access technology switching, thus to provide more efficient AKA in 5G systems.

14.4.3.2 Authentication Through Legacy Cellular System

Initially, a user needs to be associated to 3GPP access in a 5G system for seamless support from legacy cellular networks (e.g. LTE, UMTS). The initialization process is to verify

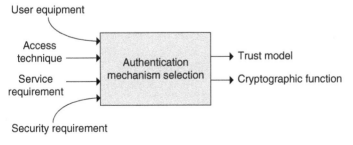

Figure 14.9 Authentication mechanism selection.

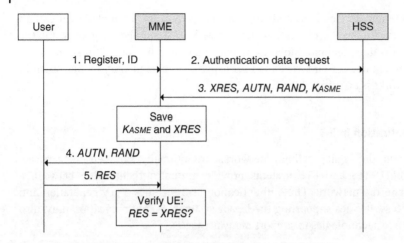

Figure 14.10 Authentication based on legacy security architecture.

identity and achieve key agreement. A user may also be initialized through a legacy system directly.

Based on the LTE security architecture, the authentication vector is generated at home subscriber server (HSS) and is then transmitted to MME, as shown in Figure 14.10. However, in the 5G security architecture, authentication vector can be generated at AMF to reduce the overhead of communications and to reduce the risk to expose the K_{ASME} and expected response (XRES). With the flexibility of network functions, AMF and UDM can be widely distributed to handle the authentication of a massive number of user devices. Nevertheless, due to the coupled control plane and user plane, MME and HSS have limited scalability. If the authentication is initialized in 5G systems through 3GPP access, the AKA process is implemented through AMF and UDM in 5G VEPC, as shown in Figure 14.11. Since AMF and UDM are both in the control plane, the overhead for information exchange between AMF and UDM is negligible compared with the overhead between entities such as MME and HSS.

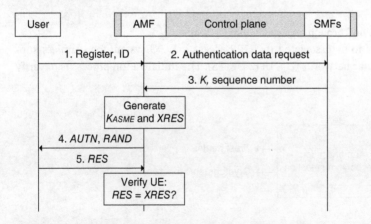

Figure 14.11 Authentication based on the 5G security architecture.

14.4.3.3 SDN Based Authentication in 5G

SDN in the core network of 5G systems not only provides enhanced data transmission performance, but also introduces new approaches to authentication. An example of SDN enabled authentication model is shown in Figure 14.12. Thanks to the optimal processing power allocation and information exchange, SDN controller can provide fast authentication to support low latency requirement of 5G systems [147]. For example, weighted secure-context-information (SCI) in SDN can be applied as a non-cryptographic security technique to improve authentication efficiency during high frequent handover in a 5G HetNet. PHY layer attributes provide unique fingerprint of the user and to simplify authentication procedure. The fingerprint can be based on multiple user-specific PHY layer attributes. The validated original attributes are obtained after a full authentication. The observations are collected through constantly sampling multiple PHY layer attributes from the received packets at the SDN controller. Both the original file and observation results contain the mean value of the attributes and variance of the chosen attributes. Then the mean attribute offset can be calculated based on the validated original attributes and observed attributes. If the attribute offset is less than a pre-determined threshold, the UE is considered legitimate. After the first full authentication in one cell, it can be readily applied in other cells with MAC address verification, which only needs local processing. Moreover, full authentication can even be done without disrupting the user communications. A valid time duration parameter can be used to flexibly adjust the secure level requirement. Compared with the digital cryptographic authentication methods, it is hard to compromise the SDN based method completely due to the user-inherent PHY layer attributes. Typically, there are more than one PHY layer characteristics used in SCI to improve the authentication reliability for applications requiring a high level of security.

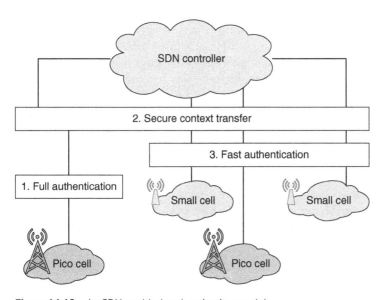

Figure 14.12 An SDN enabled authentication model.

14.4.3.4 Authentication of D2D in 5G

5G systems need to provide authentication of D2D communications. In some cases, D2D authentication utilizes simple algorithm and techniques due to limited bandwidth and energy of the devices. For example, cyclic redundancy check (CRC) based hash functions can be applied to detect double-bit errors in a given message [148]. Linear feedback shift register (LFSR) can be used to implement the CRC. The message authentication algorithm outputs an authentication tag based on a secret key and the message. It is assumed that the adversary has the family of hash functions but not the particular polynomial $g(x)$ and the pad s that are used to generate the authentication tag. The generator polynomial is changed periodically at the beginning of each session and pad s is changed for every message. The new family of cryptographic hash functions based on CRC codes with generator polynomials $g(x) = (1 + x)p(x)$ are introduced, where $p(x)$ is a primitive polynomial. The proposed CRC retains most of the implementation simplicity of cryptographically non-secure CRCs. However, the applied LFSR requires re-programmable connections.

New techniques such as a security-scoring based continuous authenticity may be applied to 5G D2D authentications [149]. The principle of legitimacy patterns is proposed to implement continuous authenticity, which enables attack detection and system security scoring measurement. For the legitimacy pattern, a redundant sequence of bits is inserted into a packet to enable the attack detection. The simulation results show the feasibility of implementing the proposed security scoring using legitimacy patterns. Moreover, legitimacy patterns considering technical perspectives and human behaviors could improve the performance.

14.4.3.5 Authentication of RFID in 5G

Radio frequency identification (RFID) can be widely integrated to applications in 5G systems. Due to hardware constraints in low-cost RFID tags, authentication mechanisms applied to RFID systems have been simple and efficient, mainly based on hash functions. The existing RFID authentication models may be applied directly to 5G use cases. However, authentication revocation process needs to be much improved because 5G use cases may involve frequency handover between different access networks as well as high-speed mobility. An example of authentication and revocation process of RFID secure applications in 5G systems is shown in Figure 14.13. The whole process is based on a typical challenge-response mechanism using hash function [150]. To authenticate or revoke a tag i, the reader generates a random number r_1 through a pseudo-random number generator (PRNG). The random number r_1 is sent together with an authentication request q to the tag. Upon receiving the request, the tag generates a random number r_2 and computes two hash values $H(ID_i\|\|r_1\|\|r_2)$ and $H(K_{i,j}\|\|r_1\|\|r_2)$, and a piece of authentication information $F = E \oplus K_{i,j}$, where E is the status flag that is used to indicate an authentication or revocation process. The reader forwards all messages to the server for verification. Depending on the status flag E, the server authenticates or revokes the tag. A hash value $H(ID_i\|\|K_{i,j}\|\|r_1, \|\|r_2)$ is computed as the final confirmation message.

> Note that (IDS, ID_i) are issued from the server to describe the public identity of tag i. Besides, a pre-shared key $K_{i,j}$ is also issued from the server for tag i. In practice, $K_{i,j}$ includes a pair of keys, i.e. $(K_{i,j}^{old}, K_{i,j}^{now})$, to be used during seamless handover.

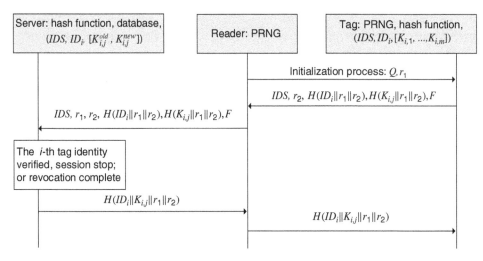

Figure 14.13 An authentication and revocation process of RFID secure applications.

14.4.4 Data Confidentiality in 5G

Data confidentiality service is commonly required to tackle eavesdropping attacks. In the legacy systems, cryptographic methods are used for implementing data confidentiality by encrypting data with secret keys. In most cases, symmetric key cryptography is adopted for data encryption, while public-key cryptography is applied to key distribution. Cryptographic approaches will continue playing an important role in the 5G wireless networks. Moreover, new methods will be implemented to ensure confidentiality. This section will focus on introducing the new implementations in PHY layer security, including power control, artificial noise, and signal processing.

14.4.4.1 Power Control

As mentioned earlier in PHY layer security, an eavesdropper relies on leaking capacity C_e from legitimate channel to recover useful information. A transmitter can set its transmit power to suppress C_e. In 5G systems, a transmission can involve more than a pair of transmitter and receiver due to its heterogeneous structure. As illustrated in Figure 14.14, a cooperator/relay can be involved in a legitimate transmission. In practice, the number

Figure 14.14 A general system model with eavesdropping attacks.

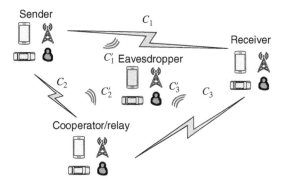

of transmitters, receivers, cooperators/relays can vary depending on use cases. With the involvement of cooperator/relay, power control is to achieve two security goals as follows:

(1) Maximize the secrecy rate for sender, i.e. the higher rate of the transmissions with/without routing through the cooperator/relay. Without loss of generality, it can be represented as:

$$C_s = \max\{C_1 - C_1', \min\{C_2 - C_2', C_3 - C_3'\}\}, \tag{14.3}$$

where C_1, C_2, C_3 are legitimate channel capacities; C_1', C_2', C_3' are eavesdropping capacities. Note that processing delay from a cooperator/relay is neglected for simplicity.

(2) Maximize the secrecy rate for cooperator/relay, i.e. the higher rate of the transmissions with/without routing through the sender. Similarly, it can be represented as:

$$C_c = \max\{C_3 - C_3', \min\{C_2 - C_2', C_1 - C_1'\}\}. \tag{14.4}$$

By selecting the route with a higher secrecy rate, the cooperator/relay may or may not be involved in a transmission. As shown in Figure 14.15, a typical 5G D2D scenario includes one base station, a number of cellular users, and one D2D link. To protect legitimate user from possible eavesdroppers, the secrecy rate based utility of each cellular user i can be computed as:

$$u_{c,i} = \log_2(1 + \theta_{c,i}) + \alpha\beta ph, \tag{14.5}$$

where $\theta_{c,i}$ indicates the signal-to-interference plus noise ratio (SINR) of cellular user i; α is the price factor; β is the scale factor; p is the transmit power of the D2D user; and h is the channel gain from the D2D user to cellular users. In this utility function, $\log_2(1 + \theta_{c,i})$ is the data rate of the cellular user, and $\alpha\beta ph$ compensates the interference from the D2D link. Similarly, the secrecy rate based utility of each D2D user can be computed as:

$$u_d = [\log_2(1 + \theta_d) - \log_2(1 + \theta_e)] - \alpha ph. \tag{14.6}$$

The utility function of D2D user includes the secrecy data rate and the payment for the interference to cellular users. A Stackelberg game can be formed to maximize the utilities for both cellular users and D2D users. The game strategy of cellular users depends on the price factor α and game strategy of D2D user depends on the transmission power P_D. The Stackelberg game is formed to maximize cellular utility function at the first stage and then the utility function of D2D user at the second stage [151].

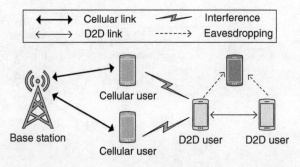

Figure 14.15 The system model with D2D link and an eavesdropper. Source: Modified from Luo et al. [151].

14.4.4.2 Artificial Noise and Signal Processing

Artificial noise can be introduced to secure the intended signal transmission. For example, an association policy can be set such that a random eavesdropper cannot reach its receiving SNR threshold, while legitimate users have maximized secrecy probability [137]. Moreover, if a sender has multiple antennas, then single-antenna eavesdroppers can be mitigated in a mm-Wave system if partial channel state information of the eavesdropper is known [152]. However, artificial noise increases power consumption and thus requires special design for energy efficiency [146]. Special signal processing may protect the system from eavesdroppers without increasing power consumption. For example, in the original symbol phase rotated secure transmission scheme [153], the base station randomly rotates the phase of the original symbols before transmission. Thus only legitimate users can correctly receive the symbols. Non-orthogonal multiple access (NOMA) is another technology that can protect a transmission system from eavesdroppers [154]. Moreover, massive MIMO introduces more possibilities in signal processing to protect the system against eavesdroppers [140].

14.4.5 Handover Procedure and Signaling Load Analysis

Handover processes are more complex in 5G systems due to the integration of 3GPP access, non-3GPP access, and new radio access. In the 5G security architecture [132], a user is associated to an AMF, which is independent from different access technologies. Therefore, a user would be able to switch between multiple access technologies handled by the same AMF. No authentication will be needed by switching to different SMF for a new session and a new IP address allocation. A general overview of handover scenarios in a two-tier HetNet model is shown in Figure 14.16.

As shown in Figure 14.17, a handover in the 5G wireless security architecture requires a session key and IP address updates through a data update process between the AMF and SMFs. The entire process is seamless since the communication latency between AMF and SMF can be neglected compared with the communication latency from MME to HSS. Besides the fast processing between AMF and SMFs, the signaling overhead is much lower in the 5G wireless security architecture compared with the legacy cellular system due to the full separation of control plane and user plane, as indicated in Figure 14.18. To satisfy certain latency requirement, the number of gateway nodes needs to be increased by a factor of 20–30 times of the current deployment [155]. On the positive side, gateways can be distributed in a flexible way because of the plane separation. Therefore, for the new core network based on control and user plane separation, the signaling load can be significantly reduced.

14.4.6 Availability in 5G

A reliable 5G system needs to enhance its availability, especially against jamming attacks in the wireless networks. There is hardly a protection fighting back jamming attacks. Nonetheless, anti-jamming schemes seek to avoid jammed channels by using frequency-hopping techniques, i.e. to hop over multiple channels for one transmission. For example, software defined radios can be used for secret adaptive frequency hopping in

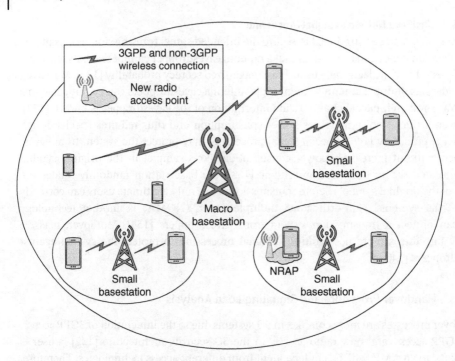

Figure 14.16 Handover scenarios in a two-tier HetNet model.

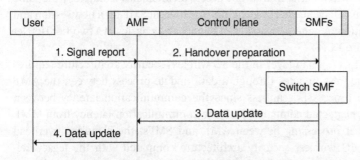

Figure 14.17 A handover procedure for changing access technologies.

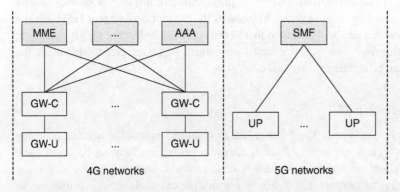

Figure 14.18 Signaling architecture comparison of 4G networks and 5G networks.

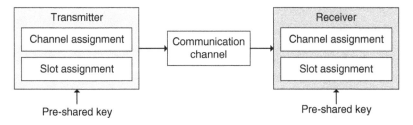

Figure 14.19 A pseudorandom time hopping system block diagram.

5G systems [156]. Bit error rate (BER) can be estimated based on PHY layer information to decide frequency blacklisting under DoS attack. However, since the frequency hopping technique requires that users have access to multiple channels, it may not work efficiently for dynamic spectrum access users due to the high switching rate and high probability of jamming. Pseudo-random time hopping may be applied as anti-jamming schemes for cognitive users in 5G systems [157], as shown in Figure 14.19. The jamming probability relates to delay performance and error probability. The jamming probability is low when the jammer lacks the access opportunities. Switching probability of time-hopping system outperforms the frequency-hopping system. With the same average symbol energy per joule, time-hopping has a lower error probability than that of frequency-hopping, and the performance gain saturates at a certain symbol energy level. The pseudorandom time-hopping technique is a strong candidate for D2D links in 5G wireless networks due to its good energy efficiency and spectrum efficiency performance as well as its capability in providing jamming resilience with a small communication overhead. However, a pre-shared key is required for the time-hopping anti-jamming technique.

Although most jamming attacks are detectable because of the active nature, some attackers may wish to hide their locations from being detected by legitimate users. In those cases, a jamming attacker is assumed to have only limited power and computing resources, similar to user nodes. To deal with such jamming attacks, an possible approach can be optimal resource allocations such that the attacker cannot allocate its power properly to jam legitimate users without being detected. For example, a fusion center may be applied to process resource allocations against such jamming attacks [158], as shown in Figure 14.20. Specifically, a non-cooperative Colonel Blotto game is formulated between the attacker

Figure 14.20 The resource allocation model.

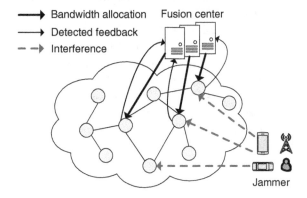

and the fusion center as an exercise in strategic resource distribution. The fusion center can allocate more bits to these nodes for reporting the measured interference. A hierarchal degree is assigned to each node based on its betweenness centrality [159]. Once the attack is detected, the fusion center will instruct the target node to increase its transmit power to maintain a proper SINR for normal communications.

14.4.7 Location and Identity Anonymity in 5G

5G wireless networks raise serious concerns on privacy leakage when supporting more and more vertical industries such as mobile-health care and smart transportations [133]. The data flows in 5G wireless networks carry extensive personal privacy information such as identity, position, and private contents. In some cases, privacy leakage may cause serious consequences. Depending on the privacy requirements of the applications, privacy protection is a big challenge in 5G wireless networks. One approach to protect location and preferences of a user in HetNets can be properly selecting access point based on a matching game framework that is established to measure the preferences of mobile users and base stations with PHY layer system parameters [160]. In particular, differentially private Gale–Shapley matching algorithm can be applied with utilities of mobile users and access points defined as packet success rate in the game framework. Identity protection in 5G systems can be extended from the temporary identity used in legacy cellular systems. For example, contextual privacy can be achieved for both data and identity protection in mobile-health [161]. In particular, the identity of the source client is encrypted by a pseudo identity of the source client with the public key of the physician using certificateless encryption mode. Meanwhile, the identity of the intended physician is also encrypted with the public key of the network manager. Through these two encryption steps, the contextual privacy can be achieved.

14.5 5G Key Management

14.5.1 3GPP 5G Key Architecture

Key management supports the establishment and maintenance of keying relationships between authorized parties, where the keying relationship is the way common data is shared between communication entities. The common data can be public or secret keys, initialization values, and other non-secret parameters. Figure 14.21 shows the 5G key hierarchy defined by 3GPP in Release 16 [162]. A long term secret key K is pre-installed in the Authentication Credential Repository and the Processing Function (ARPF) and the USIM. A pair of CK and IK is derived from K, similar to LTE and UMTS systems. The key K_{AUSF} is derived by the ARPF to secure communications between UE and ARPF. The key K_{AUSF} is passed to Authentication Server Function (AUSF) and UE to derive an anchor key K_{SEAF} for the Security Anchor Function (SEAF). The key K_{AMF} is then derived for both the AMF and UE. The AMF further derives four keys from as follows: K_{N3IWF} is used to secure communications between UE and the Non-3GPP Interworking Function (N3IWF); K_{NASint} and K_{NASenc} are two keys used to protect the integrity and confidentiality in non-access stratum (NAS); K_{gNB} is derived to secure communications between gNB and

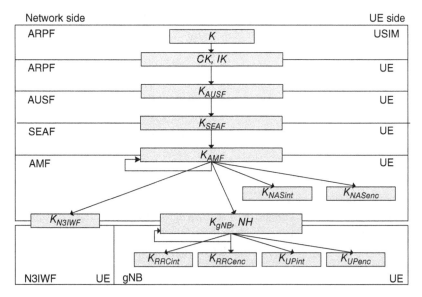

Figure 14.21 5G key hierarchy defined in 3GPP Release 16.

UE. Finally, the gNB uses K_{gNB} to derive K_{RRCint}, K_{RRCenc}, K_{UPint}, and K_{UPenc} for integrity and confidentiality on RRC and UP communications.

14.5.2 Key Management in 5G Handover

Figure 14.22 shows the handover key chaining in 5G networks. In a handover process, the source gNB (i.e. gNB_s) generates a new key $K_{NG-RAN}*$ to be used between gNB_t and UE

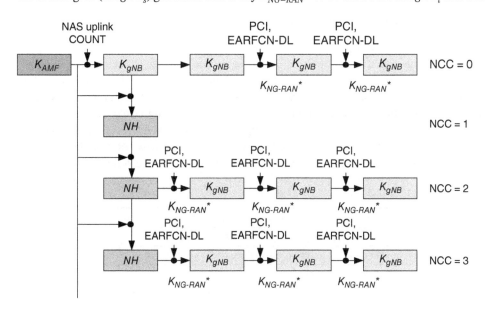

Figure 14.22 5G handover key chaining.

either through vertical key derivation or horizontal key derivation. If the gNB_s has a pair of unused Next Hop (NH) parameter and Next Hop Chain Counter (NCC), i.e. $\{NH, NCC\}$, then a vertical key derivation is performed as follows:

$$K_{NG-RAN}* = KDF(NH\|PCI\|\text{ARFCN-DL}), \tag{14.7}$$

where $KDF(\cdot)$ is a hash based key derivation function; PCI and ARFCN-DL are the PHY cell identity and absolute radio frequency channel number-down link of gNB_t, respectively. If an unused $\{NH, NCC\}$ pair is not available, then the horizontal key derivation is performed as follows:

$$K_{NG-RAN}* = KDF(K_{gNB}\|PCI\|\text{ARFCN-DL}). \tag{14.8}$$

After generating $K_{NG-RAN}*$, gNB_s sends $\{K_{NG-RAN}*, NCC\}$ pair to gNB_t. The NCC is the NH chaining counter to be included in a handover acknowledgment message and forwarded to the UE by gNB_s. Once gNB_t completes the handover, it shall send a path shift request to the AMF. Upon receiving the path shift request, the AMF increments NCC and computes a new NH as follows:

$$NH^* = KDF(K_{AMF}\|NH). \tag{14.9}$$

The new $\{NH, NCC\}$ pair is sent to gNB_t for the next handover process. Note that the initial NH is computed as follows:

$$NH = KDF(K_{AMF}\|K_{gNB}). \tag{14.10}$$

On the UE side, since it knows K_{AMF}, the first K_{gNB}, PCI, and ARFCN-DL, it can generate the $K_{NG-RAN}*$ at each handover and use the received NCC value to check if the key matches with the one used by gNB_t.

14.5.3 Key Management for D2D Users

Key management for D2D users has been researched in the recent years [161, 163, 164]. In a D2D group use cases, there are five security requirements for key management, as listed in Table 14.4. A group key management mechanism can be applied [164]. In particular, identity-based cryptography (IBC) scheme based on elliptic curve cryptography (ECC) can be applied to multicast group communications security. One example of D2D key exchange is illustrated in Figure 14.23. The key exchange process is based on Diffie–Hellman (DH) scheme. The eNodeB is a trusted party that certifies the authenticity between the two D2D users. Private dedicated channels can be established between the eNodeB and each D2D

Table 14.4 Security requirements in D2D group use cases.

Forward secrecy	Revoke access from users who have left the system
Backward secrecy	New users shall not hold old keys
Collusion freedom	Fraudulent users cannot deduce traffic encryption
Key independence	Keys in one group cannot be discovered with keys in another group
Trust relationship	Keys are not revealed to other parts in the same domain or other domain

Source: Based on GSMA Intelligence [119].

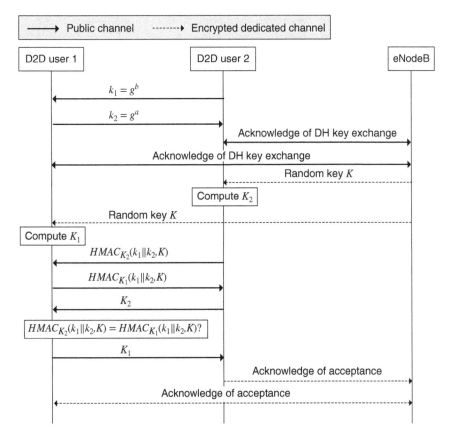

Figure 14.23 Example of D2D key exchange protocol.

user. The process can be simplified by replacing the HMAC function with other check functions with less complexity [163].

14.6 Security for New Communication Techniques in 5G

5G wireless network systems introduce several new techniques, such as HetNet, massive MIMO, D2D, SDN, and IoT. Besides security protection in the core 5G system, researchers and engineers have been working on the newly integrated communication techniques so that the 5G networks are protected as a whole system.

14.6.1 Heterogeneous Network and Massive MIMO in 5G

HetNet is a promising technique to provide blanket wireless signal coverage and high throughput in 5G wireless networks. It is a multi-tier system where different characteristics, such as transmission power, coverage size, and radio access technologies, are set for nodes depending on their tiers. With the heterogeneous characteristics, HetNet achieves higher capacity, broader coverage and better performance in energy efficiency and spectrum efficiency. In the meantime, the nature of HetNet architecture makes it more challenging in several security aspects compared with traditional single-tier cellular networks [137, 147]. For example, launching eavesdropping attacks becomes easier because of the blanket wireless signal coverage. Frequent handover between small cells in HetNet is more challenging and cannot rely on current mechanisms in legacy cellular systems. Moreover, accurate location information of a user may cause privacy leakage in a HetNet with densely deployed small cells, especially when the conventional association mechanism is applied.

Researchers have proposed to apply coordinated multi-point (CoMP) transmission to enhance communication coverage in HetNet. However, CoMP also increases the risk of eavesdropping because of the added transmission links. To counter the attack, the points chosen for transmission need to be dynamic and carefully designed to guarantee the secure coverage probability for users [165]. Moreover, artificial noise can be added to tackle the eavesdropping attacks in HetNet. However, a different mobile association policy is needed as the received signal power would always be well above noise floor due to artificial noise. For example, average received signal power (ARSP) may be applied by comparing its value to a pre-set access threshold. In this way, an association is granted only when the maximum ARSP exceeds the threshold, meaning that the user is active [137]. Such ARSP based association policy can greatly enhance secrecy throughput performance given a properly access threshold.

PHY layer association algorithms, besides data confidentiality, can also be properly designed in 5G to protect user location anonymity. On the one hand, the high density of small cells could leak more information about a user location. On the other hand, those access nodes may cooperate with each other to disguise the user from location disclosure to unauthorized users. For example, a differential private Gale–Shapley algorithm may be applied [160]. Network intrusion detection system (IDS) is a technique that can be applied to the cloud computing systems in 5G HetNet [166]. Several IDS solutions have been studied in the network domain, such as signature-based detection, anomaly-based detection, specification-based detection, stateful protocol analysis, hybrid intrusion detections. Since the cloud computing systems and core network in 5G HetNet follow typical networking structure, the existing IDS designs can be extended to support HetNet applications.

Massive MIMO can greatly enhance energy efficiency and spectrum efficiency to support a large number of users simultaneously in 5G systems [153]. Besides, massive MIMO also provides new options in security mechanisms. For example, PHY layer security such as linear zero-forcing beamforming can be used to secure downlink HetNet communications against multiple eavesdroppers [140]. However, active attackers may also utilize massive MIMO technology to compromise legitimate communications. The number of antennas and beamforming schemes at the base stations need to be carefully designed to suppress the impact from the massive MIMO equipment of attackers while maintaining high energy efficiency of the 5G systems.

14.6.2 Device-to-Device Communications in 5G

D2D devices can communicate with each other without going through base stations. Therefore, D2D communications enable efficient spectrum usage in 5G systems. Moreover,

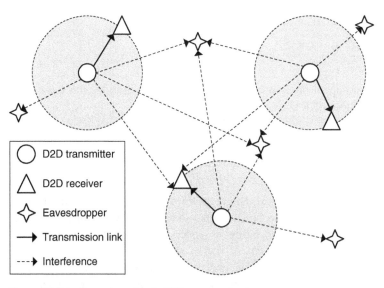

Figure 14.24 Eavesdropping in D2D communications.

D2D communications can effectively offload traffic from base stations. However, the dynamic spectrum access used in D2D links can yield security threats in large-scale deployment [149].

Eavesdropping in D2D communications can be more threatening as it can target multiple D2D communications using the same spectrum while being outside normal D2D coverage areas. In Figure 14.24, the D2D receivers on the edge of each D2D coverage area are presented as the most vulnerable nodes to eavesdropping. Thus, it is straightforward to provide the highest secrecy capacity for those users, such that,

$$\max r_t \mathcal{P}_{data} \mathcal{P}_{sec}, \tag{14.11}$$

where r_t is the transmitting rate denoted as: $r_t = W \log_2(1 + \phi_i)$. W is the bandwidth of channels; $\mathcal{P}_{data}$ is the probability of a success data link, where the SINR of the D2D receiver ϕ_i is higher than a threshold for success eavesdropping T_u; and $\mathcal{P}_{sec}$ is the security probability where the SINR of the eavesdroppers are lower than a threshold T_e. Since eavesdropping is a passive attack, for simplicity, assume that the eavesdroppers distribute as a homogeneous Poisson point process Φ_e of intensity λ_e. Then $\mathcal{P}_{data}$ can be found as:

$$\begin{aligned} \mathcal{P}_{data} &= \mathbb{P}(\phi_i \geq T_u) \\ &= \exp\left(-\frac{T_u \sigma \|x_j\|^\alpha}{p_j}\right) \mathbb{E}_{I_i}\left(\exp(-\frac{T_u I_i \|x_j\|^\alpha}{p_j})\right), \end{aligned} \tag{14.12}$$

where σ is the Gaussian noise; I_i is the interference at the D2D receiver; x_j is the location of the D2D receive; p_j is the transmit power of the D2D transmitter. If the exact locations of eavesdroppers are unknown, then $\mathcal{P}_{sec}$ can be approximated as:

$$\mathcal{P}_{sec} = \mathbb{E}_{\Phi_e}\left[\prod \mathbb{P}(\phi_e \leq T_e)\right] \approx \exp\left(-\frac{\lambda_e \text{sinc}(\delta) p_j^\delta}{\lambda_d T_e^\delta \overline{p}_{de}^\delta}\right), \tag{14.13}$$

where $\delta = 2/\alpha$; and $\bar{p}_{de}$ is calculated as:

$$\bar{p}_{de} = \frac{\sum_{d \neq j, x_d \in \Phi_d} p_d \|x_e - x_d\|^{-\alpha}}{\sum_{d \neq j, x_d \in \Phi_d} \|x_e - x_d\|^{-\alpha}}. \tag{14.14}$$

If the locations of eavesdropper are known, P'_{sec} can be found by dealing with the eavesdropper that has the highest SINR at location x_e, as follows:

$$P'_{sec} = \mathbb{P}(\phi_e \leq T_e)$$

$$= 1 - \exp\left(-\frac{T_e \sigma \|x_j - x_e\|^{\alpha}}{p_j} - \frac{\pi \lambda_d \bar{p}_{de}^{\delta} T_e^{\delta} \|x_j - x_e\|^{\alpha}}{\operatorname{sinc}(\delta) p_j^{\delta}}\right), \tag{14.15}$$

where x_e is the exact location of the most dangerous eavesdropper of R_j. In either case, the optimal SINR of D2D receivers can be computed to achieve the maximum secrecy rates [167]. However, due to the dynamic nature of wireless communication channels, the performance of PHY level security could vary.

The traditional cryptographic mechanisms may not fit D2D communications. For example, the proximity service and communication phases for D2D communications require a distributed approach for scalability. Typical public key infrastructure and key distribution center can hardly keep up with the scalability and dynamic nature of D2D communications. Researchers have proposed to apply IBC to D2D communications [164]. IBC is to generate the public key of a user based on its identity. In this way, the public keys can be generated at any node on demand or beforehand, thus to greatly reduce complexity of the system. Key revocation and update can be easily applied in IBC based mechanisms. Another way to enhance D2D key exchange involves an base station (e.g. eNodeB in legacy LTE systems). The eNodeB participates in the initial key exchange and mutual authentication of D2D users. After that, both a public channel and an encrypted dedicated channel are established for key exchange in two scenarios, one is traffic offload where D2D users are connected to the same eNodeB, and the other is social networking scenario where a D2D link is required for the applications of each D2D user [163].

Security in D2D communications requires special design in some use cases, such as m-health applications, vehicular networks, etc. For example, m-health system requires data confidentiality and integrity, mutual authentication, anonymity to anyone except intended physician, unlinkability, forward security, and contextual privacy. It also requires light weight for mobile terminals with energy and storage constraints and needs to be robust enough to fight against threats as part of the keys can be exposed. To address those issues, a certificateless public key cryptography may be applied [161]. The private key of a user is generated by both key generator center and the user, which keeps the private key unknown to the key generator center.

14.6.3 Software-Defined Network in 5G

By decoupling the control plane from the data plane, SDN simplifies centralized control in network management. SDN can provide logically centralized intelligence, programmability, and abstraction to 5G systems to improve scalability and flexibility at relatively low costs [125]. Table 14.5 lists the advantages of SDN features, the network applications,

Table 14.5 The advantages of SDN security over traditional networks.

SDN features	Application	Security use
Global view	Centralization Traffic monitoring	Network intrusion detection Switch behavior monitoring Network forensics
Self-healing mechanisms	Conditional rules Traffic statistics collection	Reactive packet dropping Reactive packet redirection
Increased control capabilities	Flow-based forwarding	Access control

and the security use. However, the advanced features of SDN and additional network applications also introduce security challenges to 5G systems. The new security issues in SDN networks are briefly summarized in Table 14.6. The forwarding plane may suffer from DoS attack on network switch due to limited forwarding table and buffering capacity. Add more resources can be a straightforward countermeasure to the switch DoS attack. Moreover, the required packet encryption and possible tunnel bypassing in the forwarding plane will hide header fields from traditional traffic analysis mechanisms.

Packet type classification may be applied to counter this issue [168, 169]. DDoS attack and compromised controller are security issues faced in the control plane mainly due to the centralized nature of SDN controller. Dynamic controller assignment and controller replication can be applied to reduce the risk of using a centralized controller. The forwarding-control link faces man-in-the-middle attack and replay attack due to unencrypted communication messages, the lack of authentication, and the lack of time stamping. It is straightforward to apply encryption and to include time stamp or nonce in 5G SDN implementation.

Table 14.6 New security issues in SDN networks and possible countermeasures.

Security issues	Causes	Possible countermeasures
Forwarding plane		
Switch DoS	Forwarding table capacity Enormous number of flows Switch buffering capacity	Proactive rule caching Rule aggregation Increase buffering capacity Reduce switch-controller COMM delay
Packet encryption Tunnel bypassing	Invisible header fields	Packet type classification
Control plane		
DDoS attack	Centralization Forwarding table capacity Enormous number of flows	Controller replication Dynamic master controller assignment Efficient controller placement
Compromised controller	Centralization	Controller replication with diversity Efficient controller assignments
Forwarding-control link		
MITM attack	Unencrypted COMM MSG Lack of authentication	Encryption Use of digital signatures
Replay attack	Unencrypted COMM MSG Lack of time stamping	Encryption Include time stamp in encrypted MSG

14.6.4 Internet-of-Things in 5G

One of the major user case of 5G wireless networks will be IoT. Due to the limited computing resources of many types of IoT nodes, those advanced security mechanisms in 5G systems cannot be utilized properly. Therefore, lightweight mechanisms and relaying/offloading need to be optimized for IoT use cases. For example, a fusion center may be involved to assist IoT nodes locating a jamming attacker [158]. In the process, each IoT node is capable of detecting its receiving interference. The decentralized interference measurements are collected at the fusion center in regular intervals on a common control channel. The fusion center measures the importance of each node with the betweenness centrality of each IoT node. A certain level threshold and aggregated received interference power level are used to determine whether a jamming attack exists or not. The fusion center can also allocate bandwidth to certain nodes to measure the interference level in order to detect the jamming attack. PHY layer security is one approach to secure IoT communication links with lower complexity. By adding a relay, the two-hop transmission can apply power allocation and codeword rate design against eavesdropping attacks [170]. Relay nodes with multiple antennas can further improve secure coverage and secrecy capacity. Moreover, IoT nodes with ultra-simplicity, e.g. RFID tags, are challenging in security protocol design. In particular, frequent authentication as well as revocation of RFID tags need to be addressed.

14.7 Challenges and Future Directions for 5G Wireless Security

With the security designs inherited from the legacy cellular networks, and many newly developed security mechanisms, there are still quite a few challenges to be addressed in the security for 5G wireless networks.

14.7.1 New Trust Models

The trust model in the legacy cellular networks involves user terminals, serving network and home network. In comparison, the trust models and authentication in 5G wireless security need an improvement to support various uses cases [128]. For some applications, there are various types of devices connected to the same network, some may be used only to gather data and some may be used only for Internet access. For different security demands, the corresponding trust model may have different security requirements. For example, a high security level demand may require both password and biometric authentication simultaneously [133]. Nonetheless, some RFID based IoT devices require efficient and frequent authentication/revocation [150].

14.7.2 New Security Attack Models

New security attacks will threaten 5G wireless networks mainly due to two factors. First, attackers in 5G systems may be equipped with advanced technologies, such as massive MIMO, cloud computing resources, etc. Those technologies will greatly increase the threats from attackers. For example, attackers with massive MIMO may eavesdrop with higher accuracy from a further distance. Moreover, attackers may even cooperate to each other, similar to legitimate user cooperation, to jam D2D transmissions without disclosing their locations. Such attacks remain to be explored in 5G systems. Second, more vulnerable points will exist in 5G systems due to the new service delivery models applied to SDN and NFV [125]. Decoupling software from hardware makes the security of software no longer depending on specific security attributes of the hardware platform [128]. Therefore, the demands on strong isolation of security for virtualization are ever increasing. Currently, network slicing is proposed to provide such isolated security [127].

14.7.3 Privacy Protection

5G wireless networks would need more attention on privacy leakage due to the open network platforms [133]. Generally speaking, privacy in 5G wireless networks include identity anonymity and location anonymity. For the identity privacy, new identity management should be considered instead of using only device-based identity management. Location privacy can be enhanced if multiple association mechanisms are applied to different use cases. In practice, encryption is widely applied to privacy protection in most systems. However, encryption may violate other service requirements such as latency and efficiency in some 5G use cases. Techniques such as data analysis and machine learning may be implemented to enhance efficiency in privacy protection. For example, before the data transmission, data analysis can be applied to find out several highly sensitive dimensions to reduce the encryption cost with privacy protection.

14.7.4 Flexibility and Efficiency

The virtualization based 5G architecture requires flexible security mechanisms to protect different applications with dynamic configurations [128]. Moreover, security setup must be customized and optimized for each specific application instead of one approach fitting all [135]. Besides the flexibility of security architecture and mechanisms, efficiency of security is another key factor in 5G wireless networks to meet both the latency and energy efficiency requirements [135, 171]. Especially for the nodes with limited computing capability and power supply in many IoT use cases. Moreover, distributed authentication nodes need to support the fast network access for massive number of devices.

14.7.5 Unified Security Management

5G wireless security needs a unified security framework with a common and essential set of security features such as access authentication, confidentiality, and privacy protection, for different services, access technologies, and devices. The basic features of these security services are similar to those in the legacy cellular networks. However, there are many new perspectives of these security features in 5G wireless networks. For example, security management across heterogeneous access need to be flexible for all access technologies. Security management of IoT applications may need to deal with burst access behavior for efficient access authentication.

14.8 Summary

In this chapter, current development, challenges, and future directions of 5G wireless network security are illustrated. 5G wireless network systems take a huge step forward comparing to the legacy LTE systems. The 5G systems will support many more use cases through 3GPP, non-3GPP, and new radio access technologies. Advanced technologies such as D2D communications, massive MIMO, and SDN will enhance both functions and security mechanism design in 5G systems. For example, PHY level security can be applied to user confidentiality as well as location anonymity; SDN supported VEPC can provide seamless handover between different access technologies. However, the new technologies also bring new challenges in security design. While regular 3GPP cellular users can rely on legacy LTE security, 5G security is still pending on advanced technologies such as D2D, non 3GPP, and new radio user cases.

15

Security in V2X Communications

Vehicle-to-everything (V2X) communications have received great attentions in recent years as a key component of the Intelligent Transportation Systems (ITS) [172]. V2X communications in general include four types of communications, i.e. vehicle-to-vehicle (V2V), vehicle-to-infrastructure (V2I), vehicle-to-network (V2N), and vehicle-to-pedestrian (V2P), as listed in Table 15.1. Vehicles and ITS entities will exchange information about traffic conditions, such as road construction, accidents and traffic jams, through V2X communication messages. Such information exchange will greatly enhance road safety for drivers and autonomous vehicle control [173]. The shared information will also support first responders in emergency. Other Internet based user applications such as routing and multimedia streaming will also benefit from the seamless vehicular networks. Similar to other wireless communication systems, a dedicated set of wireless network protocols is designed for vehicular connections within proximity. Besides, both the current LTE and the next generation cellular network technologies will support V2X communications. Due to the open nature and complexity of multiple wireless technologies in V2X communications, it is challenging to provide security protection for V2X communication networks. This chapter identifies some of the major challenges and possible solutions for security in V2X communications.

15.1 Introduction to V2X Communications

15.1.1 Generic System Architecture of V2X Communications

Entities in transportation systems connect to each other through various V2X communication techniques, as depicted in Figure 15.1. The communication capability of each vehicle is enabled through an on-board unit (OBU), which is also connected to the central computer of the vehicle for autonomous controlling. An OBU is usually paired with a tamper-proofed device (TPD) that stores credential information in securing communications. OBUs are required to send basic security related messages, such as real-time position, speed and steering information, every 300 ms. Such information can be applied to safety control and traffic control [172]. Besides vehicles, V2X communications also include multiple types of fixed infrastructure, including a trust authority (TA), road side units (RSUs), and cellular base stations [174]. The TA manages the authentication and authorization of all

Security in Wireless Communication Networks, First Edition. Yi Qian, Feng Ye, and Hsiao-Hwa Chen.
© 2022 John Wiley & Sons Ltd. Published 2022 by John Wiley & Sons Ltd.
Companion website: www.wiley.com/go/qian/sec51

Table 15.1 V2X communication types.

Type	Description	Scope and impact
V2V	Communications between vehicles within proximity	US DoT estimated that 80% of unimpaired accidents can be avoided
V2I	Communications between vehicles and road side units	The future of V2I may lead to better driver-assistance systems
V2N	Communications between vehicles and cellular network in licensed spectrum	Good for infotainment and non-mission critical applications
V2P	Communications between vehicles and pedestrians	About 20% of road fatalities in US are pedestrians

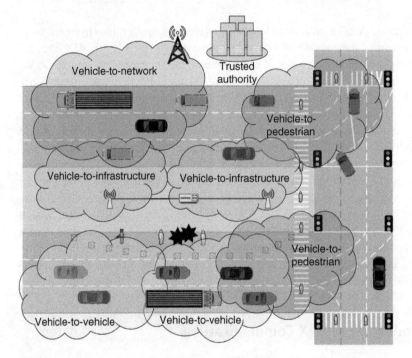

Figure 15.1 A generic architecture of V2X communications.

entities in V2X communications. Before joining V2X communications, all entities must be registered to receive a certificate at the TA [175]. In V2X communications, the TA keeps the real identities and tracks the locations of legitimate entities. Another duty of the TA is to verify suspicious behaviors from vehicles in the communications. Suspicious behaviors are generally reported from an RSU or a cellular base station. Once the report is validated by the TA, the certificate of the malicious vehicle will be revoked.

15.1.2 Dedicated Short Range Communications

The wireless technologies in V2X communications are mostly based on either dedicated short range communications (DSRC) or cellular technologies. In December

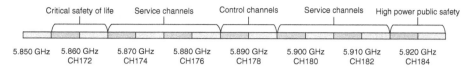

Figure 15.2 DSRC spectrum defined in the United States.

2003, the Federal Communications Commission (FCC) published a Report and Order establishing licensing and service rules for DSRC Service in ITS Radio Service in the 5.850–5.925 GHz band (also known as the 5.9 GHz band) [176–178]. The 75 MHz DSRC spectrum is further divided into multiple channels, as shown in Figure 15.2. Channel 172 is designated exclusively for V2V safety communications with high reliability and low latency for accident avoidance and mitigation, and safety of life and property applications. Channel 184 is designated exclusively for high-power, longer distance communications for public safety applications involving safety of life and property, including road intersection collision mitigation. Channel 178 is the control channel designated for broadcasting safety-related applications for vehicle safety traffic at all power levels. Channels 174, 176, 180, and 182 are service channels that can be used for both safety and non-safety communications. IEEE 802.11p standard is the basis for DSRC. As mentioned earlier, it uses channels of 10 MHz bandwidth in the 5.9 GHz band, half of which is used in IEEE 802.11a. While the DSRC is dedicated to V2X communications with low latency, however, DSRC is for short and low capacity transmissions only. It cannot support V2X communications in high density scenarios, or in large scale [179, 180].

15.1.3 Cellular Based V2X Communications

Cellular technologies, e.g. LTE and 5G, can provide larger communication coverage, lower deployment cost, and better network QoS guarantee compared with DSRC [181]. The 3GPP has made specific standards for V2X services using LTE in Releases 14 and 15 [182]. In particular, LTE-V2X focuses on enhancing V2V and V2P communications that are based on the device-to-device (D2D) communications. As part of ProSe services, D2D has a designated interface PC5 (also known as sidelink) in the physical layer. The 5G wireless networks, as the successor to LTE, have taken into consideration the support for V2X communications. 5G can be a much more promising solution to real-time V2X services because of its enhanced Mobile Broadband (eMBB) with low latency compared with IEEE 802.11p and LTE. Moreover, the HetNet structure of the 5G wireless network can satisfy different V2X communication scenarios. Many organizations from the telecommunications industry and vehicle manufacturers have contributed to 5G-V2X, such as the 5G automotive association (5GAA) [183] and 5G Communication Automotive Research (5GCAR) and innovation [184]. Besides DSRC and cellular communication technologies, some other wireless communication technologies such as Bluetooth and satellite communications are also considered for particular V2X applications. For instance, LTE-V2X (or general cellular-V2X) applies global navigation satellite system (GNSS) as its primary source of time synchronization.

15.2   Security Requirements and Possible Attacks in V2X Communications

Security and privacy are vital in V2X communications. However, due to the complex communication structures, high speed of vehicles, and low latency information exchange, it is challenging to secure V2X communications from possible attacks [185, 186]. This section briefly lists the security requirements, possible attacks, and an overview of the basic solutions.

15.2.1   Security Requirements

The primary security services in V2X communications include authentication of information, authorization of entities, privacy of users, and availability of systems [187, 188]. Table 15.2 briefly introduces the security requirements in V2X communications at the application level. As it shows, V2X security has two folds, one is to protect legitimate entities; and the other is to trace and block unauthorized entities. In LTE-V2X, the 3GPP recommends four specific security aspects [183]. First, each entity can authenticate and verify the received messages that were sent by an authorized entity. Second, data confidentiality and integrity should be guaranteed against various attacks. Third, only the fresh messages will be accepted so that the replay attack and DoS attack can be alleviated. Finally, the cryptographic algorithms should be light weighted and have high security level due to the stringent time requirement and the safety related property. 5G-V2X and other V2X technologies are expected to follow the standardization process from 3GPP.

Table 15.2   Security requirements in V2X communications.

Requirements	Description
Entity authentication	Each vehicle, RSU, eNodeB, application server, should be uniquely identified and authenticated.
Authorization	V2X services are only accessible to authorized entities.
Message authentication	All messages should be authenticated by the transmitters; and receivers can verify messages from authenticated entity.
Confidentiality	Required types of messages cannot be disclosed to unauthorized entities.
Data integrity	Modifications of V2X messages can be detected at the receiver side.
Non-repudiation	The TA can trace the real identity of a sender.
Identity anonymity	*Anonymity*: The real identity of a legitimate user cannot be disclosed to others except for the TA.
	Unlinkability: no clear relation can be found between a real identity and its corresponding pseudonyms; no clear relation can be found among multiple pseudonyms used by the same vehicle.
Location anonymity	Real location of an entity cannot be exposed to unauthorized entities.
Availability	Access to vehicular network should be granted for all legitimate entities.
Conditional traceability	The real identities of malicious (and/or unauthorized) users can be easily traced by the TA.

15.2.2 Attacks in V2X Communications

Table 15.3 lists a brief summary of possible attacks in V2X communications. V2X communications are vulnerable to various types of attacks due to open-medium wireless transmission as well as the mixture of multiple transmission technologies. Typical attacks in wireless transmission such as eavesdropping, MITM attack, timing, and relay can threat V2X communications. In the meantime, some enhanced features in V2X communications, while providing better services to users, also introduce more vulnerabilities. For example, compared with other short-range wireless technologies such as Wi-Fi, Bluetooth, etc., possible attacks on V2X (especially V2V) target more on the privacy of users. Therefore, it is straightforward to provide pseudonym to authorized users for identity privacy. Nonetheless, malicious users may take advantage of such feature and disguise their identities and locations.

Table 15.3 Attacks in V2X communications.

Attacks	Description	Targeting
Black hole Grey hole	Attackers drop all or selective messages without forwarding to targeted entities.	Availability
Bogus message	An attacker from outside or inside of a V2X system broadcasts fake/junk messages for benefits.	Authorization Authenticity
Coalition and platooning	A group of attackers within in proximity collaborate on malicious purposes.	Availability
DoS/DDoS	Attackers tend to deplete the network resources of V2X systems by injecting large volumes of messages.	Availability
Eavesdropping	An attacker taps into legitimate V2X communication links and collects information passively.	Confidentiality Privacy
Infrastructure spoofing	An attacker spoofs the IP address of a RSU, eNodeB, or Application Server and attacks.	Authorization Availability
Location tracking	Attackers track the locations of legitimate vehicles by monitoring and analyzing V2X communications.	Privacy
Masquerade	Attacker filches the identities of legitimate vehicles and sends messages on behalf.	Authenticity Privacy
Message modification	Attackers modifies an original message through deleting, adding to, changing, or reorganizing.	Authenticity Integrity
MITM	An attacker impersonates as legitimate user to both sides in V2X communications.	Availability Authenticity Integrity
Replay	A malicious vehicle maliciously or fraudulently transmits repeated messages.	Availability Authenticity
Selfish	Some V2X entities may refuse to cooperate with others to relay messages.	Availability
Sybil	An attacker joins a V2X system with multiple real and fake identities to misguide legitimate vehicles.	Authenticity authorization
Timing	A malicious vehicle delays message delivery process.	Availability Authenticity Integrity

15.2.3 Basic Solutions

Basic solutions to possible attacks in V2X communications are mainly cryptographic, trust based, and privacy oriented schemes. One or multiple schemes may target on a particular attack, as shown in Figure 15.3. Cryptographic schemes mainly protect confidentiality and message authentication from corresponding attacks. Cryptographic schemes are also the foundations in entity authentication protocols. However, cryptographic schemes alone may not be effective to some attacks in V2X communications, such as DoS/DDoS and black/gray hole. Trust based schemes introduce trust levels, e.g. by setting reputation scores, to evaluate the trustworthiness of users [189]. An attacker can be detected by checking its reputation score. Nonetheless, trust based schemes are usually incorporated with cryptographic schemes, e.g. public key infrastructure and certificates, for sender authentication and reputation score validation. Privacy oriented schemes protects user identity anonymity and location anonymity in V2X communications. Pseudonyms are normally applied to V2X communications for user identity anonymity. However, pseudonym may not hide the location for users, because a user entity can be tracked based on safety messages, such as speed, direction, previous location, etc. Moreover, V2X communications must provide robust revocation schemes to support frequent incoming/outgoing users. Most of revocation schemes are cryptographic and/or trust based.

15.3 IEEE WAVE Security Services for Applications and Management Messages

IEEE Standard 1609.2 defines security services for applications and management for Wireless Access in Vehicular Environments (WAVE) [190].

15.3.1 Overview of the WAVE Protocol Stack and Security Services

WAVE provides a communication protocol stack with two types of security services that employ both customized and general-purpose elements to be applied to V2X communication systems. An overview of WAVE protocol stack and security services are shown in

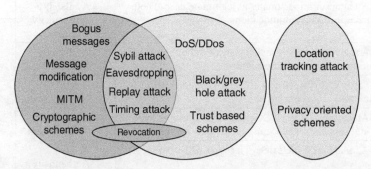

Figure 15.3 Attacks and solutions.

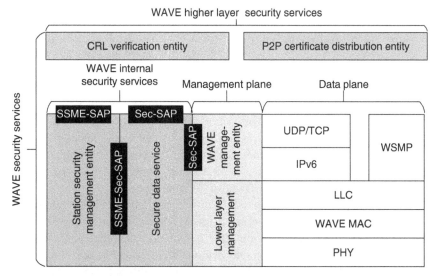

Figure 15.4 WAVE protocol stack security services.

Figure 15.4. WAVE Short Message Protocol (WSMP) supports both IP and non-IP data transfers. WAVE medium access control (MAC) is a collection of extensions to the IEEE 802.11 MAC [191]. The WAVE management entity defines the corresponding network services [192]. The Service Access Points (SAPs) are defined to support communications between WAVE security services and other entities.

The WAVE Security Services consist of internal security services and higher layer security services. The security services and corresponding functions are listed in Table 15.4. The secure data service (SDS) transforms the original protocol data units (PDUs) into secured protocol data units (SPDUs). SDS also transforms SPDUs to PDUs on reception. SDS is provided through a Secure Data Exchange Entity (SDEE). Security management is the service that manages certificates. The Certificate Revocation List (CRL) verification entity validates incoming CRLs and passes the related revocation information to the Security Service Management Entity (SSME) for storage. The Peer-to-Peer (P2P) certificate distribution entity enables P2P certificate distribution.

Table 15.4 WAVE security services.

WAVE internal security services	
Secure data service	Transforming unsecured PDUs into SPDUs
Security management	Managing information about certificates
WAVE higher layer security services	
CRL verification entity	Validates CRLs and passes the revocation information
P2P certificate distribution entity	Enables peer-to-peer certificate distribution

15.3.2 Secure Data Service and Security Service Management Entity

Figure 15.5 shows the processing flow for use of WAVE SDS. The sending SDEE invokes a sender security processing request, whose result is an SPDU from the SDS. The sender security processing request may be invoked multiple times prior to transmission of an SPDU. A similar process is operated at the receiver side to transform the SPDU into a PDU. SPDUs may be unsecured, signed, and encrypted. An unsecured SPDU provides a trivial implementation if no cryptographic protection is needed. A signed SPDU provides sender authenticity, sender authorization, message integrity, and non-repudiation to the contents. An encrypted SPDU provides confidentiality to the contents. Multiple layers of cryptographic protection can be provided to SPDUs, e.g. signed and then encrypted. An SPDU may contain another SPDU of the same or different type.

The SSME mainly stores certificates and the information that relates to both certificates for which the corresponding private key is stored by the SDS and certificates for which the corresponding private key is not stored by the SDS. In particular, the related information consists of: the last time relevant revocation information was received; the next time revocation information is expected to be received; the verification status of the certificate; and whether or not the certificate is a trust anchor. The SSME has two Service Access Points (SAPs), i.e. the Sec-SAP and the SSME-SAP. The Sec-SAP is used by higher layer entities and by the WAVE management entity. The SSME-SAP is used by certain SDS operations, e.g. invoking certain primitives. The functions of the two SSME SAPs are listed in Table 15.5.

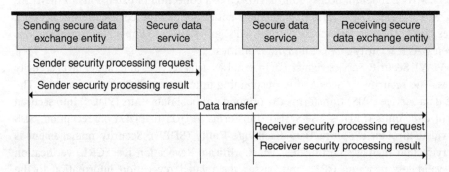

Figure 15.5 Processing flow for use of WAVE SDS.

Table 15.5 Functions of Sec-SAP and SSME-SAP.

SAP	Functions
Sec-SAP	Add information about certificates; Provide information about managed certificates; Request that a certificate is verified; Request deletion of information about a certificate; Add a certificate to its list of trust anchors; Update revocation information for a certificate; Update information about CRLs relevant to managed certificates; Provide information about CRLs relevant to managed certificates; Enable application processes associated with P2P certificate distribution;
SSME-SAP	Provide information about replayed PDUs; Provide information to enable P2P certificate distribution; Enable configuration of P2P certificate distribution

15.3.3 CRL Verification Entity and P2P Certificate Distribution Entity

WAVE defines entity revocation by recording CRL and verification through CRL verification entity. A certificate is to be revoked if an appropriately authorized entity states that the certificate is no longer trustworthy. Once revoked, all SPDUs signed by that certificate after the issue date of the revocation statement are invalid. Information about the revoked certificates is stored in CRLs by the SSME and processed by the CRL verification entity. An overview of the CRL verification entity processing flows is shown in Figure 15.6. The CRL verification entity shall verify a received CRL from the data plane. If the CRL is valid, multiple revocation requests may be invoked by the CRL verification entity to the SSME. At least one confirmation shall be received at the CRL verification entity. Moreover, the CRL verification entity shall pass the revocation information contained in the CRL to the SDEE for storage.

The P2P certificate distribution (P2PCD) is a functionality obtained by the cooperation of the P2PCD entity, the SSME, the SDS, and an appropriately behaving SDEE referred to as the trigger SDEE. The functional entities on a device that support P2PCD are illustrated in Figure 15.7. P2PCD is initiated when a device receives a signed SPDU for which WAVE Security Services are unable to construct a certificate chain due to not recognizing the issuer of the signed SPDU. The received SPDU is referred to as a trigger SPDU. The device that received the trigger SPDU uses P2PCD learning requests to request peer devices to provide the necessary certificates to complete the chain. A P2PCD learning request is a field that the SDS inserts into SPDUs when signing them on behalf of the SDEE that received the

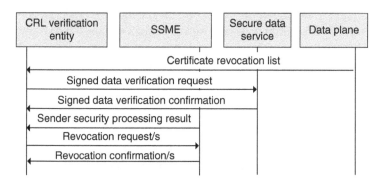

Figure 15.6 Processing flows of the CRL verification entity.

Figure 15.7 Connections of the functional entities in P2P certificate distribution.

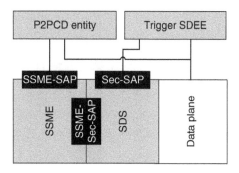

Table 15.6 Description of the functional entities in P2P certificate distribution.

Entity	Functions
P2PCD entity	Send and receive P2PCD learning responses over the data plane; Request the SSME to store the contents of received learning responses;
Trigger SDEE	Send and/or receive signed SPDUs;
Data plane	Exchange PDUs between instances of the P2PCD entity and between the instances of the trigger SDEE.
SDS	Determine if P2PCD needs to be triggered; Determine whether to include P2PCD learning requests in SPDUs;
SMEE	Provide information about incoming SPDUs to the SDS; Request the SDS to include P2PCD learning requests in SPDUs; Store certificates received via P2PCD learning response PDUs; Register and request the P2PCD entity to send P2PCD learning responses;

original SPDU. P2PCD learning responses are sent as PDUs by the P2PCD Entity to P2PCD Entities on peer devices. The design of the P2PCD service includes special mechanisms to reduce the risk of channel flooding by limiting the number of responses to a single request. Descriptions of each functional entity are given in Table 15.6.

15.4 Security in Cellular Based V2X Communications

Cellular based V2X communications are mostly based on LTE and future 5G wireless networks. The current LTE provides security to general mobile communications, thus it can be directly applied to LTE based V2I communications. However, cellular based V2V and V2P communications require special attentions. This section gives a brief introduction to the security frameworks on LTE-V2X and 5G-V2X.

15.4.1 LTE-V2X Communication Security

The 3GPP defined LTE-V2X architecture is mainly based on PC5 and LTE-Uu communication techniques, as shown Figure 15.8. User equipment (UE) defines access nodes such as

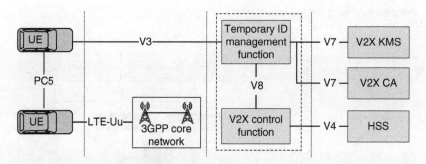

Figure 15.8 Security architecture for PC5 and LTE-Uu based V2X communications.

vehicles and pedestrians in LTE-V2X. The LTE-Uu supports V2I communications through 3GPP core network (e.g. E-UTRAN in LTE), where the traditional LTE security can be applied. UE in V2V and V2P is connected to the V2X control function (VCF), which passing corresponding security parameters to UE. The authentication and authorization of UE are actually controlled by the home subscribe server (HSS). The temporary identities and credentials of UE are distributed by the temporary identities management function (TIMF) through the V2X key management server (KMS) and the V2X certificate authority (CA).

LTE-V2X introduced three fundamental modifications to PC5 for vehicular use cases addressing high speed (up to 500 km/h) and high density (thousands of nodes). Firstly, the sub-frame structure has been modified to include four demodulation reference signal (DMRS) symbols, as illustrated in Figure 15.9a. In addition, a Tx-Rx turnaround symbol is set to the end of the sub-frame structure. The DMRS and the turnaround symbols are included for better tracking of the channel at high speed of up to 500 km/h and at the 5.9 GHz ITS band. Secondly, a new arrangement of scheduling assignment (SA) and data resources has been designed, as illustrated in Figure 15.9b. In the new design, SA or physical sidelink control channel (PSCCH) are transmitted in sub-channels using specific resource blocks (RBs) across time. Data transmission associated with the scheduling assignments is occupying adjacent RBs in the same subframe. This new design is to enhance the system level performance while meeting the requirements of high density and low latency of V2V communications. Meanwhile, another variant where scheduling assignments and associated data transmission are on nonadjacent RBs has also been standardized. Finally, a sensing with semi-persistent transmission based mechanism has been introduced for distributed scheduling. The basic idea is to estimate future congestion on a resource where congestion has been sensed. Because V2V communication traffic from a device is mostly periodic in nature, the estimation technique can optimize channel usage by enhancing resource separation between transmitters that are using overlapping resources. The design is scalable for different bandwidths including 10 MHz bandwidth.

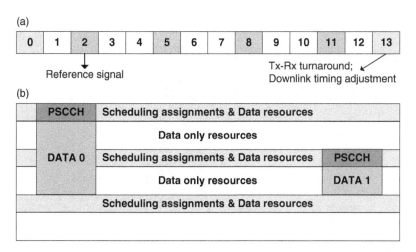

Figure 15.9 Fundamental modifications to PC5. (a) V2X sub-frame for PC5 interface. (b) New scheduling assignment and data resources.

The current LTE Authentication and Key Agreement (LTE-AKA) protocol can be applied to mutually authenticate a UE supplicant and the serving network before a UE joins a V2X communication [193]. User pseudonym assignment and air interface confidentiality can be inherited from LTE-AKA protocol for LTE-Uu communications. However, no additional privacy features beyond the regular LTE privacy features are supported for LTE-Uu communications. The 3GPP does not define security nor mandate privacy features for LTE-PC5 communications. If privacy is applied in LTE-PC5 communications, the UE shall change and randomize the source ID and IP address when the application layer identifier has changed. It is recommended to address privacy at the application layer by employing identifiers and credentials that are not linked to long-term UE or user identifiers.

15.4.2 5G-V2X Communication Security

The 5G-V2X system level architecture supports multiple operations necessary for the efficient deployment and automation of network functions of V2X communications, including multiple operators, security and privacy, smart zoning, dynamic use of multiple Radio Access Technology (RAT) and multi-links, as shown in Figure 15.10. Software Defined Networking (SDN) and Network Function Virtualization (NFV) technologies allow optimal resource allocations for most V2X use cases. Besides, it would be intuitive to integrate different wireless access technologies through the 5G framework because of the enhanced support to both 3GPP and non 3GPP accesses.

Two different solutions are proposed for 5G-V2X security and integrity check of vehicular messages. One is by applying the security checks at the application layer in the UE for a distributed manner. The other is to exploit the presence of the 5G-V2X network for robust

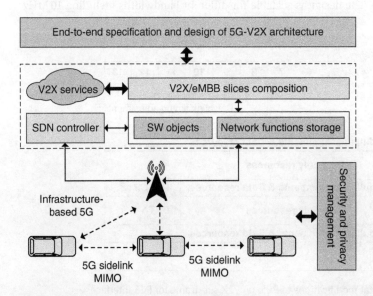

Figure 15.10 5G-V2X system level architecture.

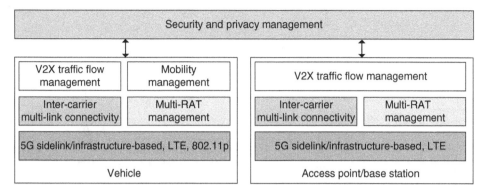

Figure 15.11 A security architecture for 5G-V2X

security. Figure 15.11 shows an overall security architecture for 5G-V2X [194]. Since the wireless communications of 5G-V2X network mainly inherits the design of LTE-V2X, security over the sidelink (i.e. PC5 or 802.11p) and LTE-Uu communications would be inherited from those of LTE-V2X. Nonetheless, the HetNet architecture, SDN and NFV technologies in 5G will greatly enhance the network performance.

15.5 Cryptography and Privacy Preservation in V2X Communications

Cryptographic schemes provide multiple security services in wireless communications. Examples of solutions have been introduced in the earlier chapters. Traditionally, symmetric key cryptography provides efficient encryption and message integrity. Public key cryptography provides non repudiation and key distributions. Cryptographic schemes also support authentication protocols and authorizations. However, typical public key infrastructure supported key distribution and certificate systems may not work for V2X communications due to high density of users and low latency requirement. This section focuses on the recent schemes targeting scalability and privacy in V2X communications.

15.5.1 Identity Based Schemes

Identity based schemes intend to simplify the public key generation and distribution so that to support large scale systems. In short, identity based schemes allow each user to generate public keys of any users with their corresponding IDs. Figure 15.12 shows an overview of how identity based schemes can be applied to V2X communications. During the initialization process, each vehicle registers at the TA with real ID (RID) and receives the secret key and other parameters needed to execute identity based schemes. Since the public keys are generated in a distributed fashion, a group head or the RSU is not needed in the process. Nonetheless, the RSU (or a central controller) is needed to detect and report abnormal behaviors from malicious vehicles to the TA.

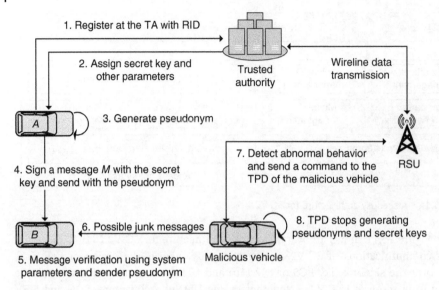

Figure 15.12 A general architecture of identity based schemes.

Introduction to bilinear pairing $\hat{e}$: $\mathbb{G}_1 \times \mathbb{G}_2 \to \mathbb{G}_T$

- Preliminaries: $\mathbb{G}_1$ and $\mathbb{G}_2$ are cyclic additive groups of prime order q, and $\mathbb{G}_T$ a cyclic multiplicative group of prime order q. In most assumptions, $\mathbb{G}_1 = \mathbb{G}_2 = \mathbb{G}$.
- Bilinearity: for all $P, Q, R \in \mathbb{G}$, and non-zero integers a, b.

$$\hat{e}(Q, P + R) = \hat{e}(P + R, Q) = \hat{e}(P, Q) \cdot \hat{e}(R, Q),$$
$$\hat{e}(aP, bP) = \hat{e}(P, bQ)^a = \hat{e}(aP, Q)^b \cdot \hat{e}(P, Q)^{ab}.$$

- Non-degeneracy: there exist $P, Q \in \mathbb{G}$ such that $\hat{e}(P, Q) \neq 1_{\mathbb{G}_T}$.
- Computability: there is an efficient algorithm to compute $\hat{e}(P, Q)$ for any $P, Q \in \mathbb{G}$.

An example in the following text provides an idea of how identity based schemes can work to provide digital signatures based on bilinear pairing [195–197]. In this example, the domain parameters can be set to:

$$(\hat{e}, q, \mathbb{G}, g, H_1, H_2, Pu),$$

where $\hat{e}$ is a bilinear pairing (as shown in the box earlier) based on $q, \mathbb{G}, \mathbb{G}_T, g; H_1 : \{0, 1\}^* \to \mathbb{Z}_q$ and $H_2 : \{0, 1\}^* \to \mathbb{G}$ are two cryptographic hash functions; $Pu = sg \in \mathbb{G}$ is the domain public key (also the public key of the TA) and $s \in \mathbb{Z}_q^*$ is the master secret of the TA. The secret key of user i is set to $Pr_i = sH_2(H_1(RID)) \in \mathbb{G}$ by the TA. With the domain secret, a vehicle user i can derive the pseudonym locally as follows:

$$VID_i = H_1(RID_i) \in \mathbb{G}. \tag{15.1}$$

The public key of user i is derived based on VID as follows:

$$Pu_i = H_2(VID). \tag{15.2}$$

Note that Pu_i can be derived locally at any user who bears VID_i. Therefore, identity based schemes can greatly enhance the performance of public key distribution in large scale V2X systems.

To sign a message M, user i first picks a random number $r \in \mathbb{Z}_q^*$ to compute $v = rg \in \mathbb{G}$. The partial signature is generated as follows:

$$\sigma_i = \hat{e}(g, H_1(M)Pr_i + rPu). \tag{15.3}$$

The overall signature includes (σ_i, v).

To verify a signature, the receiver applies both the domain public key Pu and the sender Pu_i and check if:

$$\hat{e}(Pu, H_1(M)Pu_i + v) = \sigma_i. \tag{15.4}$$

The verification works because

$$\hat{e}(Pu, H_1(M)Pu_i + v) = \hat{e}(sg, H_1(M)Pu_i + rg) = \hat{e}(g, H_1(M)sPu_i + rsg).$$

Due to the relatively high complexity, identity based schemes are mostly applied to signatures and certificates rather than encryption.

15.5.2 Group Signature Based Schemes

Group signature provides a solution for efficient processing of digital signatures within a large group by enabling every member in a V2X network to sign messages anonymously [198]. Therefore, group signature can greatly reduce the complexity in generating and distributing public keys. A typical group signature based scheme is shown in Figure 15.13.

Each vehicle is required to register at the TA with their RIDs during the initialization process. The vehicle also receives the system parameters and a valid vehicle ID (VID).

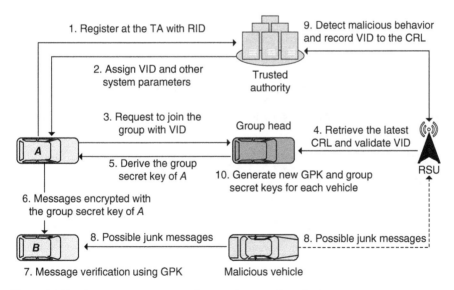

Figure 15.13 A general architecture of group signature based schemes.

System parameters are domain secret generated by the TA for further group signature cryptography. Developing group signature schemes is mostly based on bilinear paring techniques. To join a group, the VID of the applicant needs to be checked if it is on the latest CRL. If the VID is on the latest CRL, the access will not be granted. Meanwhile, the group head will report the VID to the TA, as well as generate a new group public key (GPK) and new secret keys for all legitimate vehicles. If an access is granted, the group head derives a private key for the requester and sent it through a secured channel. Practical group signature schemes can be implemented based on RSA, discrete logarithm, elliptic curve, and bilinear paring [199–202]. Group signature based V2X systems require properly distributed RSUs and trusted group head. Otherwise, the overhead of CRL update may overwhelm the RSUs.

15.5.3 Batch Verification Schemes

Batch verification is a method that can verify multiple signatures from different users in a batched fashion. Thus, batch verification schemes can be efficient in large scale V2X communications [203, 204]. Identity based cryptography is usually combined to achieve batch verification [203, 204]. In this process, a pseudo ID is used instead of the RID of a vehicle. Given the distributed nature, it can be assumed to generate pseudo IDs from the TPD of each vehicle. An example of the modules and functions in a TPD is illustrated in Figure 15.14. The authentication module first verifies the pair of RID and password (PWD). If valid, the pseudo ID generation module generates the pseudo ID pair (ID_1, ID_2). The private key generation module generates the private key pair (Pr_1, Pr_2). The process is based on a bilinear pairing $\hat{e}$ that has been introduced before. In the generation of (ID_1, ID_2), r is a random number, g is the generator of $\mathbb{G}$, and Pu_1 is one of the public keys of the TA. In the generation of (Pr_1, Pr_2), s_1 and s_2 are the secret key pair of the TA, and $H(\cdot)$ is a MapToPoint hash function [205]. Note that the TA has a pair of public keys, i.e. $Pu_1 = s_1g$ and $Pu_2 = s_2g$.

Given a message M, a signature is computed as follows:

$$\sigma = Pr_1 + h(M) \cdot Pr_2, \tag{15.5}$$

where $h(\cdot)$ is a one-way hash function. A single signature is verified as follows:

$$\hat{e}(\sigma_i, g) = \hat{e}(ID_1, Pu_1) \cdot \hat{e}(h(M))H(ID_1 \| ID_2, Pu_2). \tag{15.6}$$

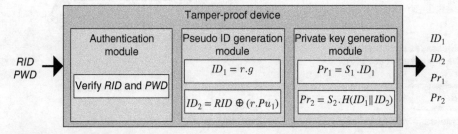

Figure 15.14 Modules and functions in a TPD.

Given a batch of signatures generated from messages $(M_1, M_2, \ldots, M_n)$, the batch verification follows:

$$\hat{e}\left(\sum_{i=1}^{n}\sigma_i, g\right) = \hat{e}\left(\sum_{i=1}^{n}ID_1^i, Pu_1\right) \cdot \hat{e}\left(\sum_{i=1}^{n}h(M_i)H(ID_1^i\|ID_2^i), Pu_2\right), \tag{15.7}$$

where ID_1^i and ID_2^i are corresponding to M_i.

15.5.4 Reputation and Trust Based Schemes

Reputation and trust management is used to complement cryptography based schemes against some attacks, e.g. DoS and black hole attacks. In general, reputation and trust based schemes check the trustworthiness of an entity based on a reputation score, which is generated based on the feedback from other users. One of the first reputation systems for vehicular networks applies a modular approach that strictly separates direct reputation handling, indirect reputation handling, and opinion generation [206]. The description of terms direct trust, indirect trust, opinion, and confidence are given in Table 15.7.

Every forwarding node evaluates on the trustworthiness of the message and appends the opinion to the message. The developers named the process *Opinion Piggybacking*, as shown Figure 15.15. An opinion O_i consists of an opinion value o_i, a source based reputation level s, and the ID of the evaluator. The opinion value o_i is computed from partial opinions $o_{i,k}$ from sender node k as follows:

$$o_{i,k} = r_k o_k + (r_{max} - o_k)(r_{max} - r_k), \tag{15.8}$$

where r_k is the direct or indirect reputation value of source node k, r_{max} is the global maximum reputation value. The source based reputation level $s = 1$ if r_k is direct, $s = 2$ if r_k is indirect, and $s = 3$ if r_k is not stored locally. Moreover, if $s = 3$, then $o_{i,k} = o_k$. The

Table 15.7 Description of the functional entities in certificate distribution.

Term	Description
Direct trust	Reputation derived from an announced event can be verified
Indirect trust	Reputation provided by nodes of which reputation information is known
Opinion	Evaluation on the trustworthiness of a message
Confidence	Decision result based on the trust-opinions and reputation levels

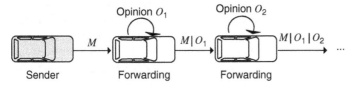

Figure 15.15 Illustration of opinion piggybacking.

forwarding node opinion o_i is a combination computed as follows:

$$o_i = \sum_{k \in \mathbf{o}_1} \frac{\alpha k}{\mathbf{o}_1} + \sum_{k \in \mathbf{o}_1} \frac{\beta k}{\mathbf{o}_2} + \sum_{k \in \mathbf{o}_3} \frac{\gamma k}{\mathbf{o}_3}, \tag{15.9}$$

where $\mathbf{o}_s$ is the set of all combined opinions with sender based reputation level s. The weights α, β, and γ are constants set according to system requirements, and $\alpha + \beta + \gamma = 1$.

A confidence decision is made by evaluating all messages related to the distinct event as follows:

$$f = |\mathbf{M}| - \left\lfloor \left| \frac{r_{\max} - \bar{r}}{a_v} - k \right| \right\rfloor, \tag{15.10}$$

where $\mathbf{M}$ is the set of messages announcing a confidence decision event, $|\mathbf{M}|$ is the number of the message, a_v is a constant factor of the reputation values, $k > 0$ is a threshold of minimum messages required to announce an event, and $\bar{r}$ is the final reputation computed similar to Eq. (15.9). It will be in the system discretion to decide if the event is prevalent.

15.5.5 Identity Anonymity Preservation

Identity anonymity includes *conditional anonymity* and *unlinkability*. The conditional anonymity requires that the real identity of a user cannot be disclosed to others except for the TA. The unlinkability requires that (i) no clear relation can be found between a real identity and its corresponding pseudonyms, and (ii) no clear relation can be found among multiple pseudonyms used by the same vehicle. The identity based schemes and the group signature based schemes described earlier have all taken into consideration the identity anonymity in their designs. Nonetheless, they take different approaches. In identity based schemes, pseudonyms are used to conceal the real identity and frequently changed to prevent tracking. Those pseudonyms are generated with some random value to provide unlinkability. In group based schemes, each legitimate group member acquires its group private key from the group head to sign messages anonymously. No regular group member can identify the real sender of a signature thus to provide anonymity within a group. Then sender is still traceable by the TA thus malicious users can be identified if abnormal activities are detected. In some cases, e.g. LTE-V2X, the TA or an RSU is responsible for generating pseudonyms. In some other cases, pseudonyms are generated locally so that they can be updated frequently in large scale mobile V2X communications [204, 207].

15.5.6 Location Anonymity Preservation

Location anonymity of a user in V2X communications can be endangered in two ways. One is by tracking the frequent update of pseudonyms and directly targeting the applications of location based services. While a pseudonym hides the identity of a vehicle, it is still linked to the broadcasting safety information, e.g. speed, traveling direction, previous location, etc. If the pseudonym updates can be tracked and linked to the same vehicle, the corresponding may be revealed to an attacker. To overcome this issue, one possible solution is updating pseudonym at a moment that the safety information of all vehicles appears indistinguishable [208, 209]. For example, the safety information for vehicles would be close to each other when the traffic light turns from red to green at a crossroad. A similar situation can be found at a parking lot, where vehicles frequently enter and leave. Whether waiting on traffic lights or parking, vehicles are in a relatively static mode, thus pseudonyms change would not be easily linked to any particular vehicle. The solution may be extended to semi-static situations. For example, vehicles platooning on a highway roughly share the same safety information. Those vehicles can form a group and update their pseudonyms based on a synchronized clock or counter.

In comparison, applications that require location based services can be more vulnerable as the location information is directly exchanged between a user and a service provider through V2X wireless communication links. Moreover, the server that handles location requests may be an untrusted entity and poses threats to user location anonymity. One way to tackle this issue is to use ambiguity to obfuscate the real location. For example, based on k-anonymity based schemes [210, 211], a user carefully picks $k - 1$ dummy locations in its vicinity area and send all k locations (including the real location) in any request. The real location will be indistinguishable among the k dummy locations from the view of an attacker. Another way to protect location anonymity if location based service is to restrict the number of requests so that an attacker cannot get enough information to fetch the exact location. For example, a cache server can be deployed to store common and frequent requests [212]. A user can request the cache server first to see if the requested service can be satisfied based on its real location. If so, location information will not be needed in the request. Otherwise, the dummy location based schemes can be applied for further request.

However, the drawback of location anonymity could be unsatisfied services due to inaccurate location information for the service provider. Moreover, dummy information processing increases the overhead in both communications and computations, as well as additional hardware support for cache servers.

15.6 Challenges and Future Research Directions

V2X communications are still under development. Therefore, many challenges and open issues still need to be solved. This section briefly highlights some of the major challenges and future research directions towards secure V2X communications.

15.6.1 Highly Efficient Authentication Schemes

V2X communications are pending highly efficient authentication schemes. Both the DSRC and IEEE 1609.2 Standard require vehicles to send safety related message at every 300 ms interval [177, 190]. Given the dynamic environment of vehicles in V2X communications, authentication and authorization must be processed highly efficient at access points (e.g. RSU and base stations) and/or locally by each vehicle in a distributed manner. In Section 15.5, the group signature and batch processing based schemes show the potential in performance improvement of authentication in V2X communications. However, while those schemes have good performance processing legitimate information, they fall short on the recovery process when verification results turn invalid. Therefore, future efforts are still needed for the enhancement.

15.6.2 Efficient Revocation Mechanisms

The IEEE WAVE standard and many recent schemes demand revocation mechanism for frequent authentication and handover of vehicles in V2X communications. While the basic ideas of revocation are clarified, the efficiency of the revocation schemes remains underdeveloped. For example, the application of CRL defined in the IEEE WAVE standard requires one-by-one processing, thus being inefficient in handling a large group of vehicles at high speed. Furthermore, management of CRL also needs enhancement for scalability of V2X communications and traceability from authorized users.

15.6.3 Advancing OBU and TPD Technologies

The core components of a vehicle in V2X communications are the OBU and TPD. Although research designs may focus on lightweight schemes that can be applied to the less powerful OBU and TPD [213], most of the developing schemes assume advanced communication capability of OBU and security capability of TPD. Therefore, OBU and TPD technology must be advanced to support the developing wireless technologies and cryptographic schemes. It is reasonable to assume that an OBU would be comparable (if not more advanced) to a smart phone or a computer in terms of communication capability to fully support V2X communications. However, TPD technologies are falling behind those have been applied to smart phones and computers.

15.6.4 Advancing Cryptography and Privacy Preservation Schemes

The cryptography and privacy preservation schemes in the regular LTE systems lay a foundation for V2X communications, especially in V2I communications. However, V2V and V2P communications that rely on sidelink of LTE and 5G, as well as other wireless technologies do not have default implementation of cryptography and privacy preservation schemes. On the positive side, new schemes are being developed, such as identity-based schemes, trust based schemes, etc., to mainly handle V2V communications. New schemes are also being developed for identity anonymity and location anonymity in V2X communications.

15.6.5 Advancing Solutions to HetNet, SDN, and NFV

Rather than being a standalone system, V2X is more likely being an integration of the Internet-of-things and other network applications, which are also supported by the 5G and beyond wireless networks. The 5G will implement HetNet, SDN, and NFV to achieve ultra-low latency and optimal resource management. Security design in 5G-V2X must take into consideration the new network structure and take advantage of the flexibility in resource management for optimality.

15.6.6 Advancing Artificial Intelligence in V2X Communication Security

Artificial intelligence (AI) will play an important role in V2X applications, e.g. for safety alert, autonomous driving, traffic monitoring, etc. [214]. AI schemes are also effective in security, especially in intrusion detection and attack identification [215, 216]. The AI designs need to be refined for better efficiency and scalability of V2X communications.

15.7 Summary

V2X communications facilitate the ITS deployment towards safety and autonomous control. Wireless technologies such as DSRC, LTE, and 5G will be applied to enable V2X communications in different applications. Security and privacy must be addressed to integrate the variety of wireless technologies and meet special requirements for V2X communications. Standards such as IEEE WAVE and LTE-V2X set a general guideline for V2X security implementations. New cryptography schemes such as group signature and trusted based schemes are being developed. There are still challenges to be addressed for security in V2X communications, including efficient schemes, hardware enhancement, integration of AI algorithms, and some other issues.

References

1 Cisco (2020). Cisco Annual Internet Report (2018–2023) White Paper.

2 Tanenbaum, A.S. and Wetherall, D. (1996). *Computer Networks*. Prentice Hall.

3 Andress, J. (2014). *The Basics of Information Security: Understanding the Fundamentals of InfoSec in Theory and Practice*. Syngress.

4 Rappaport, T.S. (2002). *Wireless Communications: Principles and Practice*. Prentice Hall PTR: Upper Saddle River, NJ

5 Jamalipour, A. (2003). *The Wireless Mobile Internet*, 368–384. Wiley.

6 Pahlavan, K. and Krishnamurthy, P. (2011). *Principles of Wireless Networks: A Unified Approach*. Prentice Hall PTR.

7 International Telecommunication Union (1991). X. 800 security architecture for Open Systems Interconnection for CCITT applications. *ITU-T (CCITT) Recommendation*.

8 Shirey, R. (2000). RFC 2828: Internet Security Glossary.

9 NIST (2017). NIST Special Publication 800-53: Security and Privacy Controls for Federal Information Systems and Organizations. *Federal Inf. Process. Stds. (NIST FIPS)*.

10 NIST (2004). NIST PUB 199: Standards for Security Categorization of Federal Information and Systems. *Federal Inf. Process. Stds. (NIST FIPS)*.

11 Washington, L.C. (2008). *Elliptic Curves: Number Theory and Cryptography*. CRC Press.

12 Scott, W.R. (2012). *Group Theory*. Courier Corporation.

13 Cohn, H. (2012). *Advanced Number Theory*. Courier Corporation.

14 Simmons, G.J. (1979). Symmetric and asymmetric encryption. *ACM Computing Surveys (CSUR)* 11 (4): 305–330.

15 Forouzan, B.A. (2007). *Cryptography & Network Security*. McGraw-Hill, Inc.

16 Goyal, K. and Kinger, S. (2013). Modified Caesar cipher for better security enhancement. *International Journal of Computer Applications* 73 (3): 0975–8887.

17 Lewand, R.E. (2000). *Cryptological Mathematics*, vol. 16. American Mathematical Society.

18 Bowers, W.M. (1959). *Digraphic Substitution: The Playfair Cipher, the Four Square Cipher*, vol. 1. American Cryptogram Association.

19 Stinson, D.R. (1995). Classical cryptography. In: *Cryptography, Theory and Practice*, 2e (ed. K.H. Rosen), 1–20. Chapman & Hall/CRC.

20 Shannon, C.E. (1949). Communication theory of secrecy systems. *The Bell System Technical Journal* 28 (4): 656–715.

Security in Wireless Communication Networks, First Edition. Yi Qian, Feng Ye, and Hsiao-Hwa Chen.
© 2022 John Wiley & Sons Ltd. Published 2022 by John Wiley & Sons Ltd.
Companion website: www.wiley.com/go/qian/sec51

21 Diffie, W. and Hellman, M.E. (1977). Special feature exhaustive cryptanalysis of the NBS data encryption standard. *Computer* 10 (6): 74–84.

22 NIST (2001). Advanced Encryption Standard. *Federal Inf. Process. Stds. (NIST FIPS)*.

23 Daemen, J. and Rijmen, V. (2013). *The Design of Rijndael: AES-The Advanced Encryption Standard*. Springer Science & Business Media.

24 NIST (1980). DES Modes of Operations. *Federal Inf. Process. Stds. (NIST FIPS)*.

25 Dworkin, M. (2001). Recommendation for Block Cipher Modes of Operation. Methods and Techniques. *Tech. Rep.* National Inst of Standards and Technology Gaithersburg MD Computer security Div.

26 IEEE Std 1619-2007 (2008). *IEEE Standard for Cryptographic Protection of Data on Block-Oriented Storage Devices*. IEEE Standard Association, pp. c1–c32. https://doi .org/10.1109/IEEESTD.2008.4493450.

27 Hellman, M. (1978). An overview of public key cryptography. *IEEE Communications Society Magazine* 16 (6): 24–32.

28 Nechvatal, J. (1991). Public-Key Cryptography. *Tech. Rep.* Gaithersburg, MD: National Computer Systems Lab.

29 Odlyzko, A.M. (1994). Public key cryptography. *AT&T Technical Journal* 73 (5): 17–23.

30 Rivest, R.L., Shamir, A., and Adleman, L. (1977). On Digital Signatures and Public-Key Cryptosystems. *Tech. Rep.* Massachusetts Inst of Tech Cambridge Lab for Computer Science.

31 Shamir, A. (1999). Factoring large numbers with the TWINKLE device. *International Workshop on Cryptographic Hardware and Embedded Systems*. Springer, pp. 2–12.

32 Jueneman, R., Matyas, S., and Meyer, C. (1985). Message authentication. *IEEE Communications Magazine* 23 (9): 29–40.

33 Wikipedia (2017). Hash function. https://en.wikipedia.org/wiki/Hash_function (accessed 08 June 2021).

34 Flajolet, P., Gardy, D., and Thimonier, L. (1992). Birthday paradox, coupon collectors, caching algorithms and self-organizing search. *Discrete Applied Mathematics* 39 (3): 207–229.

35 NIST (1985). 800-113: Computer Data Authentication. *Federal Inf. Process. Stds. (NIST FIPS)*.

36 NIST (2012). 180-4. Secure Hash Standard (SHS). *Federal Inf. Process. Stds. (NIST FIPS)*.

37 Wang, X., Yu, H., and Yin, Y.L. (2005). Efficient collision search attacks on SHA-0. *Annual International Cryptology Conference*. Springer, pp. 1–16.

38 Stevens, M., Bursztein, E., Karpman, P. et al. (2017). The first collision for full SHA-1. *Annual International Cryptology Conference*. Springer, pp. 570–596.

39 Rijmen, V. (2000). The WHIRLPOOL hash function. http://www.larc.usp.br/pbarreto/ WhirlpoolPage.html (accessed 08 June 2021).

40 IETF (1997). HMAC: Keyed-Hashing for Message Authentication.

41 Dworkin, M. (2001). Recommendation for Block Cipher Modes of Operation: The CMAC Mode for Authentication. *Tech. Rep.* National Inst of Standards and Technology Gaithersburg MD Computer security Div.

42 American Nantional Standards Institute (2005). ANSI X9.62 Public Key Cryptography for the Financial Services Industry: The Elliptic Curve Digital Signature Algorithm (ECDSA). *The elliptic curve digital signature algorithm (ECDSA)*.

43 Kerry, C.F. and Director, C.R. (2013). FIPS PUB 186-4 federal information processing standards publication digital signature standard (DSS).

44 IEEE Working Group (2000). IEEE Std 1363-2000: IEEE Standard Specifications for Public-Key Cryptography.

45 Dang, Q.H. (2015). Secure Hash Standard. *Federal Inf. Process. Stds. (NIST FIPS)*.

46 IETF (2001). Internet Message Format.

47 IETF (1997). Internet X.509 Public Key Infrastructure Certificate and Certificate Revocation List (CRL) Profile.

48 Wi-Fi Alliance. https://www.wi-fi.org (accessed 08 June 2021).

49 IEEE Standard Association (2016). IEEE Standard for Information technology—Telecommunications and information exchange between systems Local and metropolitan area networks—Specific requirements - Part 11: Wireless LAN Medium Access Control (MAC) and Physical Layer (PHY) Specifications. *IEEE Std 802.11-2016 (Revision of IEEE Std 802.11-2012)*, pp. 1–3534. https://doi.org/10.1109/IEEESTD.2016.7786995.

50 Wi-Fi Alliance (2010). Wi-Fi certified Wi-Fi direct. *White paper*.

51 Wi-Fi Alliance (2006). WPA2 security now mandatory for Wi-Fi certified products. *Press Release*.

52 Wi-Fi Alliance (2012). The State of Wi-Fi Security: Wi-Fi Certified WPA2 Delivers Advanced Security to Homes, Enterprises and Mobile Devices. *Wi-Fi Alliance, Tech. Rep.*

53 Wi-Fi Alliance 2020. WPA3 Specification Version 1.0., pp. 1–7.

54 IEEE Standard Association (2010). IEEE Standard for Local and Metropolitan Area Networks–Port-Based Network Access Control. *IEEE Std 802.1X-2010 (Revision of IEEE Std 802.1X-2004)*, pp. 1–205. https://doi.org/10.1109/IEEESTD.2010.5409813.

55 Aboba, B., Blunk, L., Vollbrecht, J. et al. (2004). RFC 3748-Extensible Authentication Protocol (EAP). *Network Working Group*.

56 Aboba, B. and Simon, D. (1999). PPP EAP TLS Authentication Protocol. *IETF RFC 2716*.

57 Pall, G. and Zorn, G. (2001). Microsoft Point-To-Point Encryption (MPPE) Protocol. *RFC3078*.

58 IEEE Std 802.11i-2004 (2004). *IEEE Standard for Information Technology-Telecommunications and Information Exchange Between Systems-Local and Metropolitan Area Networks-Specific Requirements-Part 11: Wireless LAN Medium Access Control (MAC) and Physical Layer (PHY) Specifications: Amendment 6: Medium Access Control (MAC) Security Enhancements.* IEEE Standard Association.

59 Michael, F.N. (2002). An Improved MIC for 802.11 WEP. *IEEE doc. 802.11-2/020r0.*

60 Paterson, K.G., Poettering, B., and Schuldt, J.C. (2014). Plaintext recovery attacks against WPA/TKIP. *International Workshop on Fast Software Encryption*. Springer, pp. 325–349.

61 Vanhoef, M. and Piessens, F. (2017). Key reinstallation attacks: forcing nonce reuse in WPA2. *Proceedings of the 2017 ACM SIGSAC Conference on Computer and Communications Security*, pp. 1313–1328.

62 Wi-Fi Alliance (2013). Discover Wi-Fi, Wi-Fi Direct.

63 Harkins, D. and Kumari, W. (2017). RFC 8110-Opportunistic Wireless Encryption.

64 Braeckel, P. (2011). Feeling bluetooth: from a security perspective. In: *Advances in Computers* (ed. M. Zelkowitz), vol. 81, 161–236. Elsevier.

65 Paul, M. (2010). Bluesnarfing. In: *Encyclopedia of Information Assurance-4 Volume Set (Print)* (ed. R. Herold, M.K. Rogers), 255–259. Auerbach Publications.

66 Padgette, J., Scarfone, K., and Chen, L. (2012). *Guide to Bluetooth Security: Recommendations of the National Institute of Standards and Technology (Special Publication 800-121 Revision 1)*. CreateSpace Independent Publishing Platform.

67 Laurie, A. and Herfurt, M.H.M. (2004). Hacking bluetooth enabled mobile phones and beyond–full disclosure. *21st Chaos Communication Congress, Berliner Congress Center*, Berlin, Germany.

68 Scarfone, K. and Padgette, J. (2008). *Guide to Bluetooth Security*, vol. 800, 121. NIST Special Publication.

69 Whitman, M.E., Mattord, H.J., Mackey, D., and Green, A. (2012). *Guide to Network Security*. Cengage Learning.

70 Gehrmann, C., Persson, J., and Smeets, B. (2004). *Bluetooth Security*. Artech House.

71 Massey, J.L., Khachatrian, G.H., and Kuregian, M.K. (1998). SAFER+ Cylink Corporation's submission for the advanced encryption standard. *Proceedings of the 1st Advanced Encryption Standard Candidate Conference*.

72 NIST (2009). FIPS PUB 186-3: Digital Signature Standard (DSS). *Federal Inf. Process. Stds. (NIST FIPS)*.

73 IEEE LAN/MAN Standards Committee (2003). *IEEE Standard 802.15.4, Part 15.4: Wireless Medium Access Control (MAC) and Physical Layer (PHY), Specifications for Low-Rate Wireless Personal Area Networks (LR-WPANs)*.

74 Lee, J.S., Chuang, C.C., and Shen, C.C. (2009). Applications of short-range wireless technologies to industrial automation: a ZigBee approach. *2009 5th Advanced International Conference on Telecommunications*. IEEE, pp. 15–20.

75 Gomez, C. and Paradells, J. (2010). Wireless home automation networks: a survey of architectures and technologies. *IEEE Communications Magazine* 48 (6): 92–101.

76 Zigbee Alliance. http://www.zigbee.org (accessed 08 June 2021).

77 Sastry, N. and Wagner, D. (2004). Security considerations for IEEE 802.15.4 networks. *Proceedings of the 3rd ACM Workshop on Wireless Security*. ACM, pp. 32–42.

78 Sciancalepore, S., Piro, G., Boggia, G., and Grieco, L. (2014). Application of IEEE 802.15.4 security procedures in OpenWSN protocol stack. *Standards Education e-Magazine* 4 (2): 1–9.

79 Daidone, R., Dini, G., and Anastasi, G. (2014). On evaluating the performance impact of the IEEE 802.15.4 security sub-layer. *Computer Communications* 47: 65–76.

80 Zillner, T. and Strobl, S. (2015). ZigBee exploited: the good, the bad and the ugly. *Black Hat–2015*.

81 Gascón, D. (2009). Security in 802.15.4 and ZigBee networks. 28: 1–5. https://www.libelium.com/libeliumworld/security-802-15-4-zigbee/.

82 Matyas, S.M. (1985). Generating strong one-way functions with cryptographic algorithm. *IBM Technical Disclosure Bulletin* 27: 5658–5659.

83 Karygiannis, T., Eydt, B., Barber, G. et al. (2007). Guidelines for securing radio frequency identification (RFID) systems. *NIST Special Publication* 80: 1–154.

84 Xiao, Q., Gibbons, T., and Lebrun, H. (2009). RFID technology, security vulnerabilities, and countermeasures. In: *Supply Chain the Way to Flat Organization* (eds. Y. Huo and F. Jia), 357–382. Publisher-Intech.

85 US VISIT Smart Border Alliance (2005). RFID Feasibility Study. *Final Report*. Application note.

86 Weis, S.A., Sarma, S.E., Rivest, R.L., and Engels, D.W. (2004). Security and privacy aspects of low-cost radio frequency identification systems. In: *Security in Pervasive Computing* (eds. D. Hutter, G. Müller, W. Stephan, M. Ullmann), 201–212. Springer.

87 Juels, A. (2005). Strengthening EPC tags against cloning. *Proceedings of the 4th ACM Workshop on Wireless Security*, pp. 67–76.

88 Juels, A., Rivest, R.L., and Szydlo, M. (2003). The blocker tag: selective blocking of RFID tags for consumer privacy. *Proceedings of the 10th ACM Conference on Computer and Communications Security*, pp. 103–111.

89 Hopper, N.J. and Blum, M. (2001). Secure human identification protocols. *International Conference on the Theory and Application of Cryptology and Information Security*. Springer, pp. 52–66.

90 Juels, A. and Weis, S.A. (2005). Authenticating pervasive devices with human protocols. *Annual International Cryptology Conference*. Springer, pp. 293–308.

91 Bringer, J., Chabanne, H., and Dottax, E. (2006). HB++: a lightweight authentication protocol secure against some attacks. *2nd International Workshop on Security, Privacy and Trust in Pervasive and Ubiquitous Computing (SecPerU'06)*, IEEE, pp. 28–33.

92 Regev, O. (2009). On lattices, learning with errors, random linear codes, and cryptography. *Journal of the ACM (JACM)* 56 (6): 1–40.

93 Huurdeman, A.A. (2003). *The Worldwide History of Telecommunications*. Wiley.

94 European Telecommunications Standards Institute. https://www.etsi.org/ (accessed 08 June 2021).

95 Briceno, M., Goldberg, I., and Wagner, D. (1998). An implementation of the GSM A3A8 algorithm. *Unpublished report* 5.

96 35.205, G.T. (2008). 3G Security; Specification of the MILENAGE Algorithm Set.

97 Xenakis, C. (2006). Malicious actions against the GPRS technology. *Journal in Computer Virology* 2 (2): 121–133.

98 Wagner, D. (2017). GSM Cloning.

99 Briceno, M., Goldberg, I., and Wagner, D. (1999). A pedagogical implementation of the GSM A5/1 and A5/2 "voice privacy" encryption algorithms. *Cryptome* 26.

100 International Telecommunication Union (2011). About mobile technology and IMT-2000.

101 Niemi, V. and Nyberg, K. (2006). *UMTS Security*. Wiley.

102 3GPP (1999). Technical Specifications and Technical Reports for a UTRAN-Based 3GPP System.

103 3GPP (2005). 3G Security; Security Architecture (Release 6).

104 3GPP (2020). 3rd Generation Partnership Project; Technical Specification Group Services and System Aspects; 3G Security; Security architecture (Release 16).

105 3GPP (1999). 3G Security; Specification of the 3GPP Confidentiality and Integrity Algorithms; Document 1: F8 and F9 Specification.

106 3GPP (1999). 3G Security; Specification of the 3GPP Confidentiality and Integrity Algorithms; Document 2: Kasumi Specification.

107 Cichonski, J. and Franklin, J. (2015). LTE security–how good is it?

108 Zumerle, D. (2011). 3GPP LTE Security Aspects.

109 Cao, J., Ma, M., Li, H. et al. (2014). A survey on security aspects for LTE and LTE-A networks. *IEEE Communication Surveys and Tutorials* 16 (1): 283–302.

110 3GPP (2012). 3GPP TS 33.401 V12. 5.0, 3GPP System Architecture Evolution (SAE); Security Architecture.

111 Forsberg, D., Horn, G., Moeller, W.D., and Niemi, V. (2012). *LTE Security*. Wiley.

112 Orhanou, G., Hajji, S.E., Bentaleb, Y., and Laassiri, J. (2010). EPS confidentiality and integrity mechanisms algorithmic approach. *IJCSI International Journal of Computer Science Issues* 7 (4): 15–23.

113 Orhanou, G. and El-Hajji, S. (2013). The new LTE cryptographic algorithms EEA3 and EIA3: verification, implementation and analytical evaluation. *Applied Mathematics & Information Sciences* 7 (6): 2385.

114 Dworkin, M. (2001). Recommendation for Block Cipher Modes of Operation. Methods and Techniques. *Tech. Rep.* DTIC Document.

115 3GPP (2016). 3GPP TS 35.221 V13. 0.0, 3rd Generation Partnership Project; Specification of the 3GPP Confidentiality and Integrity Algorithms EEA3 & EIA3; Document 1: EEA3 and EIA3 Specifications.

116 3GPP (2016). 3GPP TS 35.222 V13. 0.0, 3rd Generation Partnership Project; Specification of the 3GPP Confidentiality and Integrity Algorithms EEA3 & EIA3; Document 2: ZUC Spec.

117 Panwar, N., Sharma, S., and Singh, A.K. (2016). A survey on 5G: the next generation of mobile communication. *Physical Communication* 18: 64–84.

118 The 5G Infrastructure Public Private Partnership (2015). 5G Vision (The Next Generation of Communication Networks and Services).

119 GSMA Intelligence (2014). Understanding 5G: perspectives on future technological advancements in mobile. *White paper*, pp. 1–26.

120 NGMN Alliance (2015). Next generation mobile networks, white paper. *5G white paper*, vol. 1.

121 Andrews, J.G., Buzzi, S., Choi, W. et al. (2014). What will 5G be? *IEEE Journal on Selected Areas in Communications* 32 (6): 1065–1082.

122 Agiwal, M., Roy, A., and Saxena, N. (2016). Next generation 5G wireless networks: a comprehensive survey. *IEEE Communication Surveys and Tutorials* 18 (3): 1617–1655.

123 Qiao, J., Shen, X.S., Mark, J.W. et al. (2015). Enabling device-to-device communications in millimeter-wave 5G cellular networks. *IEEE Communications Magazine* 53 (1): 209–215.

124 Wei, L., Hu, R.Q., Qian, Y., and Wu, G. (2015). Energy efficiency and spectrum efficiency of multihop device-to-device communications underlaying cellular networks. *IEEE Transactions on Vehicular Technology* 65 (1): 367–380.

125 Dabbagh, M., Hamdaoui, B., Guizani, M., and Rayes, A. (2015). Software-defined networking security: pros and cons. *IEEE Communications Magazine* 53 (6): 73–79.

126 Zhang, J., Xie, W., and Yang, F. (2015). An architecture for 5G mobile network based on SDN and NFV.

127 NGMN Alliance (2016). 5G Security Recommendations Package #2: Network Slicing. *NGMN*, pp. 1–12.

128 Ericsson (2017). 5G Security - Scenarios and Solutions.

129 Global Mobile Suppliers Association (2015). *The Road to 5G: Drivers, Applications, Requirements and Technical Development*. Global Mobile Suppliers Association.

130 Qualcomm Technologies, Inc. (2016). Leading the World to 5G.

131 3GPP (2015). SA1 completes its study into 5G requirements.

132 Fang, D., Qian, Y., and Hu, R.Q. (2018). Security for 5G mobile wireless networks. *IEEE Access* 6: 4850–4874.

133 Huawei Technologies Co., Ltd (2016). 5G Security: Forward Thinking - Huawei White Paper.

134 NGMN Alliance (2016). 5G security recommendations Package #1. *White paper*.

135 Nokia (2017). Security challenges and opportunities for 5G mobile networks.

136 Liyanage, M., Abro, A.B., Ylianttila, M., and Gurtov, A. (2016). Opportunities and challenges of software-defined mobile networks in network security. *IEEE Security and Privacy* 14 (4): 34–44.

137 Wang, H.M., Zheng, T.X., Yuan, J. et al. (2016). Physical layer security in heterogeneous cellular networks. *IEEE Transactions on Communications* 64 (3): 1204–1219.

138 Chen, M., Qian, Y., Mao, S. et al. (2016). Software-defined mobile networks security. *Mobile Networks and Applications* 21 (5): 729–743.

139 Tian, F., Zhang, P., and Yan, Z. (2017). A survey on C-RAN security. *IEEE Access* 5: 13372–13386.

140 Deng, Y., Wang, L., Wong, K.K. et al. (2015). Safeguarding massive MIMO aided hetnets using physical layer security. *2015 International Conference on Wireless Communications & Signal Processing (WCSP)*, IEEE, pp. 1–5.

141 Schneider, P. and Horn, G. (2015). Towards 5G security. *2015 IEEE Trustcom/BigDataSE/ISPA*, vol. 1, IEEE, pp. 1165–1170.

142 Conti, M., Dragoni, N., and Lesyk, V. (2016). A survey of man in the middle attacks. *IEEE Communication Surveys and Tutorials* 18 (3): 2027–2051.

143 Fang, Q., WeiJie, Z., Guojun, W., and Hui, F. (2014). Unified security architecture research for 5G wireless system. *2014 11th Web Information System and Application Conference*, IEEE, pp. 91–94.

144 3GPP (2016). TR 23.799: Study on Architecture for Next Generation System.

145 Shiu, Y.S., Chang, S.Y., Wu, H.C. et al. (2011). Physical layer security in wireless networks: a tutorial. *IEEE wireless Communications* 18 (2): 66–74.

146 Zappone, A., Lin, P.H., and Jorswieck, E. (2016). Artificial-noise-assisted energy-efficient secure transmission in 5G with imperfect CSIT and antenna correlation. *2016 IEEE 17th International Workshop on Signal Processing Advances in Wireless Communications (SPAWC)*, IEEE, pp. 1–5.

147 Duan, X. and Wang, X. (2016). Fast authentication in 5G HetNet through SDN enabled weighted secure-context-information transfer. *2016 IEEE International Conference on Communications (ICC)*, IEEE, pp. 1–6.

148 Dubrova, E., Näslund, M., and Selander, G. (2015). CRC-based message authentication for 5G mobile technology. *2015 IEEE Trustcom/BigDataSE/ISPA*, vol. 1, IEEE, pp. 1186–1191.

149 Abualhaol, I. and Muegge, S. (2016). Securing D2D wireless links by continuous authenticity with legitimacy patterns. *2016 49th Hawaii International Conference on System Sciences (HICSS)*, IEEE, pp. 5763–5771.

150 Fan, K., Gong, Y., Du, Z. et al. (2015). RFID secure application revocation for IoT in 5G. *2015 IEEE Trustcom/BigDataSE/ISPA*, vol. 1, IEEE, pp. 175–181.

151 Luo, Y., Cui, L., Yang, Y., and Gao, B. (2015). Power control and channel access for physical-layer security of D2D underlay communication. *2015 International Conference on Wireless Communications & Signal Processing (WCSP)*, IEEE, pp. 1–5.

152 Ju, Y., Wang, H.M., Zheng, T.X., and Yin, Q. (2016). Secure transmission with artificial noise in millimeter wave systems. *2016 IEEE Wireless Communications and Networking Conference*, IEEE, pp. 1–6.

153 Chen, B., Zhu, C., Li, W. et al. (2016). Original symbol phase rotated secure transmission against powerful massive MIMO eavesdropper. *IEEE Access* 4: 3016–3025.

154 Qin, Z., Liu, Y., Ding, Z. et al. (2016). Physical layer security for 5G non-orthogonal multiple access in large-scale networks. *2016 IEEE International Conference on Communications (ICC)*, IEEE, pp. 1–6.

155 Huawei Technologies Co., Ltd (2016). 5G network architecture - a high-level perspective.

156 Li, Y., Kaur, B., and Andersen, B. (2011). Denial of service prevention for 5G. *Wireless Personal Communications* 57 (3): 365–376.

157 Adem, N., Hamdaoui, B., and Yavuz, A. (2015). Pseudorandom time-hopping anti-jamming technique for mobile cognitive users. *2015 IEEE Globecom Workshops (GC Wkshps)*, IEEE, pp. 1–6.

158 Labib, M., Ha, S., Saad, W., and Reed, J.H. (2015). A colonel blotto game for anti-jamming in the internet of things. *2015 IEEE Global Communications Conference (GLOBECOM)*, IEEE, pp. 1–6.

159 Linton, C. (1977). Freeman. A set of measures of centrality based on betweenness. *Sociometry* 40 (1): 35–41.

160 Farhang, S., Hayel, Y., and Zhu, Q. (2015). PHY-layer location privacy-preserving access point selection mechanism in next-generation wireless networks. *2015 IEEE Conference on Communications and Network Security (CNS)*, IEEE, pp. 263–271.

161 Zhang, A., Wang, L., Ye, X., and Lin, X. (2016). Light-weight and robust security-aware D2D-assist data transmission protocol for mobile-health systems. *IEEE Transactions on Information Forensics and Security* 12 (3): 662–675.

162 3GPP (2020). TR 33.501 V16: Security Architecture and Procedures for 5G System.

163 Sedidi, R. and Kumar, A. (2016). Key exchange protocols for secure device-to-device (D2D) communication in 5G. *2016 Wireless Days (WD)*, IEEE, pp. 1–6.

164 Abd-Elrahman, E., Ibn-Khedher, H., and Afifi, H. (2015). D2D group communications security. *2015 International Conference on Protocol Engineering (ICPE) and International Conference on New Technologies of Distributed Systems (NTDS)*, IEEE, pp. 1–6.

165 Xu, M., Tao, X., Yang, F., and Wu, H. (2016). Enhancing secured coverage with CoMP transmission in heterogeneous cellular networks. *IEEE Communications Letters* 20 (11): 2272–2275.

166 Gai, K., Qiu, M., Tao, L., and Zhu, Y. (2016). Intrusion detection techniques for mobile cloud computing in heterogeneous 5G. *Security and Communication Networks* 9 (16): 3049–3058.

167 Ying, D. and Ye, F. (2019). D2D-assisted physical-layer security in next-generation mobile network. *2019 International Conference on Computing, Networking and Communications (ICNC)*, IEEE, pp. 324–328.

168 Wang, P., Ye, F., Chen, X., and Qian, Y. (2018). DataNet: deep learning based encrypted network traffic classification in sdn home gateway. *IEEE Access* 6: 55380–55391.

169 Wang, P., Chen, X., Ye, F., and Sun, Z. (2019). A survey of techniques for mobile service encrypted traffic classification using deep learning. *IEEE Access* 7: 54024–54033.

170 Xu, Q., Ren, P., Song, H., and Du, Q. (2016). Security enhancement for IoT communications exposed to eavesdroppers with uncertain locations. *IEEE Access* 4: 2840–2853.

171 Zou, Y., Zhu, J., Wang, X., and Hanzo, L. (2016). A survey on wireless security: technical challenges, recent advances, and future trends. *Proceedings of the IEEE* 104 (9): 1727–1765.

172 MacHardy, Z., Khan, A., Obana, K., and Iwashina, S. (2018). V2X access technologies: regulation, research, and remaining challenges. *IEEE Communication Surveys and Tutorials* 20 (3): 1858–1877.

173 Shen, X., Cheng, X., Yang, L. et al. (2014). Data dissemination in VANETs: a scheduling approach. *IEEE Transactions on Intelligent Transportation Systems* 15 (5): 2213–2223.

174 Papadimitratos, P., Buttyan, L., Holczer, T. et al. (2008). Secure vehicular communication systems: design and architecture. *IEEE Communications Magazine* 46 (11): 100–109.

175 Raya, M. and Hubaux, J.P. (2005). The security of vehicular Ad Hoc networks. *Proceedings of the 3rd ACM Workshop on Security of Ad Hoc and Sensor Networks*, pp. 11–21.

176 Auto Alliance and Global Automaker (2013). 5.9 GHz DSRC connected vehicles for intelligent transportation systems.

177 IEEE 802.11p (2010). *Amendment to Standard for Information Technology-Telecommunications and Information Exchange Between Systems-Local and Metropolitan Area Networks-Specific requirements - Part 11: Wireless LAN Medium Access Control (MAC) and Physical Layer (PHY) Specifications-Amendment 7: Wireless Access in Vehicular Environment*. IEEE Standard Association.

178 Azees, M., Vijayakumar, P., and Deborah, L.J. (2016). Comprehensive survey on security services in vehicular Ad-Hoc networks. *IET Intelligent Transport Systems* 10 (6): 379–388.

179 Chen, S., Hu, J., Shi, Y. et al. (2017). Vehicle-to-everything (V2X) services supported by LTE-based systems and 5G. *IEEE Communications Standards Magazine* 1 (2): 70–76.

180 Amadeo, M., Campolo, C., and Molinaro, A. (2016). Information-centric networking for connected vehicles: a survey and future perspectives. *IEEE Communications Magazine* 54 (2): 98–104.

181 Huang, J., Fang, D., Qian, Y. et al. (2020). Recent advances and challenges in security and privacy for V2X communications. *IEEE Open Journal of Vehicular Technology* 1 (1): 244–266.

182 3GPP TR 33.885 (2017). 3rd Generation Partnership Project; Technical Specification Group Services and System Aspects; Study on Security Aspects for LTE Support of V2X Services.

183 (2016). 5G Automotive Association.

184 Fallgren, M. (2019). 5G Communication Automotive Research and Innovation.

185 Raya, M. and Hubaux, J.P. (2007). Securing vehicular Ad Hoc networks. *Journal of Computer Security* 15 (1): 39–68.

186 Qian, Y. and Moayeri, N. (2008). Design of secure and application-oriented VANETs. *VTC Spring 2008-IEEE Vehicular Technology Conference*, IEEE, pp. 2794–2799.

187 Karagiannis, G., Altintas, O., Ekici, E. et al. (2011). Vehicular networking: a survey and tutorial on requirements, architectures, challenges, standards and solutions. *IEEE Communication Surveys and Tutorials*, 13 (4): 584–616.

188 Qu, F., Wu, Z., Wang, F.Y., and Cho, W. (2015). A security and privacy review of VANETs. *IEEE Transactions on Intelligent Transportation Systems* 16 (6): 2985–2996.

189 Gyawali, S., Qian, Y., and Hu, R.Q. (2020). Machine learning and reputation based misbehavior detection in vehicular communication networks. *IEEE Transactions on Vehicular Technology* 69, 8, 8871–8885.

190 IEEE 1609 Working Group (2016). IEEE standard for wireless access in vehicular environments-security services for applications and management messages. *IEEE Std*, Volume 1609.

191 IEEE 1609 Working Group (2016). *IEEE Standard for Wireless Access in Vehicular Environments (WAVE)-Multi-Channel Operation*. IEEE Std 1609.

192 IEEE 1609 Working Group (2016). *IEEE Standard for Wireless Access in Vehicular Environments (WAVE)-Networking Services*. IEEE Std 1609.

193 Muhammad, M. and Safdar, G.A. (2018). Survey on existing authentication issues for cellular-assisted V2X communication. *Vehicular Communications* 12: 50–65.

194 Svensson, T. (2019). Tutorial 2: 5G Cellular-V2X communications-Introduction to 5GCAR, and the Role of 5G in Automotive Industry.

195 Boneh, D. and Franklin, M. (2003). Identity-based encryption from the Weil pairing. *SIAM Journal on Computing* 32 (3): 586–615.

196 Boneh, D., Lynn, B., and Shacham, H. (2004). Short signatures from the Weil pairing. *Journal of Cryptology* 17 (4): 297–319.

197 Ye, F., Qian, Y., and Hu, R.Q. (2016). Identity-based schemes for a secured big data and cloud ICT framework in smart grid system. *Security and Communication Networks* 9 (18): 5262–5277.

198 Chaum, D. and Van Heyst, E. (1991). Group signatures. *Workshop on the Theory and Application of of Cryptographic Techniques*. Springer, pp. 257–265.

199 Xia, M. and Sun, X. (2009). An efficient group signatures based on discrete logarithm. *2009 International Conference on Wireless Networks and Information Systems*, pp. 50–53.

200 Cui, S., Cheng, X., and Chan, C.W. (2006). Practical group signatures from RSA. *20th International Conference on Advanced Information Networking and Applications - Volume 1 (AINA'06)*, vol. 1, pp. 5–115.

201 Lin, X., Sun, X., Ho, P., and Shen, X. (2007). GSIS: a secure and privacy-preserving protocol for vehicular communications. *IEEE Transactions on Vehicular Technology* 56 (6): 3442–3456.

202 Zhang, C., Xue, X., Feng, L. et al. (2019). Group-signature and group session key combined safety message authentication protocol for VANETs. *IEEE Access* 7: 178310–178320.

203 Huang, J.L., Yeh, L.Y., and Chien, H.Y. (2010). ABAKA: an anonymous batch authenticated and key agreement scheme for value-added services in vehicular ad hoc networks. *IEEE Transactions on Vehicular Technology* 60 (1): 248–262.

204 Zhang, C., Lu, R., Lin, X. et al. (2008). An efficient identity-based batch verification scheme for vehicular sensor networks. *IEEE INFOCOM 2008-The 27th Conference on Computer Communications*, IEEE, pp. 246–250.

205 Boneh, D., Lynn, B., and Shacham, H. (2001). Short signatures from the Weil pairing. *International Conference on the Theory and Application of Cryptology and Information Security*. Springer, pp. 514–532.

206 Dotzer, F., Fischer, L., and Magiera, P. (2005). VARS: a vehicle ad-hoc network reputation system. *6th IEEE International Symposium on a World of Wireless Mobile and Multimedia Networks*, IEEE, pp. 454–456.

207 Huang, D., Misra, S., Verma, M., and Xue, G. (2011). PACP: an efficient pseudonymous authentication-based conditional privacy protocol for VANETs. *IEEE Transactions on Intelligent Transportation Systems* 12 (3): 736–746.

208 Lu, R., Lin, X., Luan, T.H. et al. (2012). Pseudonym changing at social spots: an effective strategy for location privacy in VANETs. *IEEE Transactions on Vehicular Technology* 61 (1): 86–96.

209 Ullah, I., Wahid, A., Shah, M.A., and Waheed, A. (2017). VBPC: velocity based pseudonym changing strategy to protect location privacy of vehicles in VANET. *2017 International Conference on Communication Technologies (ComTech)*, pp. 132–137.

210 Sweeney, L. (2002). k-anonymity: a model for protecting privacy. *International Journal of Uncertainty, Fuzziness and Knowledge-Based Systems* 10 (05): 557–570.

211 Niu, B., Li, Q., Zhu, X. et al. (2014). Achieving k-anonymity in privacy-aware location-based services. *IEEE INFOCOM 2014 - IEEE Conference on Computer Communications*, pp. 754–762.

212 Niu, B., Li, Q., Zhu, X. et al. (2015). Enhancing privacy through caching in location-based services. *2015 IEEE Conference on Computer Communications (INFOCOM)*, pp. 1017–1025.

213 Huang, J., Qian, Y., and Hu, R.Q. (2020). Secure and efficient privacy-preserving authentication scheme for 5G software defined vehicular networks. *IEEE Transactions on Vehicular Technology* 69, 8, 8542–8554.

214 Tong, W., Hussain, A., Bo, W.X., and Maharjan, S. (2019). Artificial intelligence for vehicle-to-everything: a survey. *IEEE Access* 7: 10823–10843.

215 Yang, L., Moubayed, A., Hamieh, I., and Shami, A. (2019). Tree-based intelligent intrusion detection system in internet of vehicles. *2019 IEEE Global Communications Conference (GLOBECOM)*, pp. 1–6.

216 Al-Jarrah, O.Y., Maple, C., Dianati, M. et al. (2019). Intrusion detection systems for intra-vehicle networks: a review. *IEEE Access* 7: 21266–21289.

Index

Security in Wireless Communication Networks, First Edition. Yi Qian, Feng Ye, and Hsiao-Hwa Chen.
© 2022 John Wiley & Sons Ltd. Published 2022 by John Wiley & Sons Ltd.
Companion website: www.wiley.com/go/qian/sec51